T0137244

Springer Theses

Recognizing Outstanding Ph.D. Research

Aims and Scope

The series "Springer Theses" brings together a selection of the very best Ph.D. theses from around the world and across the physical sciences. Nominated and endorsed by two recognized specialists, each published volume has been selected for its scientific excellence and the high impact of its contents for the pertinent field of research. For greater accessibility to non-specialists, the published versions include an extended introduction, as well as a foreword by the student's supervisor explaining the special relevance of the work for the field. As a whole, the series will provide a valuable resource both for newcomers to the research fields described, and for other scientists seeking detailed background information on special questions. Finally, it provides an accredited documentation of the valuable contributions made by today's younger generation of scientists.

Theses are accepted into the series by invited nomination only and must fulfill all of the following criteria

- They must be written in good English.
- The topic should fall within the confines of Chemistry, Physics, Earth Sciences, Engineering and related interdisciplinary fields such as Materials, Nanoscience, Chemical Engineering, Complex Systems and Biophysics.
- The work reported in the thesis must represent a significant scientific advance.
- If the thesis includes previously published material, permission to reproduce this must be gained from the respective copyright holder.
- They must have been examined and passed during the 12 months prior to nomination.
- Each thesis should include a foreword by the supervisor outlining the significance of its content.
- The theses should have a clearly defined structure including an introduction accessible to scientists not expert in that particular field.

More information about this series at http://www.springer.com/series/8790

Alice Fiona Charteris

^{15}N Tracing of Microbial Assimilation, Partitioning and Transport of Fertilisers in Grassland Soils

Doctoral Thesis accepted by
the University of Bristol, Bristol, UK

 Springer

Author
Dr. Alice Fiona Charteris
School of Chemistry
University of Bristol
Cantock's Close
Bristol BS8 1TS, UK

Supervisor
Prof. Richard P. Evershed
School of Chemistry
University of Bristol
Cantock's Close
Bristol BS8 1TS, UK

ISSN 2190-5053   ISSN 2190-5061 (electronic)
Springer Theses
ISBN 978-3-030-31059-2   ISBN 978-3-030-31057-8 (eBook)
https://doi.org/10.1007/978-3-030-31057-8

This Springer imprint is published by the registered company Springer Nature Switzerland AG
The registered company address is: Gewerbestrasse 11, 6330 Cham, Switzerland

Supervisor's Foreword

This book arose from Alice Charteris' doctoral research carried out between 2013 and 2016. The research is an exemplar of how, within the four-year time frame of a UK studentship, an industrious, creative and meticulous Ph.D. student can impressively progress an embryonic idea from method development, through laboratory experiments, to an ambitious field study, yielding wholly new insights into the agricultural impacted nitrogen (N) cycle. The book is beautifully written and attractively illustrated. The data richness and interpretive rigour are quite exceptional—the work is a tour de force.

The project motivation was to develop a new approach to preventing agricultural N fertiliser contamination of drinking water supplies, particularly aquifers. The novelty of the approach ultimately devised lay in recognition of the potential of the soil microbial community to trap applied fertiliser N, e.g. nitrate, ammonium and urea, in newly biosynthesised microbial organic N. The success of this new method hinged on the use of dynamic ^{15}N-stable isotope probing (^{15}N-SIP) to trace the assimilation of fertiliser N into microbial N. Assessments of the soil microbial biochemical assimilation of N fertilisers rested on the hydrolytic extraction of proteinaceous amino acids from soils then, critically, the use of gas chromatography-combustion-isotope ratio mass spectrometry (GC-C-IRMS) for N isotopic analysis. The latter allowed sensitive and specific quantitative assessments of N fertiliser assimilation rates and efficiencies by soil microbes, determinations that were impossible until now.

The seven chapters of the book begin with an introduction (Chap. 1), which reviews the N-cycle from water supply and N pollution perspectives. Modern analytical methods used to study N-cycling in the environment are woven in, with the end of the chapter offering explicit statements of the goals of the research. Chapter 2 (sites, sampling, materials and methods) gives away all the secrets as to how others can do what Alice did! The following four results chapters document the path of Alice's research, beginning with the development of the ^{15}N-SIP method (Chap. 3). This chapter is built on the peer-reviewed publication led by Alice (Charteris et al. Rapid Commun. Mass Spectrom., 2016), which provides the first glimpse of the capabilities of the new approach in action. Chapter 4 offers critical biochemical perspectives on the hitherto unattainable insights into the N-cycle based on application of the new approach to grassland soils from southern Britain. The analytical effort and volume of data generated by Alice was colossal, with the findings showing her efforts were fully justified. The results reveal for the first time how different soils and different fertilisers can elicit markedly different patterns of assimilation of fertiliser N into the microbial protein pool. Up to this point in Alice's research, all the experiments had been conducted in laboratory mesocosms; recognising the potential limitations of such experiments, she took the decision to undertake an ambitious field experiment to test theories that were beyond laboratory experiments. The experiment conducted at one of Wessex Water's aquifer sites produced such a large volume of high-quality data, and it is presented in two separate chapters (Chaps. 5 and 6). Reassuringly, the findings from the meticulously planned and executed field experiment validated the laboratory mesocosm-based microbial ^{15}N-assimilation assays. The book ends with an insightful conclusions chapter (Chap. 7), which highlights the major findings and remaining challenges, genuinely pointing the way forward for future research.

The overriding achievement of Alice's work is that she has demonstrated how ^{15}N-amino acid SIP can be used to assess soil microbial N use efficiency in different grassland systems, and that there is scope to expand this approach to any farming system which uses N fertiliser. In due course, this will likely prove to be the key to refining N-cycling models used to guide fertiliser and soil management practices to mitigate agricultural N pollution and enhance economics.

Bristol, UK   Prof. Richard P. Evershed, F.R.S.
June 2019

Abstract

Freshwaters worldwide are becoming increasingly affected by nitrate (NO_3^-) pollution, which may threaten human health via drinking water contamination and poses a considerable risk to the natural environment by altering the nutrient balance of ecosystems, reducing diversity, resilience and value (economic, social and environmental). Agriculture represents the main source of this pollution, primarily through the application of fertilisers and manures, which are necessary to feed the world's burgeoning population. The conflict between maintaining agricultural production and sufficient clean water is therefore only likely to continue and intensify. Symptomatically treating polluted waters can be effective, but is economically and environmentally unsustainable in the face of rising fuel prices and atmospheric greenhouse gas concentrations. It is clearly preferable from both a water quality and food security perspective to reduce NO_3^- pollution at source and maintain reactive nitrogen (N_r) in soils for plant uptake. Amongst other things, this requires detailed understanding of soil N-cycling to develop appropriate land management practices. In particular, the lack of knowledge regarding organic N-cycling, the transfer of N between inorganic and organic forms and the N-supplying capacity of soils have been identified as key research targets.

This work has therefore focused on developing, validating and applying a compound-specific amino acid (AA) ^{15}N stable isotope probing (^{15}N-SIP) method to assess microbial N assimilation. The approach provides valuable insights into the biochemical routing and transfer of N from the inorganic pool into a quantitatively important and active pool of soil organic N (proteinaceous N), which is likely to have considerable influence on the N-supplying capacity of the soil. The microbial assimilations of a range of ^{15}N-labelled fertiliser N compounds ($^{15}NO_3^-$, $^{15}NH_4^+$ and ^{15}N-urea) were assessed in laboratory incubation experiments with two grassland soils and found generally to increase in accordance with microbial N source preferences—i.e. $^{15}NO_3^- < {}^{15}N\text{-urea} < {}^{15}NH_4^+$. Interesting differences were revealed in the extents of assimilation of particular substrates between the two soils (1.8 vs. 6.4%, 13.4 vs. 6.3% and 5.4 vs. 4.4% of the applied $^{15}NO_3^-$, $^{15}NH_4^+$ and ^{15}N-urea, respectively), likely due to differences in soil type and/or their management histories. The representativeness of the results from these laboratory mesocosm experiments were tested in the field by repeating the incubation for $^{15}NO_3^-$ on a 1-m^2 scale under field conditions at one of the sites. The results of the laboratory and field experiments were comparable, both in terms of the biochemical routing of $^{15}NO_3^-$ into hydrolysable soil AAs, but also in terms of the percentage of the applied ^{15}N assimilated (6.4 and 6.6%, respectively). Microbial $^{15}NO_3^-$ assimilation was consistently low (1.8–6.6%) indicating other processes were quantitatively more important for NO_3^- cycling in the soil system and the physical partitioning and (vertical and lateral) transport of $^{15}NO_3^-$ was investigated in a unique plot-scale field experiment. This work highlighted the difficulties of accurately assessing N partitioning in the field and indicated that vertical ^{15}N transport predominated lateral, and transport occurred via a combination of uniform and preferential flow.

The major achievements of the work include the development, validation and use of compound-specific AA ^{15}N-SIP to investigate microbial N assimilation in mesocosm laboratory experiments and the field. New insights have been provided into the differential processing of different N fertilisers in different soils and the mechanistic links between molecular-scale processes and the observations of field-scale N fertiliser immobilisation studies elucidated. The method developed and results generated will be extremely valuable in further work aimed at improving N fertiliser recommendations and soil management practices for increased productivity and reduced N_r losses.

Parts of this thesis have been published in the following journal articles:

Charteris, A. F., Knowles, T. D. J., Michaelides, K. and Evershed, R. P. (2016) Compound-specific amino acid ^{15}N stable isotope probing of nitrogen assimilation by the soil microbial biomass using gas chromatography–combustion–isotope ratio mass spectrometry, *Rapid Communications in Mass Spectrometry*, **30**, 1846–1856, DOI: 10.1002/rcm.7612.

Acknowledgements

Firstly, I would like to thank my supervisors, Prof. Richard Evershed and Dr. Katerina Michaelides, for their advice, support and encouragement throughout my Ph.D. Their unfailing enthusiasm for research and learning has been extremely inspiring.

My Ph.D. was funded by a NERC CASE studentship between the University of Bristol and Wessex Water. The team at Wessex Water (Luke de Vial, Paul Stanfield, Jeremy Graham, Adrian Moore, Esa-Pekka Tuominen and Sean Tyrrell) have absolutely been the best collaborators I could have asked for. Always interested and willing to support my ideas and provide assistance whenever requested. I am so grateful. I am particularly indebted to Adrian Moore for his help with my field experiment (installing porous pots, building a fence, taking soil cores and sampling groundwater!); much of this work could not have been accomplished without it. Sampling and the field experiment were conducted on Longlands Dairy Farm, and I am very grateful to Tom Foot for allowing this. Jenni Dungait and Adrian Joynes are thanked for their assistance in sampling soil from the other site, Rowden Moor, North Wyke. Ian Bull and Alison Kuhl are thanked for maintaining the excellent instrument facilities we are lucky enough to have in the OGU. I am particularly grateful to Alison for all of her help with the not-so-excellent Deltas on N_2 and with amino acid fragmentation patterns. Helen Grant of the Lancaster Node of the NERC LSMSF deserves thanks for carrying out all of the bulk $\delta^{15}N$ analyses presented in this thesis (especially of the soil water residues, which were annoyingly small samples). Helen was extremely patient with these and with the many questioning emails I have sent her over the years. Thanks also to Alison Carswell and Paul Henderson for being so accommodating with the old Seal AutoAnalyser in Geography, I think we made a great little team (until we killed it...!). I would also like to thank Tim Knowles for laying the main foundations of my work in his Ph.D. and for answering all of my questions over the years. Similarly, Charlotte Lloyd has been a valuable sounding board having had the same supervisors during her Ph.D. and knowing much more about hydrology than me. I am also very grateful for her help with putting together the site maps in this thesis.

Another Helen (H. Whelton), my Ph.D. twin, deserves thanks for answering all of the questions I thought might be too silly to ask anyone else (which way up does that little thing in the GC inlet go??) and for being a sympathetic ear for my complaints and concerns. It has been an honour to sit behind the brilliant Mélanie Roffet-Salque, and I have thoroughly enjoyed all of our chats over the years. Thank you Julie Dunne, for 'getting things done/fixed' and censoring important emails for rudeness, although I think they were generally always 'almost too polite'. I am very grateful to Ron, my fellow XP sufferer for sharing this pain with me and for his friendship. Finally, a shout out to my back-lab buddies, Ili Johari and Katrina Pears, we did a good job keeping the chaos at bay, and to my fellow DWs Matt Carmichael and Jon Pemberton for understanding what working in UoB RHS is like! Everyone in the upstairs office has been subjected to my frequent rants about this and other things, and no one has seemed to mind too much, so thank you all. The people you work with make a big difference, and the OGU has been amazing.

I cannot thank my parents enough for their constant and unquestioning willingness to help out whenever necessary, be it taking soil cores, fixing fences or checking references. Even both of my brothers and my uncle, David Sandeman, have helped with field work at various times—I am so lucky to have such a wonderful family. Finally, thank you Alex for going through this with me and putting up with everything that came with that!

Contents

Abbreviations

AA	Amino acid
ANI	Added nitrogen interaction
AOA	Ammonia-oxidising archaea
AOB	Ammonia-oxidising bacteria
ATP	Adenosine triphosphate
BADC	British Atmospheric Data Centre
BBC	British Broadcasting Corporation
BNF	Biological nitrogen fixation
CBED	Concentration-Based Estimated Deposition (model)
CEH	Centre for Ecology and Hydrology
DDW	Double-distilled water
Defra	Department for Environment, Food and Rural Affairs
DGGE	Denaturing gradient gel electrophoresis
DMC	Dry matter content
DNA	Deoxyribonucleic acid
DNRA	Dissimilatory nitrate reduction to ammonium
DON	Dissolved organic nitrogen
EA-IRMS	Elemental analysis–isotope ratio mass spectrometry
EI	Electron ionisation
EU	European Union
FAO	Food and Agriculture Organisation of the United Nations
FID	Flame ionisation detector
G	Glucose
GC	Gas chromatography
GC-C-IRMS	Gas chromatography–combustion–isotope ratio mass spectrometry
GC-MS	Gas chromatography–mass spectrometry
GDH	Glutamate dehydrogenase (pathway)
GS-GOGAT	Glutamine synthetase–glutamate synthase
HDPE	High-density polyethylene
HFA	Home field advantage

IFAD	International Fund for Agricultural Development
IRM-GC-MS	Isotope-ratio-monitoring gas chromatography–mass spectrometry
IS	Internal standard
IT	Identical treatment (principle)
LMW	Low molecular weight
LSMSF	Life Sciences Mass Spectrometry Facility
MAFF	Ministry of Agriculture, Fisheries and Food
MC	Moisture content
MIT	Mineralisation–immobilization turnover
MORECS	Met Office Rainfall and Evaporation Calculation System
NERC	Natural Environment Research Council
N_r	Reactive nitrogen (all except N_2)
NUE	Nitrogen use efficiency
NVZ	Nitrate vulnerable zone
PCR	Polymerase chain reaction
PE	Priming effect
PMN	Potentially mineralisable nitrogen
PTFE	Polytetrafluoroethylene
PTV	Programmable temperature vaporisation
PWS	Public water supply
QC	Quality control
RF	Response factor
RM	Rowden Moor
RNA	Ribonucleic acid
RuBisCO	Ribulose-1,5-bisphosphate carboxylase–oxygenase
SD	Standard deviation
SE	Standard error of the mean
SIP	Stable isotope probing
SMD	Soil moisture deficit
SOM	Soil organic matter
THAA	Total hydrolysable amino acid
TN	Total nitrogen
TON	Total oxidised nitrogen
TRF	Terminal restriction fragment
TRFLP	Terminal restriction fragment length polymorphism
U	Urea
USDA	United States Department of Agriculture
USGS	United States Geological Survey
WA	Winterbourne Abbas
WFP	World Food Programme
WHC	Water holding capacity
WHO	World Health Organisation
WS-CRDS	Wavelength-scanned cavity ring-down spectroscopy
% OM	Percentage organic matter
% TC	Percentage total carbon

% TIC	Percentage total inorganic carbon
% TN	Percentage total nitrogen
ꙅ	Slurry

List of Figures

List of Tables

Chapter 1
Introduction

1.1 Freshwater Nitrate Contamination and Food Security

Despite its epithet as the 'Blue Planet' only about 1% of the water on Earth is liquid freshwater (containing <0.5 g l^{-1} dissolved salts) and thereby directly useable by humans [1, 2]. The majority of this useable water is present as groundwater [1], which is a vital, and in some cases, non-renewable or very slowly renewing resource, supplying, for example, approximately two thirds of the water for domestic use in Europe [2, 3]. Groundwater resources worldwide are however, increasingly under threat from over-exploitation and/or quality degradation by a wide variety of pollutants [2, 4]. Nitrate (NO_3^-) represents a major and growing, globally important freshwater contaminant, which is hazardous to human health and the natural environment [5–10], but is itself a valuable resource in soils/for plants.

The European Drinking Water Directive in agreement with the World Health Organisation (WHO) prescribe a limit of 11.3 mg N l^{-1} (or 50 mg NO_3^- l^{-1}; Directive 98/83/EC; [11]) to obviate the risks of NO_3^- pollution to human health [3]. Such risks include methemoglobinemia and certain cancers [12], but the true threat of these diseases as a result of drinking water NO_3^- contamination is still debated [13]. Indeed, the environmental impacts of groundwater NO_3^- pollution, which result directly from NO_3^- leaching to groundwater, as well as from antecedent and subsequent processes linked to the wider anthropogenic alteration of the nitrogen (N) cycle [5, 6, 10, 14, 15–21]; Fig. 1.1) and even from water treatment processes [22], are almost undoubtedly more threatening. It is estimated, for example, that 80% of freshwaters in Europe are contaminated with N concentrations that pose a risk to biodiversity of over 1.5 mg l^{-1} [21].

Globally, agriculture represents the main source of freshwater NO_3^- pollution, primarily due to the run-off and leaching of mobile NO_3^- ions derived from manures and synthetic fertilisers [5–7]. Reactive N (N_r; all forms of N except N_2; Fig. 1.2) is the most commonly limiting nutrient in terrestrial ecosystems [8, 19, 23, 24] and modern food production, and by extension the current world population, is dependent

© Springer Nature Switzerland AG 2019
A. F. Charteris, *15N Tracing of Microbial Assimilation, Partitioning and Transport of Fertilisers in Grassland Soils*, Springer Theses,
https://doi.org/10.1007/978-3-030-31057-8_1

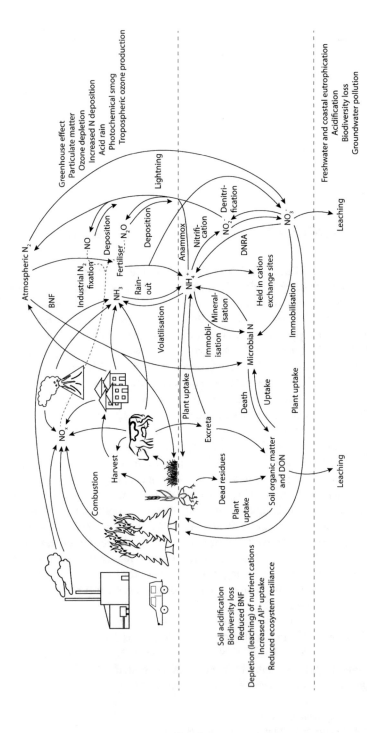

Fig. 1.1 The global (terrestrial) N cycle, including anthropogenic influences and a summary of the damaging effects of these influences in the atmosphere, terrestrial environments and freshwaters and coastal waters. BNF; biological nitrogen fixation; DON; dissolved organic nitrogen; DNRA; dissimilatory nitrate reduction to ammonium

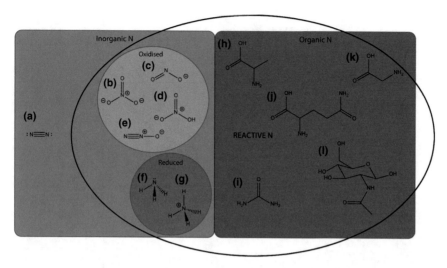

Fig. 1.2 Range of N species which may be categorised in different ways. **a** Dinitrogen gas (N_2); **b** Nitrate (NO_3^-); **c** nitrite (NO_2^-); **d** Nitric acid (HNO_3); **e** Nitrous oxide (N_2O); **f** Ammonia (NH_3); **g** Ammonium (NH_4^+); **h** Alanine (Ala, A); **i** Urea; **j** Glutamine (Gln, Q); **k** Glycine (Gly, G); and **l** N-acetylglucosamine

on added synthetic nitrogenous fertilisers [14, 25]; Fig. 1.3). These are supplied by the Haber-Bosch process and Erisman et al. [14] estimated that in 2008, 48% of the world's population was supported by Haber-Bosch N.

Yet the global distribution of nitrogenous fertilisers and therefore food is distinctly uneven [14, 17, 25, 26]. While the populations of developed nations devour once luxury products to the detriment of their health and create vast amounts of food waste, the latest Food and Agriculture Organisation (FAO) report estimates that 795 million people remain undernourished [14, 17, 25, 27]. Although fertiliser use in many developed countries, including Great Britain, for example, has decreased over the last 20–30 years (Fig. 1.4), the overall sustainability of dietary practices in developed nations is unlikely to greatly improve imminently, despite efforts (e.g. [28]). In addition, the N demands of diets in developing nations are rising [25], so N fertiliser use amongst those with already sufficient food seems set only to grow, in the near future at least. Moreover, in spite of emphasis on 'sustainable intensification', the drive to achieve global food security (requiring 70–100% more food by 2050) as world population rises to nine billion is likely to also lead to increases in N fertiliser use in regions where it has historically been low [29–32]. This is before considering further potential increases in N fertiliser demand for biofuel production, which could lead to a doubling of global N demand to 200 Tg N year^{-1} [14].

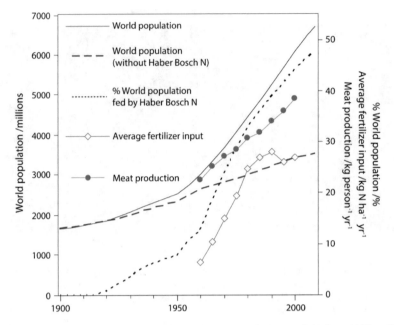

Fig. 1.3 Growth of world population and the influence of Haber-Bosch N from 1900 to 2008, including estimates of world population without Haber-Bosch N and the percentage of the world population that is sustained by Haber-Bosch N (adapted from [14]). Increases in average fertiliser use per hectare of agricultural land and per capita meat production are also shown

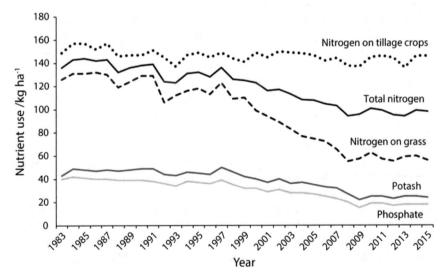

Fig. 1.4 Trends in nutrient use on all crops and grassland in Great Britain 1983–2015. Adapted from The British Survey of Fertiliser Practice, Department for Environment, Food and Rural Affairs (Defra) [33]

The conflict between maintaining agricultural production and ensuring a sufficient supply of clean water is therefore likely to continue and even intensify in the foreseeable future [7, 14, 34, 35]. Accordingly, water companies will have no choice but to manage freshwater NO_3^- pollution and continue to provide clean, safe water under increasingly challenging circumstances [7, 36]. Water treatment works represent the traditional approach, which although effective, are costly to build and are an energy- and carbon (C)-expensive solution in the face of rising fuel prices and global warming trends [7, 36]. Furthermore, water treatment only confronts the symptomatic problem not the root cause of NO_3^- contamination and to a certain extent simply results in pollution swapping, as carbon dioxide (CO_2) and nitrous oxide (N_2O) are released [37].

Reducing the amount of NO_3^- entering freshwaters in the first place is clearly preferable and more sustainable than symptomatic treatment, particularly since NO_3^- in soils is a valuable plant nutrient and much energy and money is invested the production and application of N fertilisers. Whilst simultaneously reducing waste and environmental pollution in agricultural production is clearly a somewhat idealistic solution, it is widely acknowledged that much can be done to improve N use efficiency and agricultural sustainability without impacting yields [21, 38–40]. It is estimated, for example, that on average, only half of the fertiliser N applied to crops is taken up and, under certain conditions, (soil type, season/climate, crop and management practices, especially fertiliser application rates) leaching losses can be substantial [41, 42].

1.2 Nitrate Leaching

1.2.1 Process

As an anion, NO_3^- is not held on the soil's cation exchanges sites and may be easily transported through the soil by water [39, 41, 43]. Water movement through soils is itself a complex, well-studied process, which in the most simple terms can be broken down to proceeding via some combination of slow, uniform percolation (described by Darcy's Law for laminar flow in saturated soils and Richard's equation in unsaturated soils) and preferential flow (which can occur via various processes and due to its heterogeneous nature is more challenging to model; [44–47]; Fig. 1.5). The particular combination of these two extremes will vary between sites, on a variety of scales (due to soil texture, structure, depth, organic matter content, moisture content and degree of heterogeneity as well as faunal activity and plant cover and roots) and as a result of the prevailing conditions (wet/dry and freeze/thaw cycles and rainfall intensity and duration; [41, 44–47]).

Accordingly, a selected soil volume/field is likely to contain several distinct fluxes of water (e.g. residing in particular soil structures and/or from different rainfall events) of differing chemistry, moving at different velocities and, depending largely

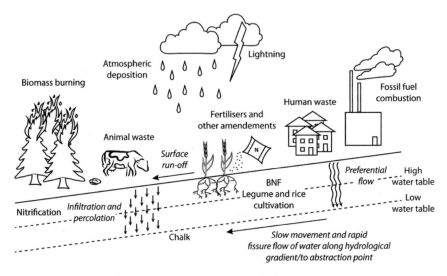

Fig. 1.5 Potential sources of NO$_3^-$ (roman text) and transport routes to groundwater (italic) in a (chalk) catchment. BNF; biological nitrogen fixation

on these velocities, mixing with one another to different extents [44, 46, 48]. Faster moving waters in larger macropores (e.g. >20 μm in diameter) flow too quickly for solute exchange with the matrix to occur and do not equilibrate with other fluxes of water en route to groundwater [49].

To add to this complexity, NO$_3^-$ is non-uniformly distributed in soils and in particular fluxes of water, the different flow paths and velocities of which will lead to hydrodynamic dispersion of NO$_3^-$ during transport [41, 44, 45]. Furthermore, NO$_3^-$ diffuses along concentration gradients within and between fluxes of water and is subject to biogeochemical processing along flowpaths and other complex and sometimes transient variations in soil structure and chemistry during leaching [41, 44–46, 48, 50]. As a result it is extremely difficult to accurately gauge, model or predict NO$_3^-$ leaching rates and concentrations in a field setting. This is not to say that the factors controlling NO$_3^-$ leaching are poorly understood and a wide range of empirical, conceptual and process-based NO$_3^-$ leaching models varying in scale and complexity are extensively used in industry and academia.

1.2.2 Controlling Factors

The two most important factors determining NO$_3^-$ leaching losses are soil NO$_3^-$ concentrations (in excess of plant N demand) and the availability of water to facilitate NO$_3^-$ transport [39, 41]. Leaching NO$_3^-$ may have been supplied to the soil directly in the form of NO$_3^-$, either as fertiliser to promote plant growth or in wet or dry deposition, as nitric acid (HNO$_3$) in acid rain, for example ([51]; Fig. 1.5).

Alternatively, it may be derived from nitrification in the soil, whereby soil microorganisms oxidise either ammonium (NH_4^+) or amino-N to produce NO_3^- under oxic conditions (Eqs. 1.1–1.3; [19, 52]; Fig. 1.5).

$$2NH_4^+ + 3O_2 \rightarrow 2NO_2^- + 2H_2O + 4H^+ + Energy \quad \Delta G^0 = -66.5 \, kcal \quad (1.1)$$

$$2NO_2^- + O_2 \rightarrow 2NO_3^- + Energy \quad \Delta G^0 = -17.5 \, kcal \quad (1.2)$$

$$\underset{-3}{RNH_2} \rightarrow \underset{-1}{RNHOH} \rightarrow \underset{+1}{RNO} \rightarrow \underset{+3}{RNO_3} \rightarrow \underset{+5}{NO_3^-} \quad (1.3)$$

The possible sources of NH_4^+ and amino-N for nitrification in soils are numerous and include: atmospheric N deposition, fertiliser, slurry, crop residue, digestate or other amendment N, animal excreta, N released by N_2-fixing plants and microorganisms and (mineralised) soil organic N (Fig. 1.5). Nitrification of NH_4^+ predominates in most soils and substrate availability, which is strongly dependent on the balance of mineralisation and immobilisation, is the most important control governing nitrification rates [19]. Nitrifiers are generally considered poor competitors for NH_4^+, so in the absence of NH_4^+ fertilisation, high nitrification rates are usually only observed when C is limited, inhibiting NH_4^+ immobilisation, or mineralisation increases, following soil disturbance, such as ploughing, for example [19, 41, 43, 53].

Soil NO_3^- available for leaching or leaching NO_3^- can also be reduced or removed by biological processes, such as uptake/immobilisation, denitrification, dissimilatory nitrate reduction to ammonium (DNRA) or non-respiratory denitrification [19]. Microbial NO_3^- assimilation is generally thought to be low, occurring only under low NH_4^+ availability [54–58], while plant NO_3^- uptake is highly dependent on plant growth rate [39–41]. Denitrification is the process whereby NO_3^- is sequentially reduced to nitrite (NO_2^-), nitric oxide (NO), N_2O and finally dinitrogen gas (N_2) under anoxic conditions (Eq. 1.4; [41, 59, 60]). As a result, denitrification in soils tends to be high where NO_3^- is available and water-logging has inhibited oxygen (O_2) diffusion leading to the development of anoxia [19, 41, 59]. In addition denitrification during/following leaching has been observed in unsaturated/vadose zones, in riparian zones between groundwater and streams and in groundwater [19, 43, 50, 61–65].

$$2NO_3^- \rightarrow 2NO_2^- \rightarrow 2NO \rightarrow N_2O \rightarrow N_2 \quad (1.4)$$

Dissimilatory nitrate reduction to ammonium and non-respiratory denitrification are generally considered more minor in soils, but are relatively poorly characterised N cycle processes [19, 59]. The process of DNRA reduces NO_3^- to NO_2^- and then NH_4^+ under anoxic conditions [19, 59], while non-respiratory denitrification produces mainly N_2O under oxic conditions [19].

Any management practices or conditions that result in soil NO_3^- concentrations in excess of plant N demand, either directly or by altering the balance of soil N cycle processes, are likely to lead to high NO_3^- leaching losses if sufficient water is available. As a result, NO_3^- leaching losses in temperate regions are commonly largest in winter when cooler conditions reduce plant uptake and evapotranspiration rates, increasing the availability of drainage water and under some circumstances, NO_3^- [39, 41]. High drainage volumes generally result in high NO_3^- leaching losses [39, 61, 66], but it has also been suggested that in some cases high drainage volumes could instead dilute leaching NO_3^-/groundwater NO_3^- concentrations. In addition, water may stimulate plant growth and thus NO_3^- uptake, while saturation encourages the development of anoxia, which could stimulate denitrification, reducing NO_3^- leaching [39, 40]. Yet on the other hand, dry conditions and weather that stunts plant growth can allow NO_3^- to build up in the soil, ready to be leached by the next high rainfall event, which can also stimulate mineralisation in rewetting the soil [39, 41]. Weather therefore plays a dominant, but variable role in NO_3^- leaching via the impact of rainfall and temperature on microbial processes, plant growth and drainage at a particular point in time [67].

Soil texture, structure and depth are also important in determining NO_3^- leaching losses by affecting drainage rates and denitrification potential [39–41]. Nitrate leaching losses are generally lowest from fine-textured, poorly draining, deep soils [39–41]. Any structures that reduce NO_3^- surface to groundwater transit times (by reducing drainage distance and/or increasing drainage speed), such as macropores (created by earthworms, plant roots, freezing and thawing cycles or wetting and drying of soils, for example), chalk fissures and artificial drainage systems have the potential to increase NO_3^- leaching losses [39, 41, 49, 68]. The increased aeration capacity of these structures also further reduces denitrification potential en route to groundwater [41]. Alternatively, however, macropores, fissures or artificial drainage systems can transport NO_3^--poor waters quickly to groundwater, bypassing a comparably NO_3^--rich matrix and thereby retarding NO_3^- leaching and diluting groundwater NO_3^- concentrations.

Given the complexity in the process of NO_3^- leaching and the range of potential NO_3^- sources (Fig. 1.5), freshwater NO_3^- pollution is generally diffuse and usually derives from a wide variety of sources in any particular catchment [8, 69]. Land use and the choice of farming system will, however, strongly impact NO_3^- leaching losses in a catchment/from particular parts of the catchment. It is difficult to rank NO_3^- losses from particular land uses and farming systems due to site-specific differences and the wide range of management practices and particular combinations thereof [41]. In general, however, in terms of land use, NO_3^- leaching losses typically increase in the order: forests < cut grassland < grazed grassland, arable crops < ploughing of pasture < horticultural crops [39, 41]. Nitrate leaching losses are also usually highest for intensive farming systems and can be low for well-managed extensive and organic farming systems, but integrated systems will often present the best compromise [40, 67].

1.3 Previous Research and Remaining Unknowns

Nitrogen has been recognised as a potential water pollutant since the 19th century [70], but it was not until the late 1960s and early 1970s that groundwater NO_3^- pollution became of particular concern [71, 72]. Since then, considerable research effort has been devoted to monitoring, understanding and predicting groundwater NO_3^- trends, as well as developing suitable methods, such as tracer techniques (e.g. by comparison with chloride [Cl^-] and tritium [^3H] profiles; [73]) and computer models (e.g. [72]). Land use and fertiliser application were relatively quickly identified as potential sources of the groundwater NO_3^- pollution observed [72] and around the same time, further concerns were raised about the wider environmental impact of agriculture and in particular, the inefficient use of N [74]. While the problem of fertiliser N losses from surface soils had been recognised well in advance of this (e.g. [75]), the combination of increasing concerns about the environment, the economics and the development of computer technology led to an acceleration of NO_3^- leaching research during the 1970s [74]. At the same time, this encouraged research into methods to assess the N supplying capacity of soils [76] and the early development of best management practices to increase N use efficiency (NUE) and reduce N loss [77].

Improving NUE without decreasing yields and developing N best management practices (for particular crops and farming systems) has since been the focus of much research [40, 67, 78]. Progress has been made in quantifying, simulating and predicting NO_3^- leaching losses at a range of scales from different soils under different land uses and management [40, 79], but the spatial and temporal variability of losses still presents a challenge, both for experimentally-derived estimates and model simulations (e.g. [80]). This work has been extended to investigate the effects and interactions of particular management practices for different soils and land uses and has enabled the development of more detailed and verified best management principles (e.g. [40, 67]).

In many countries best management advice is disseminated to farmers via various agencies (principally Defra, Natural England and the Environment Agency in the UK; e.g. [81]) and is supported by simple fertiliser recommendation systems (e.g. the Fertiliser Manual, RB209; [82]) and more complex, but user-friendly computer-based decision support models (e.g. SimUlation of Nitrogen Dynamics in Arable Land; SUNDIAL [83] and NGUAGE [84]). Achieving widespread and *wholly successful* implementation of best management advice and decision support systems *at farm scale*, however, remains a challenge [40, 78]. In addition, the potential for particular advisory practices to result in wider unintended consequences or 'pollution swapping' has more recently come to the fore—for example, buffer strips to reduce NO_3^- leaching may increase N_2O emissions [40, 42]. Research is still required to understand these interactions and ensure the development of policies that institute truly sustainable farming systems [40, 42].

Many of the major gains in crop yields in past years have resulted from the introduction of new, more disease and pest resistant varieties, as well as from improved management and crop protection (Fig. 1.6; [40, 42]). Such advancements can also positively impact NUE due to the more sustained N uptake capacity of a healthy, pest- and disease-free crop [40]. Further opportunities to select/breed for greater plant NUE in a given environment are provided by improved knowledge of the mechanistic (e.g. [85]) and genetic controls of plant N assimilation [86]. Integration of this knowledge with increased understanding (and thereby potentially greater control) of the N supplying capacity of soils is paramount to the current NUE challenge of achieving synchronicity between plant N demand and N supply (native soil N plus appropriate N fertilisation, in quantity and timing). Since soil N supply is determined largely by microbially-mediated soil N cycling, and in particular soil organic matter (SOM) turnover, developing understanding of the controls, responses and feedbacks

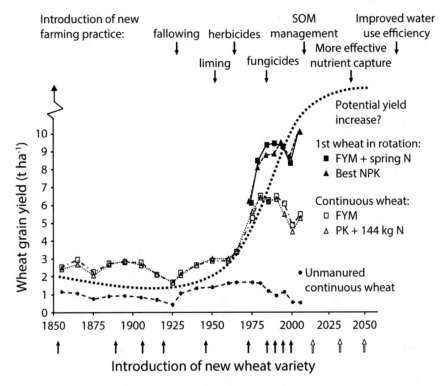

Fig. 1.6 Improvements in the yield of winter wheat (grain only, t ha^{-1}) on selected plots of the Broadbalk wheat experiment between 1850 and 2005, including the timings of introductions of new farming practices and new wheat varieties which have aided these improvements (adapted from [42, 87]). A hypothetical curve has been fitted to the wheat yield data to show the potential yield increases resulting from better SOM management, more effective nutrient capture and the development of further new varieties. FYM; farmyard manure

operating on these processes is critical (Fig. 1.6). This will necessarily also involve further investigation of the complex and interactive relationships between the cycling of N and other nutrients (especially C and P; [42, 88]).

Indeed, NO_3^- leaching losses commonly result from the turnover of SOM and specifically from the mineralisation and nitrification of soil N, rather than from fertilisers directly (except where a large excess of fertiliser is applied before rainfall). Accordingly, soil N mineralisation-immobilisation turnover (MIT) has been the focus of much research (e.g. [89–95]). This work, and wider research interrogating soil N cycling and the role of soil microorganisms, measuring NUE and developing mechanistic soil N cycling models has been facilitated by ^{15}N-labelling stable isotope techniques.

1.4 Nitrogen Isotopes as Tracers of Soil Nitrogen Cycling

1.4.1 Variation in and Use of Natural Abundance Nitrogen Isotopes

Nitrogen is one of the 61 elements known to have two (or more) stable isotopes [96]. The two stable isotopes of N have mass numbers of 14 and 15 and natural isotopic abundances of 99.6 and 0.4%, respectively [97]. The N isotopic compositions of different substances vary, however, as a result of the kinetic and equilibrium effects of the small mass difference between ^{14}N and ^{15}N in physiochemical processes leading to isotopic fractionation [96, 98–100]. Nitrogen isotopic compositions are generally reported as δ^{15}N values (in per mil, ‰), relative to the internationally agreed measurement standard, AIR-N_2 (Eq. 2.9).

Due largely to the wide range of N oxidation states (-3 to $+5$) and transformations between these states, the isotopic compositions of naturally occurring N-containing materials span over 200‰ [101]. Variations in the N isotopic compositions of particular materials and compounds have been used: (i) as source indicators and to assess the relative contributions of various inputs; (ii) to infer processes and to estimate the rates of these processes; and (iii) in developing and constraining system models [99, 102]. A few examples of such uses of natural abundance N isotopes in environmental research are shown in Table 1.1.

As Table 1.1 demonstrates, however, the spatial and temporal variability, and range and complexity of processes in which N isotopes are involved (and fractionated through) in environmental systems often introduces a degree of uncertainty in the use and interpretation of natural abundance N isotopic data. In addition, this variability and complexity usually precludes quantitative source apportionment or assessments of process importance. This has encouraged the use of ^{15}N-labelled tracers in soil and soil-plant systems research [148].

Table 1.1 Some examples of the uses of natural abundance N isotopes as source indicators and to infer processes in environmental research

Use	Theory	Caveats	References
Sources of NO_3^- in surface water, drainage and groundwater	• Nitrate in rainfall is isotopically distinct from nitrification-derived NO_3^- in soils • Fertiliser N tends to be ^{15}N-depleted relative to soil N while animal waste and sewage are ^{15}N-enriched • These differences can sometimes be used to distinguish the sources of NO_3^- in freshwater	• There may be overlap in the ranges of $\delta^{15}N$ values between these sources • The $\delta^{15}N$ values of sources can vary considerably over time and in space • Due the complexity of soil processes (and potential for isotopic fractionation) definitive source assignments are uncommon and further supporting evidence is usually required • Semi-quantitative only, although progress could be made through the application of relatively recently developed Bayesian isotopic mixing models (e.g. SIAR; Parnell et al. [103])	*Source discrimination based on $\delta^{15}N$ values possible:* Spoelstra et al. [104], Mayer et al. [105], Deutsch et al. [69], Kellman [106], Moore et al. [107], Baily et al. [108], Kelley et al. [109], Pastén-Zapata et al. [110] *Source discrimination based on $\delta^{15}N$ values not possible:* Ostrom et al. [111], Burns and Kendall [112], Kellman and Hillaire-Marcel [113], Seiler [114], Granger et al. [68], Showers et al. [115], Heaton et al. [116] *Reviews:* Heaton [117], Kendall et al. [118], Xue et al. [119]
Estimate plant N_2-fixation	• The fraction of plant N derived from N_2-fixation can be calculated by comparing plants which derive N from a mixture of sources with nearby plants which rely solely on soil N	• Direct absorption gaseous N (ammonia $[NH_3]/NO_X$) from atmosphere assumed zero/equal • Requires that species compared have similar root distributions, N uptake patterns and N preferences • Unreliable if any soil N sources have the same N isotopic composition as atmospheric N_2 • Not quantitative/semi-quantitative	*Source discrimination based on $\delta^{15}N$ values possible:* Bowman et al. [120], Cadisch et al. [121], Arndt et al. [122] *Source discrimination based on $\delta^{15}N$ values not possible:* Binkley et al. [123], Högberg [124], Spriggs et al. [125] *Reviews:* Shearer and Kohl [126, 127], Högberg [99], Marshall et al. [128]

(continued)

Table 1.1 (continued)

Use	Theory	Caveats	References
Source of N_2O	• N_2O produced by nitrification is more ^{15}N-depleted than that produced by denitrification • The intramolecular site preference ($\delta^{15}N^\alpha - \delta^{15}N^\beta$) of N_2O isotopomers is independent of the isotopic signature of the precursor and can be used to distinguish between nitrification, and denitrification as sources of N_2O • The site preference of N_2O produced by nitrification is generally higher than that of denitrification	• There may be some overlap in the site preference of autotrophic nitrification and fungal denitrification • The site preference of denitrification may increase over the course of extensive N_2O reduction. This is because reduction involves the cleavage of N–O bonds and therefore favours cleavage of ^{14}N–O bonds, resulting in increasing ^{15}N–O bonds in the residual N_2O. Modelling approaches are being developed to overcome this effect (e.g. Wu et al. [129])	Yoshida [130], Yoshida and Toyoda [131], Schmidt et al. [132], Toyoda et al. [133], Sutka et al. [134], Cardenas et al. [135], Ostrom et al. [136], Meijide et al. [137], Bergstermann et al. [138], Köster et al. [139], Wu et al. [129] *Reviews:* Stein and Yung [140]
Evidence of denitrification in groundwater	• ^{15}N-depleted N_2 is generated by kinetic isotope fractionation resulting in increasingly ^{15}N-enriched groundwater NO_3^-	• Affected by NO_3^- inputs. • Not quantitative/semi-quantitative	Vogel et al. [64], Mariotti et al. [62], Böhlke and Denver [141] *Reviews:* Heaton [117]
Indicator of N-deposition/saturation in N-limited ecosystems	• Nitrification is increased by N-deposition on N-limited ecosystems and results in ^{15}N-enriched NH_4^+ and ^{15}N-depleted NO_3^-. Plants which preferentially assimilate NH_4^+ subsequently become ^{15}N-enriched and so do litter-layers (usually ^{15}N enrichment increases with depth)	• Not always straightforward/reliable due to the complexity of soil processes and plant uptake • Not quantitative/semi-quantitative	Högberg et al. [142], Högberg et al. [143], Johannisson [144], Emmett et al. [145], Martinelli et al. [146] *Review:* Högberg [99], Garten et al. [147]

1.4.2   Use of ^{15}N-Labelled Stable Isotope Tracers

Substances in which the naturally rare stable isotope of nitrogen, ^{15}N has been artificially increased may be referred to as '^{15}N-labelled' and are widely used in soil N cycle, environmental and related research. The ^{15}N may act as physical tracer (e.g. in lysimeter leaching studies; [149]), or less commonly, as a biochemical tracer (e.g. to trace the fate of amino acid [AA] N in soil; [150]). Alternatively, ^{15}N-labelled substances (usually ^{15}NH$_4^+$ or ^{15}NO$_3^-$) can be employed in ^{15}N pool dilution experiments to measure gross rates of mineralisation and immobilisation or nitrification (e.g. to determine gross N mineralisation during RuBisCO decomposition; [151]).

The first use of a ^{15}N-labelled tracer in soil/agronomic research was reported by Norman and Werkman [152]. These authors found that only 20% of the ^{15}N applied in ^{15}N-fertiliser-labelled soybean residues was incorporated into growing plants during decomposition of the soybean residues in soil. Much similar ^{15}N tracer research following the (physical) fate of applied ^{15}N (e.g. [153–155]) and assessing the NUE of particular fertilisers by selected crops and in the context of certain soils, field rotations or other management practices (e.g. [156–160]) has taken place since. In further related work, ^{15}N-labelled fertilisers have also contributed much in studies aiming to close the elusive N balance (e.g. [161, 162]), although this has yet to be achieved, even in 'simpler' arable systems [163].

Since the 1940s, ^{15}N-labelled tracers have also been applied in more detailed soil N cycling research [164], including investigations of N mineralisation and immobilisation (e.g. [165]), priming effects (PEs; e.g. Kuzyakov [166]), microbial and plant N assimilation preferences (e.g. [167, 168]), competition between plants and microorganisms for N (e.g. [169, 170]) and in the development of mechanistic soil N cycle models (e.g. [171]). The complexity and heterogeneity of soil organic N, however, confounds the use of ^{15}N as true biochemical tracer in soil N cycling studies. As a result, much previous work has either assumed that any applied ^{15}N remaining in the soil after a reasonable period of time is 'organic ^{15}N' (e.g. [172]), or has estimated 'organic ^{15}N' by extracting/subtracting ^{15}N-enriched inorganic compounds from bulk soils/values (e.g. [173, 174]). Both methods only provide an estimate of the bulk N isotopic composition of a multifarious, unevenly ^{15}N-enriched and undefined 'organic N' pool.

The existence of differentially active sub-pools of soil organic N has been posited since the 1930s [175] and establishing a reliable measure of actively cycling or 'available' soil N remains an important goal [164]. One approach has been to assess N cycling through the soil microbial biomass (using methods based on that of Brookes et al. [176]), but a perhaps more direct focus has been on establishing methods to determine 'potentially mineralisable nitrogen' (PMN). Many methods and N availability indices have been proposed, but the relatively early incubation and leaching method of Stanford and Smith [177] is generally considered to provide the best estimate of PMN [178]. The method is time-consuming, however, and may not accurately reflect the situation in the field, or the true availability of N to plants (which can also assimilate organic N and whose N assimilation depends on complex rhizospheric interactions, for example). An urgent need remains therefore, to improve understanding of soil N cycling, particularly in relation to organic N, the transfer of N between the inorganic and organic N pools and the N suppling capacity of soils.

1.4.3 Compound-Specific ^{15}N Stable Isotope Probing of Soil Organic Nitrogen

Compound-specific ^{15}N stable isotope probing (^{15}N-SIP) of soil organic N does enable the biomolecular fate of applied N to be traced into particular organic N pools and a variety of approaches could be considered to fit under this banner. Analysis of the N isotopic composition of enriched AAs has been carried out on individually isolated and degraded AAs since Schoenheimer et al. [179] recognised their potential to study animal protein metabolism in the late 1930s [179–181]. Measurement precision of the N_2 gas produced by compound degradation/combustion improved with instrument development (e.g. [182, 183]) and on-line techniques emerged, [184, 185] with the first involving on-line separation of individual compounds for N isotopic measurements, isotope-ratio-monitoring gas chromatography-mass spectrometry (IRM-GC-MS) being reported by Matthews and Hayes in [184]. Barrie et al.'s [186] paper describes the first true gas chromatography-combustion-isotope ratio mass spectrometry (GC-C-IRMS) system [187], while Merritt and Hayes [188] developed the first GC-C-IRMS system for compound-specific N isotopic analyses.

Compound-specific gas chromatography-mass spectrometry (GC-MS) based isotope ratio techniques, developed from the early method by Sweeley et al. [185] are still in use, (e.g. He et al. [189], for ^{15}N-enriched amino sugars; Geisseler and Horwath [190], for microbial AA utilisation; Nelson et al. [191], investigating barley leaf proteins with high turnover rates) but can generally only be used precisely (± 0.01 atom %) with highly enriched compounds where enrichment is easily detectable above natural background values [187, 192, 193]. The first successful 'IRM-GC-MS' N isotopic measurements of AAs at natural abundance levels were reported in 1991 [188] and the value of quickly improving GC-C-IRMS systems for precise and accurate analyses (0.5–2.0‰; 0.0002–0.0008 atom %) was soon advocated [188, 193–195]. Further developments have included new/improved derivatisation methods for AAs, an ever-increasing range of applications exploiting the ability of GC-C-IRMS to measure AA δ^{15}N values at natural abundance levels and new methods for N isotopic analyses of other compounds (Table 1.2).

Only a handful of workers, however, have exploited the high selectivity and sensitivity of GC-C-IRMS [187, 219, 220] in compound-specific ^{15}N-SIP studies and almost half of these studies relate to mammalian physiology (e.g. [221–228]). The remainder consist of a few reports in several different research areas: N uptake in aquatic systems (e.g. [229–232]); plant N uptake (e.g. [233, 234]); plant-microbe associations (e.g. [235]); microbial cultures (e.g. [236]); and soil N partitioning (e.g. [150, 237, 238]).

Table 1.2 Some examples of developments in the use of GC-C-IRMS for N isotopic analyses

Development	References
New AA derivatisation techniques	
N-acetyl, O-n-propyl (NAP); and N-pivaloyl, O-isopropyl (NPP) AA esters	Metges et al. [193]
N-trifluoroacetyl (TFA), O-isopropyl (IP) AA esters	Macko et al. [196]
N(O)-dimethyl-t-butylsilyl; N-trifluoroacetyl, O-isopropyl; N-acetyl, O-isopropyl; and N-pivaloyl, O-isopropyl AA esters	Hofmann et al. [197]
N-pivaloyl, O-(R)-2-butyl; and N-pivaloyl, O-(S)-2-butyl AA esters	Takano et al. [198]
N-methoxycarbonyl, O-methyl AA esters	Walsh et al. [199]
New applications for natural abundance AA $\delta^{15}N$ values	
Ecotoxicological applications—hydrolysable wheat AA $\delta^{15}N$ values under ozone exposure	Hofmann et al. [200]
Free AAs in human blood plasma	Metges and Petzke [201]
Indicators of early land use practices	Simpson et al. [202, 203]
Soil AA $\delta^{15}N$ values	Bol et al. [204–206]
Hair stable isotope ratios for nutritional and metabolic status	Petzke et al. [207–209]
Modern and archaeological bone collagen and plant tissues	Styring et al. [210, 211]
Discrimination between organically and conventionally grown wheat	Paolini et al. [212]
Symbiotic trophic relationships	Sabadel et al. [213]
New methods for other compounds	
Authentication of mandarin essential oils	Faulhaber et al. [214]
3,4-methylenedioxymethamphetamine source discrimination	Palhol et al. [215, 216]
For tetrapyrroles	Chikaraishi et al. [217]
For N-containing intact polar lipids	Svensson et al. [218]

This extremely powerful tool to trace the fate of ^{15}N, applied at environmentally relevant concentrations and appropriately low enrichments, through a variety of ecosystems and into a range of N-containing products is therefore very much underexploited and the full range of applications and insights that can be gained unrealised. This may be partly due to the challenges associated with compound-specific ^{15}N analyses via GC-C-IRMS, as compared with those of ^{13}C (N is generally much less abundant than C in organic molecules; 2 N atoms are required to produce each N_2 molecule for analysis; additional reduction chemistry is required to successfully convert an N-containing molecule to N_2 for analysis; the ionisation efficiency of N_2 is only 70% that of CO_2; small leaks can be detrimental due to the high abundance of N_2 in air; and there is potential for interfering ionic species, such as $CO^{(+)}$, at m/z 28, 29 and 30; [194, 195]), which can make the technique somewhat temperamental, but may also be due to a lack of awareness regarding the potential of compound-specific ^{15}N-SIP using GC-C-IRMS.

A different and relatively recently developed compound-specific ^{15}N-SIP technique that can also be applied to investigate active microbial ecosystems is deoxyribonucleic acid (DNA) or ribonucleic acid (RNA) ^{15}N-SIP [239–242]. This culture-independent method employs isopycnic centrifugation to separate ^{15}N-enriched and un-enriched nucleic acids on the basis of the increased buoyant density ^{15}N-enrichment induces [240–242]. The enriched nucleic acids can then be amplified and analysed via a variety of fingerprinting methods (e.g. polymerase chain reaction; PCR and terminal restriction fragment length polymorphism; TRFLP or denaturing gradient gel electrophoresis; DGGE; [240, 242]). Nucleic acid ^{15}N-SIP allows ^{15}N substrates to be traced into the nucleic acids of actively assimilating microorganisms and can enable the natural environmental metabolic activity of these organisms to be linked with their taxonomic identity [240, 241]. The technique has been used in a range of applications, some examples of which are shown in Table 1.3.

As with ^{15}N-SIP using GC-C-IRMS, nucleic acid ^{15}N-SIP is also more challenging than the corresponding ^{13}C-SIP technique due to the lower N content of nucleic acids and the thus smaller potential increase in buoyant density and the higher ^{15}N enrichments required (>50 atom %; [240–242]). In addition, the potential for cross-feeding and trophic cascades to cause ^{15}N-enrichment in non-target nucleic acids should be considered, especially where long incubations are required due to slow growth rates and it may, in such cases, be useful to track the incorporation of [15]N

Table 1.3 Some examples of applications of nucleic acid ^{15}N-SIP

Application	References
To identify novel uncultivated diazotrophs in soils	Buckley et al. [243]
To identify novel uncultivated diazotrophic methanotrophs in soils	Buckley et al. [244]
To identify microorganisms that actively degrade the explosive contaminant, hexahydro-1,3,5-trinitro-1,3,5-triazine (RDX)	Roh et al. [245], Andeer et al. [246], Jayamani et al. [247], Cho et al. [248]
To investigate the assimilation of $^{15}NO_3^-$, $^{15}NH_4^+$, ^{15}N-urea, ^{15}N-glutamate or ^{15}N-amino acids by *Synechococcus* spp. and diatoms from a marine ecosystem	Wawrik et al. [249]
To assess the decomposition of organic residues by identified soil fungi and bacteria	España et al. [250, 251]
To identify bacteria utilising added N in petroleum contaminated arctic soils	Bell et al. [252]
To evaluate the autotrophic growth of ammonia oxidising bacteria (AOB) and archaea (AOA) at different NH_4^+ concentrations	Niu et al. [253]
To assess the effects of long-term fertilisation on soil AOB and AOA	Wang et al. [254]

through food webs with time [240, 243, 244]. Finally, it can also be difficult to absolutely identify the organisms that assimilated the ^{15}N label by matching terminal restriction fragments (TRFs; which may be shared by multiple organisms) with sequences from clone libraries and careful comparison of enriched and control samples is essential [240, 243].

In summary, compound-specific ^{15}N-SIP using GC-C-IRMS and nucleic acid ^{15}N-SIP are contrasting, complementary approaches that both provide valuable, but entirely disparate insights in the investigation of active microbial ecosystems, including soils. While compound-specific ^{15}N-SIP using GC-C-IRMS affords insights into the active N cycling of the whole community and the biomolecular fate of applied ^{15}N, nucleic acid ^{15}N-SIP can be used to identify the members of the community most responsible for this processing. Both techniques are, however, relatively complex and somewhat time-consuming and, partly as a result, have not been applied in conjunction.

1.5   Approach and Aims

The need to reduce atmospheric, terrestrial and freshwater N_r pollution is clear. As a major contributor to freshwater N_r contamination globally, agriculture represents a key target industry, in which improvements in NUE will not only reduce N_r pollution, benefiting the environment, but could also play an important role in achieving food security and improving the balance between food and water security. One of the challenges to progress in increasing agricultural NUE and decreasing environmental N_r losses, however, remains land owner engagement, uptake and (successful) implementation of the most up-to-date best management advice (Sect. 1.3). This is often due to the low (or even negative) short-term economic benefit of adopting recommended practices [78], many of which require additional time and money in planning and implementation, but save little in direct costs (e.g. via reduced fertiliser inputs) and do not necessarily boost yields comparably.

Since, however, the estimated costs of N_r pollution in the European Union (EU) alone lie between €70 billion and €320 billion per year [255], more than double the estimated economic benefit of synthetic nitrogenous fertilisers to EU farm incomes [21], the economic argument for reducing N_r pollution must be felt somewhere. In the case of diffuse freshwater NO_3^- contamination in the UK, much of the rising cost of dealing with this pollution falls on water companies and, by extension, on their customers. This combination of rising treatment requirements, costs, environmental impacts and the need for more sustainable N_r use (Sect. 1.1) has led to the adoption and development of a 'catchment management' approach to diffuse NO_3^- pollution by several UK water companies, with the support of local groups and other organisations including non-governmental organisations (NGOs), regulators and government. Catchment management is a more sustainable, pre-emptive strategy in which water companies work with land owners to identify and reduce NO_3^- pollution sources

[36]. Through catchment management programmes water companies raise awareness and understanding and provide best management advice, data and analytical services (e.g. soil and manure sampling and analysis) to improve soil and nutrient management plans, fertiliser spreader calibrations, assistance with agri-environment Countryside Stewardship applications and other financial support.

One such water company is Wessex Water, which supplies water and sewerage services to 2.8 million people across several counties in south-west England. The work presented in this thesis is the result of a Natural Environment Research Council (NERC) Open CASE Ph.D. Studentship between the University of Bristol and Wessex Water, supporting Wessex Water's catchment management strategy. The overarching aim of the project is to contribute further scientific understanding to the NO_3^- pollution problem which will ultimately support the development of evidence-based mitigation strategies.

As identified in Sects. 1.3 and 1.4.3, closing knowledge gaps in soil N cycling and particularly the transfer of N between the inorganic and organic N pools in soil and the N supplying capacity of soils would be extremely valuable in improving soil management practices for reduced N_r losses and better NUEs. The work presented in this thesis therefore focuses on investigating the microbial transformations of applied N fertilisers into the soil organic N pool at a biomolecular level in laboratory and field experiments and the partitioning and transport of NO_3^- in a field setting using ^{15}N stable isotope tracers. The power of ^{15}N stable isotope tracing and more specifically, compound-specific ^{15}N-SIP in tracing the biomolecular fate of fertiliser N is reviewed in Chap. 3, alongside the relevant newly developed methodology for this work. Short reviews of other relevant literature, including microbial N uptake and assimilation (Sect. 4.1.2), fertiliser immobilisation (Sect. 4.1.3), the problem of scale (Sect. 5.1), plant N uptake and assimilation (Sect. 6.1.2) and the competition for N between plants and microbes (Sect. 6.1.3) are also included in the introductions of the most applicable chapters.

The specific objectives of the work were to:

1. Develop a novel compound-specific ^{15}N-SIP approach to assess the microbial assimilation and biomolecular fate of N in soils.
2. Use this method to assess the biosynthetic routing, rates and extents of the microbial assimilation of a range of common fertiliser N forms in laboratory mesocosm experiments using two grazed grassland soils.
3. Assess whether/to what extent the microbial assimilation results of laboratory mesocosm experiments are replicated in a field environment.
4. Trace the spatial and temporal partitioning and transport of fertiliser NO_3^- in a field environment.

Two grazed grassland soils were selected for the laboratory experiments as one of these soils had previously been used in similar compound-specific ^{15}N-SIP work by a former Ph.D. student [150, 256] with which direct comparison was deemed valuable, while the other was sampled from the field adjacent to a Wessex Water public water supply (PWS) source which suffers from high and rising NO_3^- contamination (see Sect. 2.2 for more site and soil information). In addition, grasslands are globally important ecosystems covering approximately 40% of the world's land surface [257] and grazed grasslands are susceptible to high NO_3^- leaching losses due to the exceedingly concentrated patches of N generated by the random re-deposition of 85–95% of ingested N in livestock excreta [41]. The field experiment was then also carried out at the latter site for comparability.

References

1. Durand P, Breuer L, Johnes PJ, Billen G, Butturini A, Pinay G, van Grinsven H, Garnier J, Rivett M, Reay DS, Curtis C, Siemens J, Maberly S, Kaste Ø, Humborg C, Loeb R, de Klein J, Hejzlar J, Skoulikidis N, Kortelainen P, Lepistö A, Wright R (2011) Nitrogen processes in aquatic ecosystems. In: Sutton MA, Howard CM, Erisman JW, Billen G, Bleeker A, Grennfelt P, van Grinsven H, Grizzetti B (eds) The European nitrogen assessment: sources, effects and policy perspectives. Cambridge University Press, UK, pp 126–146
2. Vitousek PM, Mooney HA, Lubchenco J, Melilla JM (1997) Human domination of earth's ecosystems. Science 277:494–499. https://doi.org/10.1126/science.277.5325.494
3. Grizzetti B, Bouraoui F, Billen G, van Grinsven H, Cardoso AC, Thieu V, Garnier J, Curtis C, Howarth R, Johnes P (2011) Nitrogen as a threat to European water quality. In: Sutton MA, Howard CM, Erisman JW, Billen G, Bleeker A, Grennfelt P, van Grinsven H, Grizzetti B (eds) The European nitrogen assessment: sources, effects and policy perspectives. Cambridge University Press, UK, pp 379–404
4. Refsgaard JC, Thorsen M, Jensen JB, Kleeschulte S, Hansen S (1999) Large scale modelling of groundwater contamination from nitrate leaching. J Hydrol 221:117–140
5. Galloway JN, Aber JD, Erisman JW, Seitzinger SP, Howarth RW, Cowling EB, Cosby BJ (2003) The nitrogen cascade. Bioscience 53:341–356
6. Galloway JN, Dentener FJ, Capone DG, Boyer EW, Howarth RW, Seitzinger SP, Asner GP, Cleveland CC, Green PA, Holland EA, Karl DM, Michaels AF, Porter JH, Townsend AR, Vörösmarty CJ (2004) Nitrogen cycles: past, preserichnt, and future. Biogeochemistry 70:153–226
7. Howden NJK, Burt TP, Worrall F, Mathias SA, Whelan MJ (2013) Farming for water quality: balancing food security and nitrate pollution in UK river basins. Ann Assoc Am Geogr 103:397–407. https://doi.org/10.1080/00045608.2013.754672
8. Smith VH, Tilman GD, Nekola JC (1999) Eutrophication: impacts of excess nutrient inputs on freshwater, marine, and terrestrial ecosystems. Environ Pollut 100:179–196
9. Spalding RF, Exner ME (1993) Occurrence of nitrate in groundwater—a review. J Environ Qual 22:392–402
10. Vitousek PM, Aber JD, Howarth RW, Likens GE, Matson PA, Schindler DW, Schlesinger WH, Tilman DG (1997) Human alteration of the global carbon cycle: sources and consequences. Ecol Appl 7:737–750
11. WHO (2011) Guidelines for drinking-water quality, 4th edn. WHO Press, Switzerland
12. Ward MH, deKok TM, Levallois P, Brender J, Gulis G, Nolan BT, VanDerslice J (2005) Workgroup report: drinking-water nitrate and health—recent findings and research needs. Environ Health Perspect 113:1607–1614
13. Powlson DS, Addiscott TM, Benjamin N, Cassman KG, de Kok TM, van Grinsven H, L'hirondel J-L, Avery AA, Van Kessel C (2008) When does nitrate become a risk for humans?. J Environ Qual 37:291–295. https://doi.org/10.2134/jeq2007.0177
14. Erisman JW, Sutton MA, Galloway J, Klimont Z, Winiwarter W (2008) How a century of ammonia synthesis changed the world. Nat Geosci 1:636–639

15. Fowler D, Coyle M, Skiba U, Sutton MA, Cape JN, Reis S, Sheppard LJ, Jenkins A, Grizzetti B, Galloway JN, Vitousek P, Leach A, Bouwman AF, Butterbach-Bahl K, Dentener F, Stevenson D, Amann M, Voss M (2013) The global nitrogen cycle in the twenty-first century. Philos Trans R Soc B 368:20130164

16. Galloway JN (1998) The global nitrogen cycle: changes and consequences. Environ Pollut 102:15–24

17. Galloway JN, Townsend AR, Erisman JW, Bekunda M, Cai Z, Freney JR, Martinelli LA, Seitzinger SP, Sutton MA (2008) Transformation of the nitrogen cycle: recent trends, questions, and potential solutions. Science 320:889–892

18. Nosengo N (2003) Fertilized to death. Nature 425:894–895

19. Paul EA (2007) Soil microbiology, ecology and biochemistry, 3rd edn. Academic Press, UK

20. Rosswall T (1982) Microbiological regulation of the biogeochemical nitrogen cycle. Plant Soil 67:15–34

21. Sutton MA, Oenema O, Erisman JW, Leip A, van Grinsven H, Winiwarter W (2011) Too much of a good thing. Nature 427:159–161

22. Redmile-Gordon MA, Armenise E, Hirsch PR, Brookes PC (2014) Biodiesel co-product (BCP) decreases soil nitrogen (N) losses to groundwater. Water, Air Soil Pollut 225. https://doi.org/10.1007/s11270-013-1831-7

23. Erisman JW, van Grinsven H, Grizzetti B, Bouraoui F, Powlson D, Sutton MA, Bleeker A, Reis S (2011) The European nitrogen problem in a global perspective. In: Sutton MA, Howard CM, Erisman JW, Billen G, Bleeker A, Grennfelt P, van Grinsven H, Grizzetti B (eds) The European nitrogen assessment: sources, effects and policy perspectives. Cambridge University Press, UK, pp 9–31

24. von Liebig J (1855) Principles of agricultural chemistry with special reference to the late researches made in England (ed Gregory W), Walton & Maberly, London, UK

25. Smil V (2002) Nitrogen and food production: proteins for human diets. Ambio 31:126–131

26. Vitousek PM, Naylor R, Crews T, David MB, Drinkwater LE, Holland E, Johnes PJ, Ketzenberger J, Martinelli LA, Matson PA, Nziguheba G, Ojima D, Palm CA, Robertson GP, Sanchez PA, Townsend AR, Zhang FS (2009) Nutrient imbalances in agricultural development. Science 324:1519–1520

27. FAO, IFAD, WFP (2015) The state of food insecurity in the world 2015. Meeting the 2015 international hunger targets: taking stock of uneven progress, FAO, Rome, Italy

28. Leach AM, Galloway JN, Bleeker A, Erisman JW, Kohn R, Kitzes J (2012) A nitrogen footprint model to help consumers understand their role in nitrogen losses to the environment. Environ Dev 1:40–66

29. Godfrey HCJ, Beddington JR, Crute IR, Haddad L, Lawrence D, Muir JF, Pretty J, Robinson S, Thomas SM, Toulmin C (2010) Food security: the challenge of feeding 9 billion people. Science 327:812–818. https://doi.org/10.1126/science.1185383

30. Mueller ND, Gerber JS, Johnston M, Ray DK, Ramankutty N, Foley JA (2012) Closing yield gaps through nutrient and water management. Nature 490:254–257. https://doi.org/10.1038/nature11420

31. Sánchez PA (2010) Tripling crop yields in tropical Africa. Nat Geosci 3:299–300

32. Tscharntke T, Clough Y, Wanger TC, Jackson L, Motzke I, Perfecto I, Vandermeer J, Whitbread A (2012) Global food security, biodiversity conservation and the future of agricultural intensification. Biol Cons 151:53–59

33. Defra (2016) The British survey of fertiliser practice 2015: statistical notice. Report available from https://www.gov.uk/government/collections/fertiliser-usage. Accessed 27 Jan 2017

34. Carvalho FP (2006) Agriculture, pesticides, food security and food safety. Environ Sci Policy 9:685–692

35. Dalin C, Qiu H, Hanasaki N, Mauzerall DL, Rodriguez-Iturbe I (2015) Balancing water resource conservation and food security in China. Proc Natl Acad Sci USA. https://doi.org/10.1073/pnas.1504345112

36. de Vial L, Bowles F, Dennis PJ (2010) Protecting water resources and health by protecting the environment: a case study. Issues Environ Sci Technol 30:122–139. https://doi.org/10.1039/9781849731058-00122

37. Gong Y-K, Peng Y-Z, Yang Q, Wu W-M, Wang S-Y (2012) Formation of nitrous oxide in a gradient of oxygenation and nitrogen loading rate during denitrification of nitrite and nitrate. J Hazard Mater 227–228:453–460
38. Cassman KG, Doberman A, Walters DT (2002) Agroecosystems, nitrogen-use efficiency, and nitrogen management. Ambio 31:132–140
39. Di HJ, Cameron KC (2002) Nitrate leaching in temperate agroecosystems: sources, factors and mitigating strategies. Nutr Cycl Agroecosyst 46:237–256
40. Goulding K, Jarvis S, Whitmore A (2008) Optimizing nutrient management for farm systems. Philos Trans R Soc B 363:667–680. https://doi.org/10.1098/rstb.2007.2177
41. Cameron KC, Di HJ, Moir JL (2013) Nitrogen losses from the soil/plant system: a review. Ann Appl Biol 126:145–173. https://doi.org/10.1111/aab.12014
42. Dungait JAJ, Cardenas LM, Blackwell MSA, Wu L, Withers PJA, Chadwick DR, Bol R, Murray PJ, Macdonald AJ, Whitmore AP, Goulding KWT (2012) Advances in the understanding of nutrient dynamics and management in UK agriculture. Sci Total Environ 434:39–50
43. Nila Rekha P, Kanwar RS, Nayak AK, Hoang CK, Pederson CH (2011) Nitrate leaching to shallow groundwater systems from agricultural fields with different management practices. J Environ Monit 13:2550–2558
44. Armstrong AC, Leeds-Harrison PB, Harris GL, Catt JA (1999) Measurement of solute fluxes in microporous soils: techniques, problems and precision. Soil Use Manag 15:240–246
45. Bachmair S, Weiler M, Nützmann G (2009) Controls of land use and soil structure on water movement: lessons for pollutant transfer through the unsaturated zone. J Hydrol 369:241–252
46. Jarvis NJ (2007) A review of non-equilibrium water flow and solute transport in soil macropores: principles, controlling factors and consequences for water quality. Eur J Soil Sci 58:523–546
47. Gazis C, Feng X (2004) A stable isotope study of soil water: evidence for mixing and preferential flow paths. Geoderma 119(1–2):97–111
48. Asano Y, Compton JE, Robbins Church M (2006) Hydrologic flow paths influence inorganic and organic nutrient leaching in a forest soil. Biochemistry 81:191–204
49. Barraclough D, Gardner CMK, Wellings SR, Cooper JD (1994) A tracer investigation into the importance of fissure flow in the unsaturated zone of the British Upper Chalk. J Hydrol 156:459–469
50. Sudduth EB, Perakis SS, Bernhardt ES (2013) Nitrate in watersheds: straight from soils to streams? J Geophys Res: Biogeosciences 118:291–302
51. Dentener F, Drevet J, Lamarque JF, Bey I, Eickhout B, Fiore AM, Hauglustaine D, Horowitz LW, Krol M, Kulshrestha UC, Lawrence M, Gay-Lacaux C, Rast S, Shindell D, Stevenson D, Van Noije T, Atherton C, Bell N, Bergman D, Butler T, Cofala J, Collins B, Doherty R, Ellingsen K, Galloway J, Gauss M, Montanaro V, Muller JF, Pitari G, Rodrigues J, Sanderson M, Solmon F, Strahan S, Schultz MSK, Szopa, S, Wild O (2006) Nitrogen and sulfur deposition on regional and global scales: a multimodel evaluation. Glob Biogeochem Cycles 20:GB4003. https://doi.org/10.1029/2005gb002672
52. Barraclough D, Puri G (1995) The use of ^{15}N pool dilution and enrichment to separate the heterotrophic and autotrophic pathways of nitrification. Soil Biol Biochem 27:17–22
53. Tahovská K, Kaňa J, Bárta J, Oulehle F, Richter A, Šantrůčková H (2013) Microbial N immobilization is of great importance in acidified mountain spruce forest soils. Soil Biol Biochem 59:58–71
54. Geisseler D, Horwath WR, Joergensen RG, Ludwig B (2010) Pathways of nitrogen utilization by soil microorganisms—a review. Soil Biol Biochem 42:2058–2067. https://doi.org/10.1016/j.soilbio.2010.08.021
55. Jackson LE, Schimel JP, Firestone MK (1989) Short-term partitioning of ammonium and nitrate between plants and microbes in an annual grassland. Soil Biol Biochem 21:409–415
56. Recous S, Mary B, Faurie G (1990) Microbial immobilisation of ammonium and nitrate in cultivated soils. Soil Biol Biochem 22:913–922
57. Recous S, Machet JM, Mary B (1992) The partitioning of fertilizer-N between soil and crop: comparison of ammonium and nitrate applications. Plant Soil 144:101–111

58. Rice CW, Tiedje JM (1989) Regulation of nitrate assimilation by ammonium in soils and in isolated soil microorganisms. Soil Biol Biochem 21:597–602

59. Giles M, Morley N, Baggs EM, Daniell TJ (2012) Soil nitrate reducing processes—drivers, mechanisms for spatial variation, and significance for nitrous oxide production. Front Microbiol 3. https://doi.org/10.3389/fmicb.2012.00407

60. Thangarajan R, Bolan NS, Tian G, Naidu R, Kunhikrishnan A (2013) Role of organic amendment application on greenhouse gas emission from soil. Sci Total Environ 465:72–96. https://doi.org/10.1016/j.scitotenv.2013.01.031

61. Almasri MN, Kaluarachichi JJ (2004) Assessment and management of long-term nitrate pollution of groundwater in agriculture-dominated watersheds. J Hydrol 295:225–245

62. Mariotti A, Landreau A, Simon B (1988) N isotope biochemistry and natural denitrification process in groundwater: application to the chalk aquifer of northern France. Geochim Cosmochim Acta 52:1869–1878

63. Thomas S, Waterland H, Dann R, Close M, Francis G, Cook F (2012) Nitrous oxide dynamics in a deep soil-alluvial gravel vadose zone following nitrate leaching. Soil Sci Soc Am J 76:1333–1346

64. Vogel JC, Talma AS, Heaton THE (1981) Gaseous nitrogen as evidence for identification in groundwater. J Hydrol 50:191–200

65. Yuan L, Pang Z, Huang T (2012) Integrated assessment on groundwater nitrate by unsaturated zone probing and aquifer sampling with environmental tracers. Environ Pollut 171:226–233

66. Goulding KWT, Poulton PR, Webster CP, Howe MT (2000) Nitrate leaching from the broadbalk wheat experiment, Rothamsted, UK, as influenced by fertilizer and manure inputs and the weather. Soil Use and Management 16:244–250

67. Goulding K (2000) Nitrate leaching from arable and horticultural land. Soil Use Manag 16:145–151

68. Granger SJ, Heaton THE, Bol R, Bilotta GS, Bulter P, Haygarth PM, Owens PN (2008) Using $\delta^{15}N$ and $\delta^{18}O$ to evaluate the sources and pathways of NO_3^- in rainfall event discharge from drained agricultural grassland lysimeters at high temporal resolutions. Rapid Commun Mass Spectrom 22:1681–1689

69. Deutsch B, Liskow I, Kahle P, Voss M (2005) Variations in the $\delta^{15}N$ and $\delta^{18}O$ values of nitrate in drainage water of two fertilized fields in Mecklenburg-Vorpommern (Germany). Aquat Sci 67:156–165

70. Harremoës P (1977) Introduction to the proceedings of the conference on nitrogen as a water pollutant, Copenhagen, August 18–20 1975. Prog Water Technol 8:1–2

71. Lehr JH (1971) Forward to the proceedings of the national ground water quality symposium. Denver, Colorado, 25–27 Aug

72. Young CP, Oakes DB, Wilkinson WB (1976) Prediction of future nitrate concentrations in ground water. Ground Water 14:426–438

73. Smith DB, Wearn PL, Richards HJ, Rowe PC (1970) Water movement in the unsaturated zone of high and low permeability strata by measuring natural tritium, *Isotope Hydrology*. In: Proceedings of a symposium on the use of isotopes in hydrology held by the international atomic energy agency in co-operation with the United Nations education and scientific and cultural organisation in Vienna 9–13 Mar 1970, pp 73–87

74. Nielsen DR, Biggar JW, Wierenga PJ (1982) Nitrogen transport processes in soil. In: Stevenson FJ (ed) Nitrogen in agricultural soils. American Society of Agronomy, USA, pp 423–448

75. Allison FE (1955) The enigma of soil nitrogen balance sheets. Adv Agron 7:213–250

76. Stanford G (1982) Assessment of soil nitrogen availability. In: Stevenson FJ (ed) Nitrogen in agricultural soils. American Society of Agronomy, USA, pp 651–688

77. Keeney DR (1982) Nitrogen management for maximum efficiency and minimum pollution. In: Stevenson FJ (ed) Nitrogen in agricultural soils. American Society of Agronomy, USA, pp 605–649

78. Dobermann A, Cassman KC (2004) Environmental dimensions of fertilizer nitrogen: what can be done to increase nitrogen use efficiency and ensure global food security? In: Mosier AR, Syers JK, Freney JR (eds) Agriculture and the nitrogen cycle: assessing the impacts of fertilizer use on food production and the environment, SCOPE 65, Island Press, Washington, DC, USA pp 261–278
79. Gooday R, Anthony S, Fawcett L (2008) A field scale model of soil drainage and nitrate leaching for application in nitrate vulnerable zones. Environ Model Softw 23:1045–1055
80. Vogel H-J, Cousin I, Ippisch O, Bastian P (2006) The dominant role of structure for solute transport in soil: experimental evidence and modelling of structure and transport in a field experiment. Hydrol Earth Syst Sci 10:495–506
81. Defra (2009) Protecting our water, soil and air: a code of good agricultural practice for farmers, growers and land managers. The Stationary Office, UK
82. Defra (2010) Fertiliser manual (RB209), p 60, 65. Report available from http://www.ahdb.org.uk/projects/CropNutrition.aspx. Accessed 26 Mar 2014
83. Smith JU, Bradbury NJ, Addiscott TM (1996) SUNDIAL: a PC-based system for simulating nitrogen dynamics in arable land. Agron J 88:38–43. https://doi.org/10.2134/agronj1996.00021962008800010008x
84. Brown L, Scholefield D, Jewkes EC, Lockyer DR, del Prado A (2005) NGAUGE: a decision support system to optimise N fertilisation of British grassland for economic and environmental goals. Agr Ecosyst Environ 109:20–39. https://doi.org/10.1016/j.agee.2005.02.021
85. Garnett T, Conn V, Plett D, Conn S, Zanghellini J, Mackenzie N, Enju A, Francis K, Holtham L, Roessner U, Boughton B, Bacic A, Shirley N, Rafalski A, Dhugga K, Tester M, Kaiser BN (2013) The response of the maize nitrate transport system to nitrogen demand and supply across the lifecycle. New Phytol. https://doi.org/10.1111/nph.12166
86. Hirel B, Le Gouis J, Ney B, Gallais A (2007) The challenge of improving nitrogen use efficiency in crop plants: towards a more central role for genetic variability and quantitative genetics within integrated approaches. J Exp Bot 58:2369–2387. https://doi.org/10.1093/jxb/erm097
87. Rasmussen PE, Keith WT, Brown JR, Grace PR, Janzen HH, Körschens M (1998) Long-term agroecosystem experiments: assessing agricultural sustainability and global change. Science 282:893–896
88. Gärdenäs AI, Ågren GI, Bird JA, Clarholm M, Hallin S, Ineson P, Kätterer T, Knicker H, Nilsson SI, Näsholm T, Ogle S, Paustian K, Persson T, Stendahl J (2011) Knowledge gaps in soil carbon and nitrogen interactions—from molecular to global scale. Soil Biol Biochem 43:702–717. https://doi.org/10.1016/j.soilbio.2010.04.006
89. Barraclough D (1995) ^{15}N isotope dilution techniques to study soil nitrogen transformations and plant uptake. Fertiliser Research 42:185–192
90. Barraclough D (1997) The direct or MIT route for nitrogen immobilisation: a ^{15}N mirror image study with leucine and glycine. Soil Biol Biochem 29:101–108
91. Booth MS, Stark JM, Rastetter E (2005) Controls on nitrogen cycling in terrestrial ecosystems: a synthetic analysis of literature data. Ecol Monogr 75:139–157
92. Davidson EA, Hart SC, Shanks CA, Firestone MK (1991) Measuring gross nitrogen mineralisation, immobilisation, and nitrification by ^{15}N isotopic pool dilution in intact soil cores. J Soil Sci 42:335–349
93. Jansson SL, Persson J (1982) Mineralisation and immobilisation of soil nitrogen. In: Stevenson FJ (ed) Nitrogen in agricultural soils. American Society of Agronomy, USA, pp 229–252
94. Kirkham D, Bartholomew WV (1955) Equations for following nitrogen transformations in soil utilizing tracer data: II. Soil Sci Soc Am Proc 19:189–192. https://doi.org/10.2136/sssaj1955.03615995001900020020x
95. Bjarnason S (1988) Calculation of gross nitrogen immobilization and mineralization in soil. J Soil Sci 39(3):393–406
96. Brand WA, Coplen TB (2012) Stable isotope deltas: tiny, yet robust signatures in nature. Isot Environ Health Stud 48:393–409

97. Brand WA, Coplen TB, Vogl J, Rosner M, Prohaska T (2014) Assessment of international reference materials for isotope-ratio analysis (IUPAC technical report). Pure Appl Chem 86:425–467. https://doi.org/10.1515/pac-2013-1023
98. Hoering T (1955) Variation of nitrogen-15 abundance in naturally occurring substances. Science 122:1233–1234
99. Högberg P (1997) ^{15}N natural abundance in soil-plant systems. New Phytol 137:179–203
100. Urey HC (1947) The thermodynamic properties of isotopic substances, Liversidge Lecture, Delivered before the Chemical Society in the Royal Institution on December 18th, 1946, pp 562–581
101. Coplen TB, Hopple JA, Böhlke JK, Peiser HS, Rieder SE, Krouse HR, Rosman KJR, Ding T, Vocke RD Jr, Révész KM, Lamberty A, Taylor P, De Bièvre P (2002) Compilation of minimum and maximum isotope ratios of selected elements in naturally occurring terrestrial materials and reagents. U.S. Geological Survey Water Resources Investigations Report 01-4222, U.S. Geological Survey, Denver, USA
102. Sulzman EW (2007) Stable isotope chemistry and measurement: a primer. In: Michener R, Lajtha K (eds) Stable isotopes in ecology and environmental science, 2nd edn. Ecological methods and concepts series. Blackwell Publishing, UK, pp 1–21
103. Parnell AC, Inger R, Bearhop S, Jackson AL (2010) Source partitioning using stable isotopes: coping with too much variation. PLoS ONE 5:e9672. https://doi.org/10.1371/journal.pone.0009672
104. Spoelstra J, Schiff SL, Elgood RJ, Semkin RG, Jeffries DS (2001) Tracing the sources of exported nitrate in the Turkey Lakes watershed using ^{15}N/^{14}N and ^{18}O/^{16}O isotopic ratios. Ecosystems 4:536–544
105. Mayer B, Boyer EW, Goodale C, Jaworski NA, van Breemen N, Howarth RW, Seitzinger S, Billen G, Lajtha K, Nadelhoffer K, van Dam D, Hetling LJ, Nosal M, Paustian K (2002) Sources of nitrate in rivers draining sixteen watersheds in the northeastern U.S.: isotopic constraints. Biogeochemistry 57:171–197
106. Kellman LM (2005) A study of tile drain nitrate—δ^{15}N values as a tool for assessing nitrate sources in an agricultural region. Nutr Cycl Agroecosyst 71:131–137. https://doi.org/10.1007/s10705-004-1925-0
107. Moore KB, Ekwurzel B, Esser BK, Hudson GB, Moran JE (2006) Sources of groundwater nitrate revealed using residence time and isotope methods. Appl Geochem 21:1016–1029
108. Baily A, Rock L, Watson CJ, Fenton O (2011) Spatial and temporal variations in groundwater nitrate at an intensive dairy farm in south-east Ireland: insights from stable isotope data. Agr Ecosyst Environ 144:308–318
109. Kelley CJ, Keller CK, Evans RD, Orr CH, Smith JL, Harlow BA (2013) Nitrate-nitrogen and oxygen isotope ratios for identification of nitrate sources and dominant nitrogen cycle processes in a tile-drained dryland agricultural field. Soil Biol Biochem 57:731–738
110. Pastén-Zapata E, Ledesma-Ruiz R, Harter T, Ramírez AI, Mahlknecht J (2014) Assessment of sources and fate of nitrate in shallow groundwater of an agricultural area by using a multi-tracer approach. Sci Total Environ 470–471:855–864. https://doi.org/10.1016/j.scitotenv.2013.10.043
111. Ostrom NE, Knoke KE, Hedin LO, Robertson GP, Smucker AJM (1998) Temporal trends in nitrogen isotope values of nitrate leaching from an agricultural soil. Chem Geol 146:219–227
112. Burns DA, Kendall C (2002) Analysis of δ^{15}N and δ^{18}O to differentiate NO_3^- sources in runoff at two watersheds in the Catskill Mountains of New York. Water Resour Res 38:1051. https://doi.org/10.1029/2001WR000292
113. Kellman LM, Hillaire-Marcel C (2003) Evaluation of nitrogen isotopes as indicators of nitrate contamination sources in an agricultural watershed. Agr Ecosyst Environ 95:87–102
114. Seiler RL (2005) Combined use of ^{15}N and ^{18}O of nitrate and ^{11}B to evaluate nitrate contamination in groundwater. Appl Geochem 20:1626–1636
115. Showers WJ, Genna B, McDade T, Bolich R, Fountain JC (2008) Nitrate contamination in groundwater on an urbanized dairy farm. Environ Sci Technol 42:4683–4688

116. Heaton THE, Stuart ME, Sapiano M, Sultana MM (2012) An isotope study of the sources of nitrate in Malta's groundwater. J Hydrol 414–415:244–254

117. Heaton THE (1986) Isotopic studies of nitrogen pollution in the hydrosphere and atmosphere: a review. Chem Geol (Isot Geosci Sect) 59:87–102

118. Kendall C, Elliott EM, Wankel SD (2007) Tracing anthropogenic inputs of nitrogen to ecosystems. In: Michener R, Lajtha K (eds) Stable isotopes in ecology and environmental science, 2nd edn. Ecological methods and concepts series. Blackwell Publishing, UK, pp 375–449

119. Xue D, Botte J, De Baets B, Accoe F, Nestler A, Taylor P, Van Cleemput O, Berglund M, Boeckx P (2009) Present limitations and future prospects of stable isotope methods for nitrate source identification in surface- and groundwater. Water Res 43:1159–1170

120. Bowman WD, Schardt JC, Schmidt SK (1996) Symbiotic N_2-fixation in alpine tundra: ecosystem input and variation in fixation rates amongst communities. Oecologia 108:345–350

121. Cadisch G, Hairiah K, Giller KE (2000) Applicability of the natural ^{15}N abundance technique to measure N_2 fixation in *Arachis hypogaea* grown on an Ultisol. Neth J Agric Sci 48:31–45

122. Arndt SK, Kahmen A, Arampatsis C, Popp M, Adams M (2004) Nitrogen fixation and metabolism by groundwater-dependent perennial plants in a hyperarid desert. Oecologia 141:385–394

123. Binkley D, Sollins P, McGill WB (1985) Natural abundance of nitrogen-15 as a tool for tracing alder-fixed nitrogen. Soil Sci Soc Am J 49:444–447

124. Högberg P (1986) Nitrogen-fixation and nutrient relations in savanna woodland trees (Tanzania). J Appl Ecol 23:675–688

125. Spriggs AC, Stock WD, Dakora FD (2003) Influence of mycorrhizal associations on foliar $\delta^{15}N$ values of legume and non-legume shrubs and trees in the fynbos of South Africa: implications for estimating N_2 fixation using the ^{15}N natural abundance method. Plant Soil 255:495–502

126. Shearer G, Kohl DH (1986) N_2-fixation in field settings: estimations based on natural ^{15}N abundance. Aust J Plant Physiol 13:699–756

127. Shearer G, Kohl DH (1988) Natural ^{15}N abundance as a method of estimating the contribution of biologically fixed nitrogen to N_2-fixing systems: potential for non-legumes. Plant Soil 110:317–327

128. Marshall JD, Brooks JR, Lajtha K (2007) Sources of variation in the stable isotopic composition of plants. In: Michener R, Lajtha K (eds) Stable isotopes in ecology and environmental science, 2nd edn. Ecological methods and concepts series. Blackwell Publishing, UK, pp 22–60

129. Wu D, Köster JR, Cárdenas LM, Brüggemann N, Lewicka-Szczebak D, Bol R (2016) N_2O source partitioning in soils using ^{15}N site preference values corrected for the N_2O reduction effect. Rapid Commun Mass Spectrom 30:620–626. https://doi.org/10.1002/rcm.7493

130. Yoshida N (1988) ^{15}N-depleted N_2O as a product of nitrification. Nature 335:528–529

131. Yoshida N, Toyoda S (2000) Constraining the atmospheric N_2O budget from intramolecular site preference in N_2O isotopomers. Nature 405:330–334

132. Schmidt H-L, Werner RA, Yoshida N, Well R (2004) Is the isotopic composition of nitrous oxide an indicator for its origin from nitrification or denitrification? A theoretical approach from referred data and microbiological and enzyme kinetic aspects. Rapid Commun Mass Spectrom 18:2036–2040. https://doi.org/10.1002/rcm.1586

133. Toyoda S, Mutobe H, Yamagishi H, Yoshida N, Tanji Y (2005) Fractionation of N_2O isotopomers during production by denitrifier. Soil Biol Biochem 37:1535–1545. https://doi.org/10.1016/j.soilbio.2005.01.009

134. Sutka RL, Ostrom NE, Ostrom PH, Breznak JA, Gandhi H, Pitt AJ, Li F (2006) Distinguishing nitrous oxide production from nitrification and denitrification on the basis of isotopomer abundances. Appl Environ Microbiol 72:638–644. https://doi.org/10.1128/AEM.72.1.638-644.2006

135. Cardenas LM, Chadwick D, Scholefield D, Fychan R, Marley CL, Jones R, Bol R, Well R, Vallejo A (2007) The effect of diet manipulation on nitrous oxide and methane emissions from manure application to incubated grassland soils. Atmos Environ 41:7096–7107

136. Ostrom NE, Pitt A, Sutka R, Ostrom PH, Grandy AS, Huizinga KM, Robertson GP (2007) Isotopologue effects during N_2O reduction in soils and in pure cultures of denitrifiers. J Geophys Res 112:G02005. https://doi.org/10.1029/2006JG000287

137. Meijide A, Cardenas LM, Bol R, Bergstermann A, Goulding K, Well R, Vallejo A, Scholefield D (2010) Dual isotope and isotopomer measurements for the understanding of N_2O production and consumption during denitrification in an arable soil. Eur J Soil Sci 61:364–374. https://doi.org/10.1111/j.1365-2389.2010.01233.x

138. Bergstermann A, Cárdenas L, Bol R, Gilliam L, Goulding K, Meijide A, Scholefield D, Vallejo A, Well R (2011) Effects of antecedent soil moisture conditions on emissions and isotopologue distribution of N_2O during denitrification. Soil Biol Biochem 43:240–250

139. Köster JR, Cárdenas LM, Bol R, Lewicka-Szczebak D, Senbayram M, Well R, Giesemann A, Dittert K (2015) Anaerobic digestates lower N_2O emissions compared to cattle slurry by affecting rate and product stoichiometry of denitrification—An N_2O isotopomer case study. Soil Biol Biochem 84:65–74

140. Stein LY, Yung YL (2003) Production, isotopic composition, and atmospheric fate of biologically produced nitrous oxide. Annu Rev Earth Planet Sci 31:329–356. https://doi.org/10.1146/annurev.earth.31.110502.080901

141. Böhlke JK, Denver JM (1995) Combined use of groundwater dating, chemical, and isotopic analysis to resolve the history and fate of nitrate contamination in two agricultural watersheds, Atlantic coastal plain, Maryland. Water Resour Res 31:2319–2339

142. Högberg P (1991) Development of ^{15}N enrichment in a nitrogen-fertilised forest soil-plant system. Soil Biol Biochem 23:335–338

143. Högberg P, Högbom L, Schinkel H, Högberg M, Johannisson C, Wallmark H (1996) ^{15}N of surface soils, roots and mycorrhizas in profiles of European forest soils. Oecologia 108:207–214

144. Johannisson C (1996) ^{15}N abundance as an indicator of N-saturation of coniferous forest, Ph.D. Thesis, SLU, Umea, Sweden

145. Emmett BA, Kjønass OJ, Gundersen P, Koopmans C, Tietema A, Sleep D (1998) Natural abundance of ^{15}N in forests across a nitrogen deposition gradient. For Ecol Manage 101:9–18

146. Martinelli LA, Piccolo MC, Townsend AR, Vitousek PM, Cuevas E, McDowell W, Robertson GP, Santos OC, Treseder K (1999) Nitrogen stable isotopic composition of leaves and soil: tropical versus temperate forests. Biogeochemistry 46:45–65

147. Garten CT Jr, Hansen PJ, Todd DE Jr, Lu BB, Brice DJ (2007) Sources of variation in the stable isotopic composition of plants. In: Michener R, Lajtha K (eds) Stable isotopes in ecology and environmental science, 2nd edn. Ecological methods and concepts series. Blackwell Publishing, UK, pp 61–82

148. Evans RD (2007) Soil nitrogen isotope composition. In: Michener R, Lajtha K (eds) Stable isotopes in ecology and environmental science, 2nd edn. Ecological Methods and Concepts Series, Blackwell Publishing, Oxford, UK, pp 83–98

149. Barraclough D, Geens EL, Maggs JM (1984) Fate of fertiliser nitrogen applied to grassland. II. Nitrogen-15 leaching results. J Soil Sci 35:191–199

150. Knowles TDJ, Chadwick DR, Bol R, Evershed RP (2010) Tracing the rate and extent of N and C flow from $^{13}C,^{15}N$-glycine and glutamate into individual de novo synthesised soil amino acids. Org Geochem 41:1259–1268. https://doi.org/10.1016/j.orggeochem.2010.09.003

151. Gibbs P, Barraclough D (1998) Gross mineralisation of nitrogen during the decomposition of leaf protein I (Ribulose 1,5-diphosphate carboxylase) in the presence or absence of sucrose. Soil Biol Biochem 30:1821–1827

152. Norman AG, Werkman CH (1943) The use of the nitrogen isotope N^{15} in determining nitrogen recovery from plant materials decomposing in soil. J Am Soc Agron 35:1023–1025

153. Glendining MJ, Poulton PR, Powlson DS, Jenkinson DS (1997) Fate of ^{15}N-labelled fertilizer applied to spring barley grown on soils of contrasting nutrient status. Plant Soil 195:83–98

154. Hancock JM, McNeill AM, McDonald GK, Holloway RE (2011) Fate of fertiliser N applied to wheat on a coarse textured highly calcareous soil under simulated semi-arid conditions. Plant Soil 348:139–153. https://doi.org/10.1007/s11104-011-0917-5

155. Jenson LS, Pedersen IS, Hansen TB, Nielsen NE (2000) Turnover and fate of ^{15}N-labelled cattle slurry ammonium-N applied in the autumn to winter wheat. Eur J Agron 12:23–35

156. Bosshard C, Sørensen P, Frossard E, Dubois D, Mäder P, Nanzer S, Oberson A (2009) Nitrogen use efficiency of ^{15}N-labelled sheep manure and mineral fertiliser applied to microplots in long-term organic and conventional cropping systems. Nutr Cycl Agroecosyst 83:271–287. https://doi.org/10.1007/s10705-008-9218-7

157. Glendining MJ, Poulton PR, Powlson DS, MacDonald AJ, Jenkinson DS (2001) Availability of the residual nitrogen from a single application of ^{15}N-labelled fertiliser to subsequent crops in a long-term continuous barley experiment. Plant Soil 233:231–239

158. Kumar K, Goh KM (2002) Recovery of ^{15}N-labelled fertiliser applied to winter wheat and perennial ryegrass crops and residual ^{15}N recovery by succeeding wheat crops under different crop residue management practices. Nutr Cycl Agroecosyst 62:123–130

159. Macdonald AJ, Poulton PR, Powlson DS, Jenkinson DS (1997) Effects of season, soil type and cropping on recoveries, residues and losses of ^{15}N-labelled fertilizer applied to arable crops in spring. J Agric Sci 129:125–154

160. Nario A, Pino I, Zapata F, Paz Albornoz M, Baherle P (2003) Nitrogen (^{15}N) fertiliser use efficiency in peach (*Prunus persica* L.) cv Goldencrest trees in Chile. Sci Hortic 97:279–287

161. Goulding KWT, Webster CP, Powlson DS, Poulton PR (1993) Denitrification losses of nitrogen fertiliser applied to winter wheat following ley and arable rotations as estimated by acetylene inhibition and ^{15}N balance. J Soil Sci 44:63–72

162. Powlson DS, Hart PBS, Poulton PR, Johnston AE, Jenkinson DS (1992) Influence of soil type, crop management and weather on the recovery of ^{15}N-labelled fertilizer applied to winter wheat in spring. J Agric Sci 118:83–100

163. Jenkinson DS (2001) The impact of humans on the nitrogen cycle, with focus on temperate arable agriculture. Plant Soil 228:3–15

164. Nannipieri P, Eldor P (2009) The chemical and functional characterization of soil N and its biotic components. Soil Biol Biochem 41:2357–2369

165. Watkins N, Barraclough D (1996) Gross rates of N mineralization associated with the decomposition of plant residues. Soil Biol Biochem 28:169–175

166. Kuzyakov Y, Friedel JK, Stahr K (2000) Review of mechanisms and quantification of priming effects. Soil Biol Biochem 32:1485–1498

167. Arkoun M, Sarda X, Jannin L, Laîne P, Etienne P, Garcia-Mina J, Yvin J, Ourry A (2012) Hydroponics versus field lysimeter studies of urea, ammonium, and nitrate uptake by oilseed rape (*Brassica napus* L.). J Exp Bot 63:5245–5258

168. Puri G, Ashman MR (1999) Microbial immobilization of ^{15}N-labelled ammonium and nitrate in a temperate woodland soil. Soil Biol Biochem 31:929–931

169. Inselsbacher E, Hinko-Najera Umana N, Stange FC, Gorfer M, Schüller E, Ripka K, Zechmeister-Boltenstern S, Hood-Novotny R, Strauss J, Wanek W (2010) Short-term competition between crop plants and soil microbes for inorganic N fertilizer. Soil Biol Biochem 42:360–372. https://doi.org/10.1016/j.soilbio.2009.11.019

170. Kaštovská E, Šantrůčková H (2011) Comparison of uptake of different N forms by soil microorganisms and two wet-grassland plants: a pot study. Soil Biol Biochem 43:1285–1291. https://doi.org/10.1016/j.soilbio.2011.02.021

171. Inselsbacher E, Wanek W, Strauss J, Zechmeister-Boltenstern S, Müller C (2013) A novel ^{15}N tracer model reveals: plant nitrate uptake governs nitrogen transformation rates in agricultural soils. Soil Biol Biochem 57:301–310

172. Sebilo M, Mayer B, Nicolardot B, Pinay G, Mariotti A (2013) Long-term fate of nitrate fertilizer in agricultural soils. Proc Natl Acad Sci USA 110:18185–18189. https://doi.org/10.1073/pnas.1305372110

173. Ladd JN, Amato M (1986) The fate of nitrogen from legume and fertiliser sources in soils successively cropped with wheat under field conditions. Soil Biol Biochem 18:417–425

174. Pilbeam CJ, Hutchison D (1998) Fate of nitrogen applied in different fertilisers to the surface of a calcareous soil in Syria. Nutr Cycl Agroecosyst 52:55–60

175. Gainey PL (1936) Total nitrogen as a factor influencing nitrate accumulation in soils. Soil Sci 42:157–163

176. Brookes PC, Landman A, Pruden G, Jenkinson DS (1985) Chloroform fumigation and the release of soil nitrogen: a rapid direct extraction method to measure microbial biomass nitrogen in soil. Soil Biol Biochem 17:837–842

177. Stanford G, Smith SJ (1972) Nitrogen mineralization potentials of soils. Soil Sci Soc Am J 36:465–472

178. Sharifi M, Zebarth BJ, Burton DL, Grant CA, Cooper JM (2007) Evaluation of some indices of potentially mineralisable nitrogen in soil. Soil Sci Soc Am J 71:1233–1239. https://doi.org/10.2136/sssaj2006.0265

179. Schoenheimer R, Rittenberg D, Foster GL, Keston AS, Ratner S (1938) The application of the nitrogen isotope N^{15} for the study of protein metabolism. Science 88:599–600

180. Rittenberg D, Keston AS, Rosebury F, Schoenheimer R (1939) Studies in protein metabolism: II. The determination of nitrogen isotopes in organic compounds. J Biol Chem 127:291–299

181. Schoenheimer R, Rittenberg D (1939) Studies in protein metabolism: I. General considerations in the application of isotopes to the study of the protein metabolism. The normal abundance of nitrogen isotopes in amino acids. J Biol Chem 127:285–290

182. McKinney CR, McCrea JM, Epstein S, Allen HA, Urey HC (1950) Improvements in mass spectrometers for the measurement of small differences in isotope abundance ratios. Rev Sci Instrum 21:724–730

183. Nier AO (1950) A redetermination of the relative abundances of the isotopes of carbon, nitrogen, oxygen, argon, and potassium. Phys Rev 77:789–793

184. Matthews DE, Hayes JM (1978) Isotope-ratio-monitoring gas chromatography-mass spectrometry. Anal Chem 50:1465–1473

185. Sweeley CC, Elliott WH, Fries I, Ryhage R (1966) Mass spectrometric determination of unresolved components in gas chromatographic effluents. Anal Chem 38:1549–1553

186. Barrie A, Bricout J, Koziet J (1984) Gas chromatography-stable isotope ratio analysis at natural abundance levels. Biomed Mass Spectrom 11:583–588

187. Meier-Augenstein W (1999) Applied gas chromatography coupled to isotope ratio mass spectrometry. J Chromatogr A 842:351–371

188. Merritt DA, Hayes JM (1994) Nitrogen isotopic analyses by isotope-ratio-monitoring gas chromatography/mass spectrometry. J Am Soc Mass Spectrom 5:387–397

189. He HB, Li XB, Zhang W, Zhang XD (2011) Differentiating the dynamics of native and newly immobilized amino sugars in soil frequently amended with inorganic nitrogen and glucose. Eur J Soil Sci 62:144–151. https://doi.org/10.1111/j.1365-2389.2010.01324.x

190. Geisseler D, Horwath WR (2014) Investigating amino acid utilisation by soil microorganisms using compound specific stable isotope analysis. Soil Biol Biochem 74:100–105. https://doi.org/10.1016/j.soilbio.2014.02.024

191. Nelson CJ, Alexova R, Jacoby RP, Millar AH (2014) Proteins with high turnover rate in barley leaves estimated by proteome analysis combined with in planta isotope labelling. Plant Physiol 166:91–108. https://doi.org/10.1104/pp.114.243014

192. Lee TA, Forrest TM, Wilson GE, Hardy JK (1990) The use of multiple mass spectral line pairs for enhanced precision in isotope enrichment studies of ^{15}N-labeled amino acids. Anal Biochem 185:24–28

193. Metges CC, Petzke K-J, Hennig U (1996) Gas chromatography/combustion/isotope ratio mass spectrometric comparison of N-acetyl- and N-pivaloyl amino acid esters to measure ^{15}N isotopic abundances in physiological samples: a pilot study on amino acid synthesis in the upper gastro-intestinal tract of minipigs. J Mass Spectrom 31:367–376

194. Brand WA, Tegtmeyer AR, Hilkert A (1994) Compound-specific isotope analysis: extending toward $^{15}N/^{14}N$ and $^{18}O/^{16}O$. Org Geochem 21:585–594

195. Brenna JT (1994) High-precision gas isotope ratio mass spectrometry: recent advances in instrumentation and biomedical applications. Acc Chem Res 27:340–346

196. Macko SA, Uhle ME, Engel MH, Andrusevich V (1997) Stable nitrogen isotope analysis of amino acid enantiomers by gas chromatography/combustion/isotope ratio mass spectrometry. Anal Chem 69:926–929

197. Hofmann D, Gehre M, Jung K (2003) Sample preparation techniques for the determination of natural ^{15}N/^{14}N variations in amino acids by gas chromatography-combustion-isotope ratio mass spectrometry (GC-C-IRMS). Isot Environ Health Stud 39:233–244. https://doi.org/10.1080/1025601031000147630

198. Takano Y, Chikaraishi Y, Ogawa NO, Kitazato H, Ohkouchi N (2009) Compound-specific nitrogen isotope analysis of D-alanine, L-alanine, and valine: application of diastereomer separation to δ^{15}N and microbial peptidoglycan studies. Anal Chem 81:394–399

199. Walsh RG, He S, Yarnes CT (2014) Compound-specific δ^{13}C and δ^{15}N analysis of amino acids: a rapid, chloroformate-based method for ecological studies. Rapid Commun Mass Spectrom 28:96–108. https://doi.org/10.1002/rcm.6761

200. Hofmann D, Jung K, Segschneider H-J, Gehre M, Schüürman G (1995) ^{15}N/^{14}N analysis of amino acids with GC-C-IRMS—methodical investigations and ecotoxicological applications. Isot Environ Health Stud 31:367–375. https://doi.org/10.1080/10256019508036284

201. Metges CC, Petzke K-J (1997) Measurement of ^{15}N/^{14}N isotopic combustion in individual plasma free amino acids of human adults at natural abundance by gas chromatography-combustion-isotope ratio mass spectrometry. Anal Biochem 247:158–164

202. Simpson IA, Bol R, Dockrill SJ, Petzke K-J, Evershed RP (1997) Compound-specific δ^{15}N amino acid signals in palaeosols as indicators of early land use: a preliminary study. Archaeol Prospect 4:147–152

203. Simpson IA, Bol R, Bull ID, Evershed RP, Petzke K-J, Dockrill SJ (1999) Interpreting early land management through compound specific stable isotope analyses of archaeological soils. Rapid Commun Mass Spectrom 13:1315–1319

204. Bol R, Ostle NJ, Petzke KJ, Watson A, Cockburn J (1998) Amino acid ^{15}N/^{14}N analysis at natural abundances: a new tool for soil organic matter studies in agricultural systems. Isot Environ Health Stud 34:87–93. https://doi.org/10.1080/10256019708036336

205. Bol R, Ostle NJ, Chenu CC, Petzke K-J, Werner RA, Balesdent J (2004) Long term changes in the distribution and δ^{15}N values of individual soil amino acids in the absence of plant and fertiliser inputs. Isot Environ Health Stud 40:243–256. https://doi.org/10.1080/10256010412331305607

206. Bol R, Ostle NJ, Petzke K-J, Chenu CC, Balesdent J (2008) Amino acid ^{15}N in long-term bare fallow soils: influence of annual N fertiliser and manure applications. Eur J Soil Sci 59:617–629. https://doi.org/10.1111/j.1365-2389.2008.01013.x

207. Petzke K-J, Boeing H, Klaus S, Metges CC (2005) Carbon and nitrogen stable isotopic composition of hair protein and amino acids can be used as biomarkers for animal-derived dietary protein intake in humans. J Nutr 135:1515–1520

208. Petzke K-J, Boeing H, Metges CC (2005) Choice of dietary protein of vegetarians and omnivores is reflected in their hair protein ^{13}C and ^{15}N abundance. Rapid Commun Mass Spectrom 19:1392–1400. https://doi.org/10.1002/rcm.1925

209. Petzke KJ, Fuller BT, Metges CC (2010) Advances in natural stable isotope ratio analysis of human hair to determine nutritional and metabolic status. Curr Opin Clin Nutr Metablc Care 13:532–540. https://doi.org/10.1097/MCO.0b013e32833c3c84

210. Styring AK, Sealy JC, Evershed RP (2010) Resolving the bulk δ^{15}N values of ancient human and animal bone collagen via compound-specific nitrogen isotope analysis of constituent amino acids. Geochim Cosmochim Acta 74:241–251. https://doi.org/10.1016/j.gca.2009.09.022

211. Styring AK, Kuhl A, Knowles TDJ, Fraser RA, Bogaard A, Evershed RP (2012) Practical considerations in the determination of compound-specific amino acid δ^{15}N values in animal and plant tissues by gas chromatography-combustion-isotope ratio mass spectrometry, following derivatisation to their N-acetylisopropyl esters. Rapid Commun Mass Spectrom 26:2328–2334. https://doi.org/10.1002/rcm.6322

212. Paolini M, Ziller L, Laursen KH, Husted S, Camin F (2015) Compound-specific δ^{15}N and δ^{13}C analyses of amino acids for potential discrimination between organically and conventionally grown wheat. J Agric Food Chem 63:5841–5850. https://doi.org/10.1021/acs.jafc.5b00662

213. Sabadel AJM, Woodward EMS, Van Hale R, Frew RD (2016) Compound-specific isotope analysis of amino acids: a tool to unravel complex symbiotic trophic relationships. Food Webs 6:9–18. https://doi.org/10.1016/j.fooweb.2015.12.003
214. Faulhaber S, Hener U, Mosandl A (1997) GC/IRMS analysis of mandarin essential oils. 1. $\delta^{13}C_{PDB}$ and $\delta^{15}N_{AIR}$ values of methyl N-methylanthranilate. J Agric Food Chem 45:2579–2583
215. Palhol F, Lamoureux C, Naulet N (2003) ^{15}N isotopic analyses: a powerful tool to establish links between seized 3,4-methylenedioxymethamphetamine (MDMA) tablets. Anal Bioanal Chem 376:486–490. https://doi.org/10.1007/s00216-003-1849-6
216. Palhol F, Lamoureux C, Chabrillat M, Naulet N (2004) $^{15}N/^{14}N$ isotopic ratio and statistical analysis: an efficient was of linking seized ecstasy tablets. Anal Chim Acta 510:1–8. https://doi.org/10.1016/j.aca.2003.12.069
217. Chickaraishi Y, Kashiyama Y, Ogawa NO, Kitazato H, Satoh M, Nomoto S, Ohkouchi N (2008) A compound-specific isotope method for measuring the stable nitrogen isotopic composition of tetrapyrroles. Org Geochem 39:510–520. https://doi.org/10.1016/j.orggeochem.2007.08.010
218. Svensson E, Schouten S, Stam A, Middelburg JJ, Sinninghe Damsté JS (2015) Compound-specific stable isotope analysis of nitrogen-containing intact polar lipids. Rapid Commun Mass Spectrom 29:2263–2271. https://doi.org/10.1002/rcm.7393
219. Maxfield PJ, Dildar N, Hornibrook ERC, Stott AW, Evershed RP (2012) Stable isotope switching (SIS): a new stable isotope probing (SIP) approach to determine carbon flow in the soil food web and dynamics in organic matter pools. Rapid Commun Mass Spectrom 26:997–1004. https://doi.org/10.1002/rcm.6172
220. Petzke KJ, Metges CC (2012) Practical recommendations for the reduction of memory effects in compound-specific $^{15}N/^{14}N$-ratio analysis of enriched amino acids by gas chromatography/combustion/isotope ratio mass spectrometry. Rapid Commun Mass Spectrom 26:195–204. https://doi.org/10.1002/rcm.5319
221. Bos C, Metges CC, Gaudichon C, Petzke KJ, Pueyo ME, Morens C, Everwand J, Benamouzig R, Tomé D (2003) Postprandial kinetics of dietary amino acids are the main determinant of their metabolism after soy or milk protein ingestion in humans. J Nutr 133:1308–1315
222. Dänicke S, Nieto R, Lobley GE, Fuller MF, Brown DS, Milne E, Calder AG, Chen S, Grant I, Böttcher W (1999) Responses in the absorptive phase in muscle and liver protein synthesis rates of growing rats. Archiv für Tierernaehrung 52:41–52. https://doi.org/10.1080/17450399909386150
223. Dänicke S, Böttcher W, Simon O, Jeroch H (2001) The measurement of muscle protein synthesis in broilers with a flooding dose technique: use of ^{15}N-labelled phenylalanine, GC-MS and GC-C-IRMS. Isot Environ Health Stud 37:213–225. https://doi.org/10.1080/10256010108033297
224. Fromentin C, Sanders P, Nau F, Anton M, Fromentin G, Tomé D, Thibault J-N, Gaudichon C (2012) A pilot study for the intrinsic labeling of egg proteins with ^{15}N and ^{13}C. Rapid Commun Mass Spectrom 26:43–48. https://doi.org/10.1002/rcm.5291
225. Mariotti F, Petzke KJ, Bonnet D, Szezepanski I, Bos C, Huneau J-F, Fouillet H (2013) Kinetics of the utilization of dietary arginine for nitric oxide and urea synthesis: insight into the arginine–nitric oxide metabolic system in humans. Am J Clin Nutr 97:972–979. https://doi.org/10.3945/ajcn.112.048025
226. Metges CC, Petzke KJ, El-Khoury AE, Henneman L, Grant I, Bedri S, Regan MM, Fuller MF, Young VR (1999) Incorporation of urea and ammonia nitrogen into ileal and fecal microbial proteins and plasma free amino acids in normal men and ileostomates. Am J Clin Nutr 70:1046–1058
227. Petzke KJ, Korkushko OV, Semesko TM, Metges CC (1997) N-isotopic composition in human plasma protein amino acids at natural abundance level and after a single [$^{15}N_2$] urea administration measured by GC-C-IRMS. Isot Environ Health Stud 33:267–275. https://doi.org/10.1080/10256019708234037

228. Petzke KJ, Grigorov JG, Korkushko OV, Kovalenko NK, Semesko TG, Metges CC (1998) Incorporation of urea nitrogen into fecal protein and plasma protein amino acids in elderly human volunteers after ingestion of lactic acid bacteria. Z Für Ernährungswissenschaft 37:368–375

229. Tobias C, Giblin A, McClelland J, Tucker J, Peterson B (2003) Sediment DIN fluxes and preferential recycling of benthic microalgal nitrogen in a shallow macrotidal estuary. Mar Ecol Progess Ser 257:25–36

230. Van Engeland T, Bouma TJ, Morris EP, Brun FG, Peralta G, Lara M, Hendriks IE, van Rijswijk P, Veuger B, Soetaert K, Middelburg JJ (2013) Dissolved organic matter uptake in a temperate seagrass ecosystem. Mar Ecol Prog Ser 478:87–100. https://doi.org/10.3354/meps10183

231. Veuger B, Middelburg JJ, Boschker HTS, Houtekamer M (2005) Analysis of ^{15}N incorporation into D-alanine: a new method for tracing nitrogen uptake by bacteria. Limnol Ocean: Methods 3:230–240

232. Veuger B, Middelburg JJ (2007) Incorporation of nitrogen from amino acids and urea by benthic microbes: role of bacteria versus algae and coupled incorporation of carbon. Aquat Microb Ecol 48:35–46

233. Sauheitl L, Glaser B, Weigelt A (2009) Advantages of compound-specific stable isotope measurements over bulk measurements in studies on plant uptake of intact amino acids. Rapid Commun Mass Spectrom 23:3333–3342. https://doi.org/10.1002/rcm.4255

234. Segschneider H-J, Hofmann D, Schmidt G, Russow R (1995) Incorporation of $^{15}NO_2$ nitrogen into individual amino acids by sunflowers using Gc-C-Irms. Isot Environ Health Stud 31:315–325. https://doi.org/10.1080/10256019508036277

235. Molero G, Aranjuelo I, Teixidor P, Araus JL, Nogués S (2011) Measurement of ^{13}C and ^{15}N isotope labeling by gas chromatography/combustion/isotope ratio mass spectrometry to study amino acid fluxes in a plant-microbe symbiotic association. Rapid Commun Mass Spectrom 25:599–607. https://doi.org/10.1002/rcm.4895

236. Arndt K, Hofmann D, Gehre M, Krumbiegel P (1998) ^{15}N investigation into the effect of a pollutant on the nitrogen metabolism of *Tetrahymena pyriformis* as a model for environmental medical research. Environ Health Perspect 106:493–497

237. Dörr N, Kaiser K, Sauheitl L, Lamersdorf N, Stange CF, Guggenberger G (2012) Fate of ammonium ^{15}N in a Norway spruce forest under long-term reduction in atmospheric N deposition. Biogeochemistry 107:409–422. https://doi.org/10.1007/s10533-010-9561-z

238. Redmile-Gordon MA, Evershed RP, Hirsch PR, White RP, Goulding KWT (2015) Soil organic matter and the extracellular microbial matrix show contrasting responses to C and N availability. Soil Biol Biochem 88:257–267. https://doi.org/10.1016/j.soilbio.2015.05.025

239. Addison SL, McDonald IR, Lloyd-Jones G (2010) Stable isotope probing: technical considerations when resolving ^{15}N-labeled RNA in gradients. J Microbiol Methods 80:70–75. https://doi.org/10.1016/j.mimet.2009.11.002

240. Buckley DH, Huangyutitham V, Hsu S-F, Nelson TA (2007) Stable isotope probing with ^{15}N achieved by disentangling the effects of genome G + C content and isotope enrichment on DNA density. Appl Environ Microbiol 73:3189–3195. https://doi.org/10.1128/AEM.02609-06

241. Cadisch G, Espana M, Causey R, Richter M, Shaw E, Morgan JAW, Rahn C, Bending GD (2005) Technical considerations for the use of ^{15}N-DNA stable-isotope probing for functional microbial activity in soils. Rapid Commun Mass Spectrom 19:1424–1428. https://doi.org/10.1002/rcm.1908

242. Cupples AM, Shaffer EA, Chee-Sanford JC, Sims GK (2007) DNA buoyant density shifts during ^{15}N-DNA stable isotope probing. Microbiol Res 162:328–334. https://doi.org/10.1016/j.micres.2006.01.016

243. Buckley DH, Huangyutitham V, Hsu S-F, Nelson TA (2007) Stable isotope probing with $^{15}N_2$ reveals novel noncultivated diazotrophs in soil. Appl Environ Microbiol 73:3196–3204. https://doi.org/10.1128/AEM.02610-06

244. Buckley DH, Huangyutitham V, Hsu S-F, Nelson TA (2008) $^{15}N_2$-DNA-stable isotope probing of diazotrophic methanotrophs in soil. Soil Biol Biochem 40:1272–1283. https://doi.org/10.1016/j.soilbio.2007.05.006

245. Roh H, Yu C-P, Fuller ME, Chu K-H (2009) Identification of hexahydro-1,3,5-trinitro-1,3,5-triazine-degrading microorganisms via ^{15}N-stable isotope probing. Environ Sci Technol 43:2505–2511

246. Andeer P, Stahl DA, Lillis L, Strand SE (2013) Identification of microbial populations assimilating nitrogen from RDX in munitions contaminated military training range soils by high sensitivity stable isotope probing. Environ Sci Technol 47:10356–10363. https://doi.org/10.1021/es401729c

247. Jayamani I, Manzella MP, Cupples AM (2013) RDX degradation potential in soils previously unexposed to RDX and the identification of RDX-degrading species in one agricultural soil using stable isotope probing. Water Air Soil Pollut 224:1745. https://doi.org/10.1007/s11270-013-1745-4

248. Cho K-C, Lee DG, Fuller ME, Hatzinger PB, Condee CW, Chu K-H (2015) Application of ^{13}C and ^{15}N stable isotope probing to characterize RDX degrading microbial communities under different electron-accepting conditions. J Hazard Mater 297:42–51. https://doi.org/10.1016/j.jhazmat.2015.04.059

249. Wawrik B, Callaghan AV, Bronk DA (2009) Use of inorganic and organic nitrogen by *Synechococcus* spp. and diatoms on the West Florida Shelf as measured using stable isotope probing. Appl Environ Microbiol 75:6662–6670. https://doi.org/10.1128/AEM.01002-09

250. España M, Rasche F, Kandeler E, Brune T, Rodriguez B, Bending GD, Cadisch G (2011) Assessing the effect of organic residue quality on active decomposing fungi in a tropical Vertisol using ^{15}N-DNA stable isotope probing. Fungal Ecol 4:115–119. https://doi.org/10.1016/j.funeco.2010.09.005

251. España M, Rasche F, Kandeler E, Brune T, Rodriguez B, Bending GD, Cadisch G (2011) Identification of active bacteria involved in decomposition of complex maize and soybean residues in a tropical Vertisol using ^{15}N-DNA stable isotope probing. Pedobiologia 54:187–193. https://doi.org/10.1016/j.pedobi.2011.03.001

252. Bell TH, Yergeau E, Martineau C, Juck D, Whyte LG, Greer CW (2011) Identification of nitrogen-incorporating bacteria in petroleum-contaminated Arctic soils by using [^{15}N] DNA-based stable isotope probing and pyrosequencing. Appl Environ Microbiol 77:4163–4171. https://doi.org/10.1128/AEM.00172-11

253. Niu J, Kasuga I, Kurisu F, Furumai H, Shigeeda T (2013) Evaluation of autotrophic growth of ammonia-oxidizers associated with granular activated carbon used for drinking water purification by DNA-stable isotope probing. Water Res 47:7053–7065. https://doi.org/10.1016/j.watres.2013.07.056

254. Wang X, Han C, Zhang J, Huang Q, Deng H, Deng Y, Zhong W (2015) Long-term fertilization effects on active ammonia oxidizers in an acidic upland soil in China. Soil Biol Biochem 84:28–37. https://doi.org/10.1016/j.soilbio.2015.02.013

255. Brink C, van Grinsven H, Jacobsen BH, Rabl A, Gren I-M, Holland M, Klimont Z, Hicks K, Brouwer R, Dickens R, Willems J, Termansen M, Velthof G, Alkemade R, van Oorschot M, Webb J (2011) Costs and benefits of nitrogen in the environment. In: Sutton MA, Howard CM, Erisman JW, Billen G, Bleeker A, Grennfelt P, van Grinsven H, Grizzetti B (eds) The European nitrogen assessment: sources, effects and policy perspectives. Cambridge University Press, UK, pp 513–540

256. Knowles TDJ (2009) Following the fate of proteinaceous material in soil using a compound-specific ^{13}C- and ^{15}N-labelled tracer approach, Unpublished Ph.D. Thesis, University of Bristol, Bristol, UK

257. White R, Murray S, Rohweder M (2000) Pilot analysis of global ecosystems: grassland ecosystems. World Resources Institute, USA

Chapter 2
Sites, Sampling, Materials and Methods

2.1 Overview

The overall methodological approach of this thesis is to combine ^{15}N-SIP mesocosm experiments with a field plot ^{15}N tracing experiment in order to gain a better understanding of the microbial processing of fertiliser N in soils at a biomolecular level and relate this to the situation in a field environment in which physical partitioning and transport of N plays an important role. The ^{15}N-SIP approach developed in order to achieve this is described in Chap. 3 and a range of N fertilisers are investigated in mesocosm experiments in Chap. 4. Chapter 5 compares the results of the mesocosm experiment for ^{15}NO$_3^-$ with those obtained in a field environment, while Chap. 6 focuses on the physical partitioning and transport of ^{15}NO$_3^-$ in soils.

This chapter describes the two sites from which soil was sampled for the laboratory incubation experiments and provides details of sampling procedures, soil handling and the general laboratory incubation experiment method which is used in the work presented in Chaps. 3 and 4. Experimental protocols, instrumental analyses and equations used throughout this thesis are then described. Further details of the plot-scale field experiment, which was also conducted at the Winterbourne Abbas (WA) site (Sect. 2.2.2) are provided in Chap. 5. Any experimental procedures/analytical techniques only applied in one part of the work are included in the relevant chapter only.

Parts of this chapter have been adapted from Charteris et al. [1] (published by Rapid Communications in Mass Spectrometry under a CC BY 4.0 licence).

2.2 Site Descriptions

2.2.1 Rowden Moor Experimental Site at North Wyke Research Station

The Rowden Moor (RM) Experimental Site is part of North Wyke Research Station near Okehampton, Devon, south-west England (50°46′ N, 3°54′ W; Fig. 2.1) which belongs to Rothamsted Research. The mean annual temperature at the site is 10.5 °C and mean annual precipitation is 1035 mm [2]. Rowden Moor is underlain by the Carboniferous Crackington Formation (comprising clay shales and thin sandstone bands) and the soil has been characterised as a clayey non-calcareous Pelostagnogley of the Hallsworth series (British Classification), a Stagni-vertic cambisol under the (FAO) scheme or a Typic haplaquept by the United States Department of Agriculture (USDA; [3]). The RM Experimental Site has been under grassland for over 40 years and the dominant species is *Lolium* spp. interspersed with *Cynosurus, Festuca, Agrostis, Holcus* and *Dactylis* spp. [4]. The plot sampled for this work (plot six) has been grazed by cattle for almost 25 years (except in 2001 due to the foot and mouth outbreak) and has received approximately 200–250 kg N ha^{-1} year^{-1} as cattle slurry over this period. The site was selected as it was used in previous, related work by Knowles [5, 6].

2.2.2 Winterbourne Abbas Public Water Supply Source, Longlands Dairy Farm

The WA PWS source is located in Well Bottom Woods (50°42′ N, −2°34′ W), adjacent to Longlands Dairy Farm approximately 2 km west of the village of Winterbourne Abbas in Dorset, south-west England (Fig. 2.2). The mean minimum and maximum annual air temperature at the site is 8.5 ± 0.3 °C and 14.2 ± 0.3 °C (based on recordings from 2009 to 2014; British Atmospheric Data Centre BADC [7]) and mean annual precipitation approximately 900 mm. The site is underlain by the Coniacian/Santonian Seaford Chalk Formation (Geological Map Data © NERC 2016) and a chalk aquifer from which the supply borehole draws.

The soil is a lime rich clay loam of variable depth (0.3–0.8 m). The field sampled in this work, 'Little Broadheath' belongs to Longlands Dairy Farm and borders Well Bottom Woods to the west (Fig. 2.2). Historically, spring crops were grown in Little Broadheath, but following a tenancy change in September 2011, it was converted to grass ley (*Lolium perenne* and *Trifolium repens*) and dairying using a mobile milking parlour began. The field is fertilised with 40 kg N ha^{-1} (previously as ammonium sulfate; $(NH_4)_2SO_4$ and more recently as sulfur-coated urea; CH_4N_2) every 40 days from spring until the start of the 'closed period' on 15th September which prohibits N fertiliser application on grasslands in Nitrate Vulnerable Zones (NVZs; [8]). The site

Fig. 2.1 Map showing the location of the RM sampling site. © Crown Copyright and Database Right [01/02/2016]. Ordnance Survey (Digimap Licence)

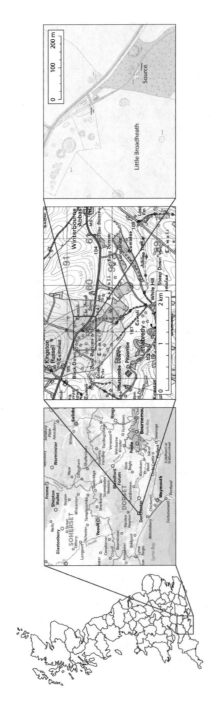

Fig. 2.2 Map showing the location of the 'Little Broadheath' sampling site, including the WA PWS source and associated source protection zones (smallest loop = SPZ 1; middle = SPZ 2; and largest = SPZ 3). © Crown Copyright and Database Right [01/02/2016]. Ordnance Survey (Digimap Licence)

was selected partly because the underlying chalk aquifer suffers from high and rising NO_3^- pollution and partly due to the presence of established monitoring activities and facilities within the catchment and the availability of historical data.

2.3 Sampling, Soil Storage and Preparation

All glassware and equipment was washed thoroughly with Decon 90 and water, rinsed with acetone and oven dried. All non-volumetric glassware was then furnaced at 450 °C for a minimum of 4 h, whilst volumetric glassware and other equipment was solvent-rinsed with dichloromethane (DCM), DCM and methanol (MeOH; 2:1 *v/v* mixture) and then MeOH. All solvents were of HPLC grade and were supplied by Rathburn Chemicals Ltd. Double-distilled water (DDW) was produced using a Bibby Aquatron still. Where not applied to living soil, DDW was extracted with DCM prior to use. This DCM extraction was performed in order to remove dissolved organic contaminants possibly present in the water, but which could not have been eliminated by double distilling.

2.3.1 Sampling

Sampling of plot six of Rowden Moor Experimental Site was undertaken on 12th February 2013. Soil was sampled randomly from the plot along two perpendicular 'W' transects to a depth of *ca.* 15 cm using a plant root corer. In total nine samples were collected. The soil had a percentage total carbon (% TC) content of 7.02 ± 0.07%, percentage total inorganic carbon (% TIC) contents of 0.22 ± 0.02% and percentage total nitrogen (% TN) contents of 0.63 ± 0.01%.

Initial sampling at the WA site was undertaken on 31st January 2013 along a random 'W' transect in Little Broadheath. Soil was taken from a depth of 2–15 cm below the surface and in total nine samples were collected. Further sampling was undertaken at WA on 21st October 2014 along a random 'W' transect focused on the centre of the field. This soil was also sampled to a depth of 2–15 cm below the surface. Average soil characteristics and chemistry are shown in Table 2.1.

2.3.2 Soil Storage and Preparation

Soils were sampled into and stored in plastic bags in a refrigerator at 5 °C until preparation in order to slow, but minimise disruption of, the microbial community [10, 11]. In addition, Jones and Shannon [12] found that storage of soil for up to 40 days in a refrigerator did not significantly affect the mineralisation rate of AAs. Aerobic conditions were maintained by placing paper towel in the opening of the bag and securing it in place with a loose elastic band.

The nine samples of each soil were combined in equal weights and homogenised to produce a pooled soil sample for each site. These pooled samples were gently

Table 2.1 Average soil characteristics and chemistry for soil from Little Broadheath, Longlands Dairy Farm, WA

	pH	P/mg l^{-1}	K/mg l^{-1}	Mg/mg l^{-1}	% OM	% TC	% TIC	% TN
Mean	6.4	24	339	99	9.16	7.46	3.29	0.45
SE	0.3	3	35	7	0.51	0.06	0.17	0.01

Several of the analyses in the table were carried out by Hill Court Farm Research as follows: soil pH was measured in deionised water; values for phosphorus (P), potassium (K) and magnesium (Mg) represent 'available' P, K and Mg contents and were obtained by 30 min soil extractions (0.5 M sodium bicarbonate [NaHCO$_3$], pH 8.5, for P and 1 M ammonium nitrate [NH$_4$NO$_3$] for K and Mg); an estimate of percentage organic matter (% OM) contents was obtained by loss on ignition (Ministry of Agriculture, Fisheries and Food [MAFF] and ADAS [9])

spread out (using two pairs of tweezers to gently divide moist clumps) in foil lined trays and air-dried so as to allow the soil to pass through a clean 2 mm sieve with as little smearing as possible. At the same time, as much flora and fauna (mainly grass and worms) as possible was removed from the soils to further aid sieving and also prevent the microbial decomposition of these components during soil preparation. The upper sections of grassland soils consist almost entirely of rhizosphere or mycorrhizosphere soil, however, so, as well as difficulties in removing all tiny roots, larger roots that were successfully removed are likely to have left some impression on the remaining soil (e.g. via root exudates). During soil spreading, care was taken to avoid compressing and smearing the soil in order to prevent the destruction of its microstructure and the soil was not allowed to over-dry which could endanger the survival of the soil microbial community [13, 14]. Once sieved, the soils were briefly stored at 5 °C in foil lined trays fitted with a foil lid punctured with small air holes to prevent further drying whilst maintaining aerobic conditions.

2.3.3   Dry Matter Content

Dry matter content (DMC) determinations were carried out in triplicate. Each soil (3 × 10 g) was weighed into pre-weighed Petri-dishes. The Petri-dishes were placed on a metal tray in a pre-heated oven at 105 °C for 24 h and then allowed to cool in a desiccator. The oven-dry weights were recorded and the DMCs (as a percentage of fresh weight) and moisture contents (MC) of the soils were calculated using Eqs. (2.1) and (2.2).

$$\% \, DMC = \left(\frac{dry \, weight \, of \, soil}{fresh \, weight \, of \, soil} \right) \times 100 \qquad (2.1)$$

$$\% \, MC = 100 - DMC \qquad (2.2)$$

2.3.4 Water Holding Capacity

Soil water holding capacity (WHC) may be determined volumetrically or gravimetrically. In this work, it was determined volumetrically. A short length of rubber tubing (~8 cm) was attached to the stem of a glass funnel and a clip attached to the bottom of the tube to enable it to be sealed shut. Glass wool (0.25 g) was weighed out, rolled into a compact cylinder, placed in the funnel at the top of the stem and tamped down firmly with a narrow spatula. First, the WHC of the funnel containing only glass wool was determined. The tube was clipped shut and 10 ml of DDW was added to the funnel via a 10 ml measuring cylinder. The funnel was left for 30 min and then the tube unclipped and the water allowed to drain into a dry 10 ml measuring cylinder for 30 min. The volume of water collected in the measuring cylinder was recorded. This was performed three times. In order to determine the WHC of each of the soils, the process was repeated, in triplicate for each, with 10 g of soil being added to the funnel prior to the addition of DDW. The WHCs of the soils were calculated using Eqs. (2.3), (2.4) and (2.5).

$$10 - (ml\ water\ retained\ by\ glass\ wool + ml\ of\ water\ collected)$$
$$= ml\ of\ water\ retained\ by\ soil\ (A) \tag{2.3}$$

$$(10 \times A) + \% MC = WHC (ml\ per\ 100\ g\ fresh\ soil;\ B) \tag{2.4}$$

$$\left(\frac{B}{\% DMC} \right) \times 100 = WHC (ml\ per\ 100\ g\ oven\ dry\ soil) \tag{2.5}$$

Using these results, the soils were re-wet to 50% WHC by the addition of DDW from an atomiser.

2.4 Incubation Method

Soil incubations were carried out as described by Knowles [5] and Knowles et al. [6] in small glass tubes (10 cm high × 2 cm diameter) which were open at the top and tapered to a small hole at the base to maintain aerobic conditions and prevent water-logging (Fig. 2.3). Wads of Soxhlet extracted (DCM:MeOH 2:1 v/v) glass wool were used to plug the base of the tubes. Sieved soils (10 g) at 50% WHC were weighed into each tube and compacted slightly. Soils were maintained at 50% WHC throughout incubations by weight and the addition of DDW via an atomiser. Drying and re-wetting of soils can affect the rates of NH_4^+ turnover and nitrification [15], so efforts were made to keep moisture levels as constant as possible. The tubes were fitted with furnaced and pierced aluminium foil lids to minimise volatile losses and water loss by evaporation.

Fig. 2.3 Incubation tube, including glass wool plug and pierced foil lid

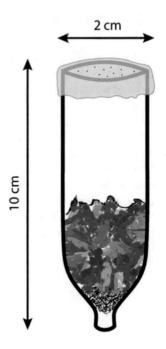

The incubation tubes were held in test tube racks with each tube sitting over a small glass vial to catch any water run-off during re-wetting of the soil (this did not occur). Incubations were carried out in a darkened temperature-controlled room at 20 °C. Since soil disturbances, such as sieving, for example, are known to increase mineralisation rates [10, 12, 16], soils were pre-incubated under the incubation conditions for a period of 4 days prior to treatment addition to allow for equilibration to the new conditions.

Substrates were introduced by injection and the needle was drawn up through the soil as the plunger was depressed in order to achieve a good distribution [17]. During application, the liquid treatment was observed to reach the inner wall of the glass tube indicating that the treatment did reach a good proportion of the soil. Substrate ^{15}N-labellings of *ca.* 10 atom % were chosen to ensure clear tracing of applied substrates whilst obviating the effects of very high labellings on changes to ^{15}N discrimination and isotopic fractionation in biological mechanisms [18, 19]. All incubations were carried out in triplicate so there were three tubes for each time point of each treatment of each soil. The incubation experiments were halted abruptly by immersion in liquid N_2 after periods of 1.5, 3, 6 and 12 h and 1, 2, 4, 8, 16, 32 and 64 days and stored at −20 °C until freeze-drying. Freeze-dried soils were finely ground using a pestle and mortar and stored in sealed 28 ml vials at −20 °C.

2.5 Extraction, Isolation and Derivatisation of Hydrolysable Amino Acids

2.5.1 Internal and External Standards

An 800 μg ml^{-1} solution of norleucine (Nle) in 0.1 M hydrochloric acid (HCl; reagent grade) was prepared for use as an internal standard (IS) for hydrolysable AA analyses. The external standard comprised a 1 mg ml^{-1} mixed solution of 14 AAs (Ala, aspartic acid [Asp], glutamic acid [Glu], Gly, hydroxyproline [Hyp], leucine [Leu], lysine [Lys], Nle, phenylalanine [Phe], proline [Pro], serine [Ser], threonine [Thr], tyrosine [Tyr] and valine [Val]) in 0.1 M HCl. The N isotopic composition of these AAs (with the exception of Hyp and Nle) had been determined previously off-line by elemental analysis-isotope ratio mass spectrometry (EA-IRMS) by Thermo Fisher Scientific in Bremen, Germany and the Merlewood/Lancaster node of the NERC Centre for Ecology and Hydrology (CEH) using the primary reference material NIST 8547 IAEA-N-1 ammonium sulfate ((NH_4)$_2SO_4$), $\delta^{15}N$ + 0.4‰). The external standard was used to: monitor instrument function; aid in the identification of sample AAs by comparison of retention times; calculate AA response factors for quantification; and as a quality control (QC) standard to monitor the performance of the gas chromatograph-combustion-isotope ratio mass spectrometer.

2.5.2 Hydrolysis and Extraction of Amino Acids from Soils

Freeze-dried, ground soils (100 mg) were weighed into culture tubes and 100 μl of Nle (800 μg ml^{-1} in 0.1 M HCl) was added to each sample as an IS. Hydrolysis with 5 ml of 6 M HCl was carried out at 100 °C for 24 h under an atmosphere of N_2 in culture tubes sealed with polytetrafluoroethylene (PTFE) tape (Fig. 2.4).

Acid hydrolysis extracts both free and proteinaceous AAs as well as catalysing the breakdown of living microbial biomass [20]. The relatively harsh conditions are necessary for the cleavage of peptide bonds between hydrophobic residues (e.g. isoleucine [Ile], Leu and Val), but also result in the deamination of asparagine (Asn) to Asp and glutamine (Gln) to Glu and the complete destruction of cysteine (Cys) and tryptophan (Trp; [20, 21]). The technique may also partially destroy Ser (ca. 10% loss), Thr (ca. 5% loss) and Tyr (loss depends on level of trace impurities in hydrolysis agent; [21] and has the potential to hydrolyse AA chains from non-proteinaceous sources, such as peptidoglycan, resulting in an over-estimation of some AAs, mostly Ala, Glu, Gly and Lys [20]. The technique is, however, considered the most reliable method for determining the total protein content of soils [20] and as such, it is reasonable to equate total hydrolysable AA concentrations to the size of the soil protein pool. The hydrolysis is performed under N_2 as the presence of O_2 can induce the thermal breakdown of hydroxyl- and sulfur-containing AAs (e.g. Ser, Thr, Tyr and methionine [Met]; [20]).

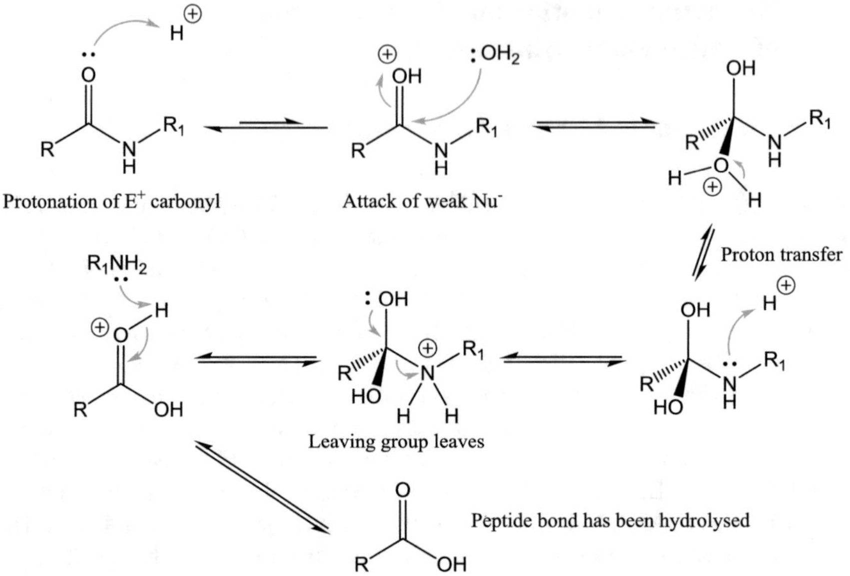

Fig. 2.4 Mechanism for the acid catalysed hydrolysis of peptide bonds. E^+; electrophile; Nu^-; nucleophile

After cooling, the culture tubes were centrifuged (3000 rpm, 10 min) and the supernatants transferred to 28 ml vials. Centrifugation of the remaining residues with 2 ml 0.1 M HCl was carried out twice and the supernatants were added to the corresponding 28 ml vials. Samples of combined supernatants were blown down at 60 °C under high N_2 flow and stored in a freezer at −20 °C under 1 ml 0.1 M HCl.

2.5.3 Preparation of Dowex Resin

Saturation of all cation exchange sites in the Dowex 50WX8 200–400 mesh ion exchange resin by protons was ensured by soaking the resin in sodium hydroxide (NaOH, 3 M) for 12 h, rising it five times with DDW (shaking the bottle and decanting the DDW water each time) and then soaking it for at least 24 h in 6 M HCl. The prepared resin was also stored under 6 M HCl until required.

2.5.4 Isolation of Hydrolysed Amino Acids

Amino acids were isolated from the hydrolysates by cation exchange column chromatography using acidified Dowex resin. The resin (1 ml) was added to glass columns and washed to neutral with DDW (3 × 2 ml). The samples (soil hydrolysates in 1 ml 0.1 M HCl) were then added to the columns and salts eluted with DDW (2 × 2 ml). Amino acids were eluted with 2 M ammonium hydroxide (NH$_4$OH; 4 × 2 ml) into fresh culture tubes and blown down under N$_2$ at 60 °C.

2.5.5 N-*Acetyl*, O-*Isopropyl Derivatisation*

Amino acids are non-volatile, polyfunctional compounds that require derivatisation to enhance their volatility [22]. Isopropylation was achieved by the addition of 0.25 ml of a fresh 4:1 *v/v* mixture of isopropanol (IP) and acetyl chloride (AC; puriss. p.a. grade; prepared by dropwise addition of AC to ice-cold IP) to the hydrolysed, purified and dried AA extracts (Fig. 2.5). The culture tubes were sealed with PTFE tape and heated at 100 °C for 1 h and the reaction quenched by freezing at −20 °C. Excess reagents were evaporated at 40 °C under a gentle stream of N$_2$ and once dry, DCM (2 × 0.25 ml) was added and gently evaporated to facilitate further egress of excess reagents.

Acetylation was carried out with 1 ml of a fresh 5:2:1 *v/v/v* mixture of acetone, triethylamine (Et$_3$N; ≥ 99.50% purity) and acetic anhydride (Ac$_2$O; ReagentPlus®® grade) heated in PTFE sealed culture tubes at 60 °C for 10 min (Fig. 2.6). Excess reagents were evaporated under a gentle stream of N$_2$ at room temperature and AAs

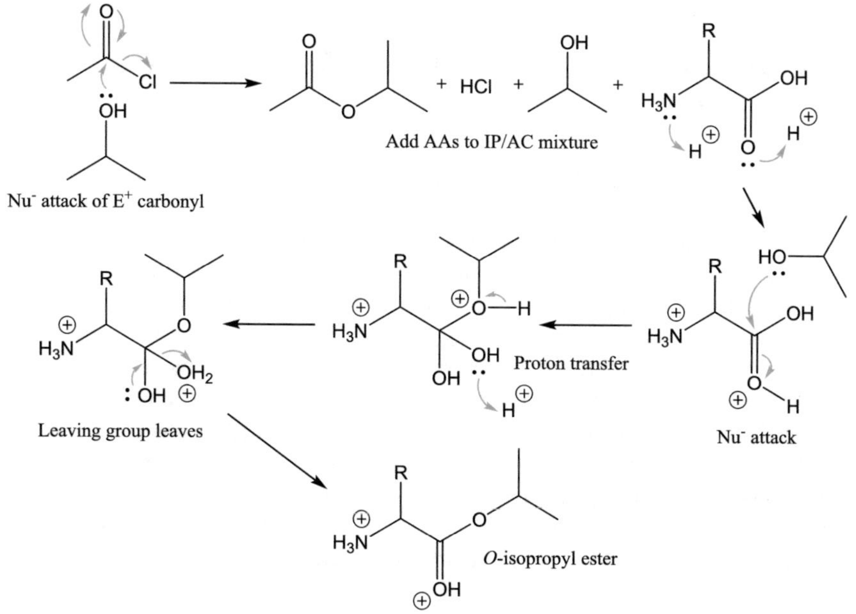

Fig. 2.5 Mechanism for the O-isopropylation step of the AA derivatisation protocol

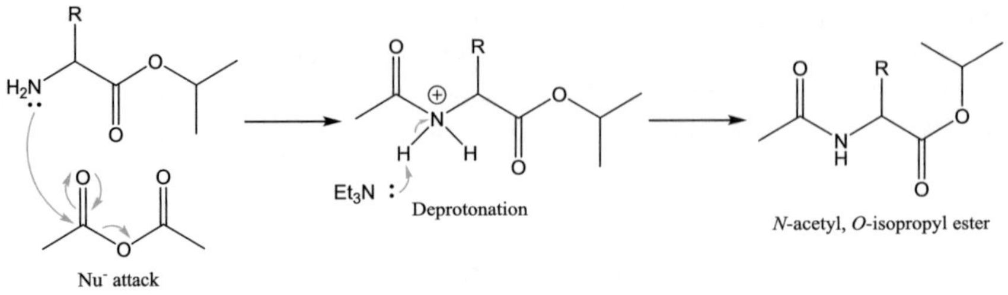

Fig. 2.6 Mechanism for the *N*-acetylation step of the AA derivatisation protocol

extracted into ethyl acetate (EtOAc; 2 ml) by vortex mixing with a saturated sodium chloride solution (NaCl; 1 ml) to affect phase separation. After allowing the layers to settle, the organic phases were transferred to 7 ml vials and the remaining NaCl washed with 1 ml EtOAc which was added to the appropriate 7 ml vial after vortex mixing and phase separation. The organic phases were dried at room temperature under a very gentle stream of N_2 and then vials were placed in an ice bath and DCM (3×1 ml) added and evaporated very gently under N_2 in order to aid evaporation of residual water. Derivatised AA samples were sealed and stored in a freezer at -20 °C until analysis.

2.6   Instrumental Analyses

2.6.1   Bulk Soil Total Carbon and Total Inorganic Carbon Content

Bulk soil % TC analyses were carried out on a Eurovector EA3000 elemental analyser. A weighed sample (*ca.* 9 mg) was sealed in a tin capsule with a combustion aid (a mixture of vanadium pentoxide [V_2O_5] and tungsten oxide [WO_3]) and introduced into a combustion tube at 1016 °C containing pure O_2. Helium (He) carrier gas then carried the combustion products over heated copper (Cu) wire to remove excess O_2 and reduce any N oxides. The resulting N_2, CO_2 and water was passed through a separation column and then measured using a thermal conductivity detector.

Bulk soil % TIC analyses were carried out using a modified Strohlein Coulomat 702 Analyser. Perchloric acid ($HClO_4$) was added to a weighed amount of sample (*ca.* 40 mg) and the resultant CO_2 flushed into a coulometric cell with a pre-set and continuously monitored pH. The CO_2 causes a change in the pH of the cell and the current required to return the pH to its original value is measured and used to determine the amount of CO_2 evolved and hence the % TIC content.

2.6.2 Bulk Sample Total Nitrogen Content and Nitrogen Isotopic Composition

Bulk sample % TN and δ^{15}N analyses were determined by EA-IRMS at the Lancaster Node of the NERC Life Sciences Mass Spectrometry Facility (LSMSF). Samples (sufficient masses to yield 50–100 µg N) were weighed into tin capsules, combusted and subsequently reduced over heated Cu wires in the elemental analyser before the resultant N_2 was passed into the isotope ratio mass spectrometer for determination of % TN contents and δ^{15}N values.

2.6.3 Gas Chromatography-Flame Ionisation Detection

A Hewlett Packard 5890 Series II gas chromatograph fitted with a *DB-35* (35%-phenyl)-methylpolysiloxane, mid-polarity coated capillary column (60 m × 0.32 mm i.d., 0.5 µm phase thickness; Agilent Technologies) and flame ionisation detector (FID) was used for quantification of individual AAs as their *N*-acetyl, *O*-isopropyl derivatives by comparison with an IS, *N*-acetyl, *O*-isopropyl Nle. The *N*-acetyl, *O*-isopropyl AAs were identified primarily by their known elution order [22] and by comparison with standards. Where necessary, peak identities were confirmed by co-injection with single *N*-acetyl, *O*-isopropyl AA standards. The carrier gas was He and the gas chromatograph was operated under constant pressure. The temperature programme utilised was: 70 °C (2 min) to 150 °C at 15 °C min^{-1}, then to 210 °C at 2 °C min^{-1} and finally to 270 °C (5 min) at 8 °C min^{-1}. Data were acquired and analysed using Clarity chromatographic station for Windows by DataApex.

2.6.4 Gas Chromatography-Mass Spectrometry

Gas chromatography-mass spectrometry analyses were performed on a Thermo Scientific Trace 1300 gas chromatograph interfaced with a Thermo Scientific ISQ single quadrupole mass spectrometer via a heated transfer line. Sample introduction was either manual or automated using a Thermo Scientific AS 1310 autosampler and via a split-splitless inlet set to splitless mode. The gas chromatograph was fitted with a *DB-35* coated capillary column (60 m × 0.32 mm i.d., 0.5 µm phase thickness; Agilent Technologies) and the carrier gas was He. A selection of samples were analysed in order to confirm AA identities using the same temperature programmes employed for GC and GC-C-IRMS analyses. The mass spectrometer was operated in electron ionisation (EI) mode. Data were acquired and analysed using Xcalibur 3.0 (Thermo Fischer Scientific). *N*-acetyl, *O*-isopropyl AAs were identified using molecular ions ($M^{+\bullet}$) and by the loss of characteristic fragments, for example, the isopropyl-derivatised –OH group $[M - 59]^+$.

2.6.5   Gas Chromatography-Combustion-Isotope Ratio Mass Spectrometry

The $\delta^{15}N$ values of individual AAs as their N-acetyl, O-isopropyl derivatives were determined using a ThermoFinnigan Trace 2000 gas chromatograph coupled with a ThermoFinnigan Delta^{Plus} XP isotope ratio monitoring mass spectrometer via a ThermoFinnigan Combustion III interface (Thermo Electron Corporation). Samples were either introduced manually or using a CTC Analytics GC Pal autosampler and via a programmable temperature vaporisation (PTV) inlet. The carrier gas was He at a flow rate of 1.4 ml min^{-1} and the gas chromatograph was fitted with a *DB-35* coated capillary column (30 m × 0.32 mm i.d. × 0.5 μm stationary phase thickness; Agilent Technologies). The temperature programme utilised was: 40 °C (5 min) to 120 °C at 15 °C min^{-1}, to 180 °C at 3 °C min^{-1}, then to 210 °C at 1.5 °C min^{-1} and finally to 270 °C at 5 °C min^{-1}. The oxidation and reduction reactor consisted of high purity Cu and nickel (Ni) wires (OEA Laboratories) and was held at 1030 °C. The mass spectrometer was operated in EI mode with three Faraday cup collectors for *m/z* 28, 29 and 30.

Amino acid $\delta^{15}N$ values were determined relative to that of a monitoring gas of known (previously determined using in-house AA standards; Sect. 2.5.1) N isotopic composition introduced directly into the ion source via an open split in four pulses at the beginning and end of each run. In order to adhere to the identical treatment (IT) principle and ensure the GC-C-IRMS system was functioning properly, each duplicated sample was bracketed by the in-house AA standard mixture of known $\delta^{15}N$ values and sample AA $\delta^{15}N$ values accepted only when at least 75% of the AAs in the standard mixture run either side of the sample were within ±1‰ and the others were within ±1.5‰, and when this was also true on average over the course of the run. Data were acquired and analysed using Isodat NT 3.0 (Thermo Electron Corporation). Figure 2.7 shows a typical *N*-acetyl, *O*-isopropyl derivatised hydrolysable soil AA sample chromatogram including the ion current signals for each *m/z* value recorded.

2.7   Data Processing

2.7.1   Quantification of Amino Acids

Amino acid quantification was carried out by comparison of sample AA peak areas with that of the IS, Nle (Eq. 2.6). Due to structural variations between AAs, however, inequivalent GC-FID responses are obtained when equal concentrations of different AAs are analysed [23]. It is therefore necessary to calculate response factors (RFs) using AA standard mixtures of known (equivalent) concentrations (Eq. 2.7) in order to determine sample AA concentrations (Eq. 2.8).

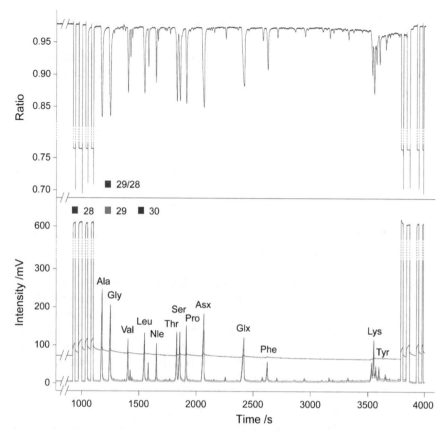

Fig. 2.7 Typical GC-C-IRMS chromatogram of an *N*-acetyl, *O*-isopropyl derivatised hydrolysable soil AA sample showing the ion current signals recorded by the GC-C-IRMS operating for N_2 (*m/z* 28, 29 and 30) and the ratio of *m/z* 28 to 29 which is used to generate $^{15}N/^{14}N$ isotope ratios

$$\text{Sample AA apparent mass} = \left(\frac{\text{Sample AA peak area}}{\text{Sample IS peak area}} \right) \times \text{mass IS} \qquad (2.6)$$

$$\text{AA RF} = \frac{\text{Standard AA peak area}}{\text{Standard IS peak area}} \qquad (2.7)$$

$$\text{Sample AA actual mass} = \text{Sample AA apparent mass} \times \text{AA RF} \qquad (2.8)$$

2.7.2 Equations Relating to Nitrogen Isotopic Compositions

The N isotopic compositions of substances are usually reported as relative difference isotope ratios using the δ notation, as $\delta^{15}N$ values (in ‰), relative to the internationally agreed measurement standard, AIR-N_2 [24]; Eq. 2.9).

$$\delta^{15}N_{P/AIR-N_2} = \left(\frac{R(^{15}N/^{14}N)_P}{R(^{15}N/^{14}N)_{AIR-N_2}} - 1 \right) \times 1000 \qquad (2.9)$$

where $R(^{15}N/^{14}N)_P$ is the isotope ratio of the sample and $R(^{15}N/^{14}N)_{AIR-N2}$ is the absolute isotope ratio of AIR-N_2 the internationally agreed measurement standard for N, which acts as a scale-anchor. The recommended value for the absolute isotope ratio of AIR-N_2 for calculations is $0.0036765 \pm 4 \times 10^{-6}$ [25–27]. While Eq. (2.9) implies that the value of the absolute isotope ratio of AIR-N_2 is important in the calculation of sample $\delta^{15}N_{AIR-N2}$ values, this is not actually the case as the values are determined in dual inlet mass spectrometers by comparison of sample and standard/reference material ion currents and ion current ratios, and AIR-N_2 is assigned a value of 0‰, regardless of its absolute isotope ratio [28]. The $\delta^{15}N_{AIR-N2}$ value of the sample should then always be the same 'distance' from 0‰. In order to ensure that this is the case and improve comparability between instruments and laboratories, the $\delta^{15}N_{AIR-N2}$ scale may be standardised by scale normalisation (the processes of stretching or shrinking the δ scale by defining the δ value of a selected secondary reference material relative to the primary standard; [29–31]). The selected secondary reference material for N is USGS32 (potassium nitrate), which has a $\delta^{15}N_{AIR-N2}$ value of 180‰.

The deliberate use of a range of further secondary reference materials, in preference to AIR-N_2 is also recommended as relative difference isotope ratio δ values are likely to be more reliable where the reference materials employed are as chemically and physically similar to the samples as possible [25, 28]. This is because the selection of appropriate reference materials enables adherence to the IT principle whereby samples and reference materials are prepared, processed and converted to final measurement gases in the same way and, as a result, isotopic fractionations 'cancel' and any other preparatory or systematic biases are removed [25, 28]. A range of 'representative' secondary reference materials for $\delta^{15}N$ analyses exist, but the availability of organic N reference materials is limited (e.g. USGS40 and USGS41, L-glutamic acid with $\delta^{15}N_{AIR-N2}$ values of -4.52 ± 0.06 and 47.57 ± 0.1‰ respectively, are out of stock) and, as the variety of available materials is insufficient, more (particularly organic) secondary reference materials are being, and need to be, developed [31].

Ideally, two-point normalisations using two reference materials of similar physical and chemical nature to the sample, but contrasting isotopic composition to one another should be used to anchor and normalise sample δ values [31]. In the work carried out in this thesis, however, this would require two secondary/tertiary reference standards composed of a mixture of AAs and of contrasting N isotopic compositions that bracket the sample AA N isotopic compositions. Such reference materials are not available, even for natural abundance samples and it not yet known whether such two-point scale normalisations will be effective with on-line techniques such as GC-C-IRMS. Moreover, in the case of ^{15}N-enriched samples, the ^{15}N enrichments of the upper bracketing standard may not fall on the 0–180‰ two-point normalised δ^{15}N$_{AIR-N2}$ scale and could be difficult to produce accurately. Thus the compound-specific AA δ^{15}N$_{AIR-N2}$ data presented in this thesis was generated via the most accurate available method, as described in Sect. 2.6.5. The need for scale-anchoring can also be further reduced by presentation of sample δ^{15}N$_{AIR-N2}$ values less control sample δ^{15}N$_{AIR-N2}$ values.

The major disadvantage of isotope δ values is that they do not vary linearly with isotopic abundance and, while deviations from linearity are negligible at natural abundance, where samples are highly enriched in ^{15}N, R(^{15}N/^{14}N)$_P$ becomes large and δ values tend to infinity [32]. In such cases the use of phi (φ) values based on atom (isotope) fractions has been suggested [33, 34], but use has not been extensive. Instead, atom % values are commonly used (despite the disadvantages of their requirement for the absolute isotopic abundances of internationally agreed measurement standards and the fact that this term has been deprecated [35]). Atom % values are based on the atom or isotope-amount fraction, $x(^{15}$N)$_P$, of ^{15}N relative to total N:

$$x(^{15}\text{N})_P = \frac{n(^{15}\text{N})_p}{n(^{15}\text{N})_p + n(^{14}\text{N})_p} = \frac{r\left(\frac{^{15}\text{N}}{^{14}\text{N}}\right)_P}{r\left(\frac{^{15}\text{N}}{^{14}\text{N}}\right)_P + 1} = \frac{R\left(\frac{^{15}\text{N}}{^{14}\text{N}}\right)_P}{R\left(\frac{^{15}\text{N}}{^{14}\text{N}}\right)_P + 1} \qquad (2.10)$$

where $n(^{15}$N)$_P$ is the amount of ^{15}N isotopes in substance P, $r(^{15}$N/^{14}N)$_P$ is the isotope-amount ratio and $R(^{15}$N/^{14}N)$_P$ is the isotope-number ratio of substance P [35]. $R(^{15}$N/^{14}N)$_P$ can be obtained by re-arranging Eq. (2.10) as follows:

$$R\left(\frac{^{15}\text{N}}{^{14}\text{N}}\right)_P = R\left(\frac{^{15}\text{N}}{^{14}\text{N}}\right)_{AIR-N_2} \left(\frac{\delta^{15}\text{N}_{P/AIR-N_2}}{1000} + 1\right) \qquad (2.11)$$

and

$$\text{Atom \%} = x(^{15}\text{N})_P \times 100 \qquad (2.12)$$

2.7.3 Nitrogen Content and Isotopic Composition Mixing Calculations

When mixing N-containing substances, both the amount of N, n, and N isotopic composition, $x(^{15}$N), of the starting substances (P and Q) are important in determining the eventual N content and N isotopic composition of the mixed substance (F; Eq. 2.13).

$$n_{\mathrm{P}}x(^{15}\mathrm{N})_{\mathrm{P}} + n_{\mathrm{Q}}x(^{15}\mathrm{N})_{\mathrm{Q}} = (n_{\mathrm{P}} + n_{\mathrm{Q}})x(^{15}\mathrm{N})_{\mathrm{F}} \tag{2.13}$$

2.7.4 Bulk Nitrogen Isotopic Compositions and Percentage Retentions

The total amount of $^{15}\mathrm{N}$ present in moles per gram of bulk sample can be calculated as follows (Knowles et al. [6]):

$$n(^{15}\mathrm{N})_{\mathrm{P}} = x(^{15}\mathrm{N})_{\mathrm{P}}\left(\frac{\%\mathrm{N}}{1400}\right) \tag{2.14}$$

The excess moles of $^{15}\mathrm{N}$ present/retained per gram in a bulk enriched sample, compared with control samples, $n^{\mathrm{E}}(^{15}\mathrm{N})_{\mathrm{P/C}}$, is given by:

$$n^{\mathrm{E}}\left(^{15}\mathrm{N}\right)_{\mathrm{P/C}} = x^{\mathrm{E}}\left(^{15}\mathrm{N}\right)_{\mathrm{P/C}}\left(\frac{\%\mathrm{N}}{1400}\right) \tag{2.15}$$

where $x^{\mathrm{E}}(^{15}\mathrm{N})_{\mathrm{P/C}}$ is excess atom fraction of $^{15}\mathrm{N}$ in the bulk enriched soil, following incubation, compared with control:

$$x^{\mathrm{E}}\left(^{15}\mathrm{N}\right)_{\mathrm{P/C}} = x(^{15}\mathrm{N})_{\mathrm{P}} - x(^{15}\mathrm{N})_{\mathrm{C}} \tag{2.16}$$

The percentage retention of applied $^{15}\mathrm{N}$ may then be calculated using:

$$\%\ \mathrm{Retention}\ ^{15}\mathrm{N} = \left(\frac{n^{\mathrm{E}}\left(^{15}\mathrm{N}\right)_{\mathrm{P/C}}}{n^{\mathrm{E}}\left(^{15}\mathrm{N}\right)_{\mathrm{A}}}\right) \times 100 \tag{2.17}$$

where $n^{\mathrm{E}}(^{15}\mathrm{N})_{\mathrm{A}}$ is the number of moles of excess $^{15}\mathrm{N}$ applied (above natural abundance values for the substrate).

2.7.5 Percentage Incorporations of ^{15}N into Amino Acids

The $^{15}\mathrm{N}$ enrichment, E of an AA following application of a $^{15}\mathrm{N}$-labelled substrate may be expressed as the number of moles of $^{15}\mathrm{N}$ derived from the applied substrate that are present in that AA in the soil:

$$E = n \times x^{\mathrm{E}}\left(^{15}\mathrm{N}\right)_{\mathrm{P/C}} \tag{2.18}$$

where n is the number of moles of N in the AA (i.e. if the molecular structure of the AA contains only one N atom, n is the same as the number of moles of the AA in the soil, but twice this if the AA structure consists of two N atoms and so on) and $x^E(^{15}N)_{P/C}$ is excess atom fraction of the AA after incubation compared with the control (Eq. 2.16). The atom fraction, $x(^{15}N)_P$, of ^{15}N in the AA can be calculated from the AA's $\delta^{15}N$ value as follows [6]:

$$x(^{15}N)_P = \frac{R\left(\frac{^{15}N}{^{14}N}\right)_{AIR-N_2}\left(\frac{\delta^{15}N_{P/AIR-N_2}}{1000}+1\right)}{1+\left(\frac{\delta^{15}N_{P/AIR-N_2}}{1000}+1\right)} \qquad (2.19)$$

which is equivalent to Eq. (2.10). The data may also be expressed in terms of the percentage of the applied ^{15}N incorporated into each AA, as in Knowles et al. [6]:

$$\% \ ^{15}N \ \text{Incorporation}_A = \left(\frac{E}{n^E\left(^{15}N\right)_A}\right) \times 100 \qquad (2.20)$$

Percentage incorporations reflect both the concentration and enrichment ($\delta^{15}N$ value) of the AA (i.e. how much was incorporated if the AA at x concentration was enriched by x ‰) and the percentage of applied ^{15}N incorporated into newly synthesised soil protein is determined by summing these results for individual AAs. Note that these percentage incorporation data will be affected by the conservation/retention of applied ^{15}N in the system, thus if ^{15}N is lost from the system (e.g. over time), AA percentage incorporations may become skewed as there is less ^{15}N available for incorporation than expected. The incubation experimental design described aims to limit any ^{15}N losses from the system in order to obviate this issue, but it is equally possible to calculate percentage incorporations at time, t, based on the moles of applied ^{15}N retained ($n^E(^{15}N)_{P/C}$; above natural abundance/control soil values) in the system at time, t (Eq. 2.15), if bulk recovery of the applied ^{15}N is low or decreases with time, i.e.:

$$\% \ ^{15}N \ \text{Incorporation}_R = \left(\frac{E}{n^E\left(^{15}N\right)_{P/C}}\right) \times 100 \qquad (2.21)$$

It can also be argued, however, that loss from the system is just another process competing against AA biosynthesis for N, so should not be discounted in this way. Both of these approaches are valid and the most appropriate one will depend on the specifics of the experimental design and desired outcomes.

2.7.6 Statistical Analyses

Statistical testing for significant differences between parameters was performed using IBM SSPS Statistics 23 (International Business Machines Corporation). In order to obtain quantitative information about the rates of change of AA δ^{15}N values and % ^{15}N incorporations, linear and non-linear regressions were fitted to time course plots as follows:

(i) Linear regressions of the form:

$$\Delta^{15}\text{N} = k_0 t \tag{2.22}$$

were fitted to early time course data in order to determine the initial rates at which Δ^{15}N values (or % ^{15}N incorporations, using an analogous equation) changed; k_0 is the initial zero-order rate constant of the reaction.

(ii) Non-linear single first-order exponential assimilation regressions of the form:

$$\Delta^{15}\text{N} = \Delta^{15}\text{N}_0 + P\left(1 - e^{-k_1 t}\right) \tag{2.23}$$

were fitted to data over the time course in order to obtain a quantitative measure of the overall rate of at which Δ^{15}N values (or % ^{15}N incorporations, using an analogous equation) changed and the overall/plateau change in AA Δ^{15}N values/% ^{15}N incorporations at 'steady state'; Δ^{15}N$_0$ (or % ^{15}N$_0$) is the AA Δ^{15}N value (or % ^{15}N incorporation) at $t = 0$ (which is 0 by definition), P is the plateau level reached and k_1 is the first-order rate constant.

Linear regressions were performed using Microsoft Excel 2013 (Microsoft Corporation), while non-linear regressions were fitted using the curve-fit tool in SigmaPlot 13 (Systat Software, Inc.).

References

1. Charteris AF, Knowles TDJ, Michaelides K, Evershed RP (2016) Compound-specific amino acid ^{15}N stable isotope probing of nitrogen assimilation by the soil microbial biomass using gas chromatography/combustion/isotope ratio mass spectrometry. Rapid Commun Mass Spectrom 30:1846–1856. https://doi.org/10.1002/rcm.7612
2. Tyson KC, Hawkins JMB, Stone AC (1993) Final report of the AFRC–ADAS drainage experiment 1982–1993. Institute of Grassland and Environmental Research, North Wyke, Okehampton, UK
3. Harrod TR, Hogan DV (2008) The soils of North Wyke and Rowden, Revised edition of original report by Harrod, T. R. 1981, Soil Survey of England and Wales (now the National Soil Resources Institute, Cranfield University, UK). Report available from Rothamstead Research, http://www.rothamsted.ac.uk/northwyke. Accessed 05 June 2015
4. Bol R, Amelung W, Friedrich C (2004) Role of aggregate surface and core fraction in the sequestration of carbon from dung in a temperate grassland soil. Eur J Soil Sci 55:71–77. https://doi.org/10.1046/j.1365-2389.2003.00582.x

5. Knowles TDJ (2009) Following the fate of proteinaceous material in soil using a compound-specific ^{13}C- and ^{15}N-labelled tracer approach, Unpublished Ph.D. Thesis, University of Bristol, Bristol, UK

6. Knowles TDJ, Chadwick DR, Bol R, Evershed RP (2010) Tracing the rate and extent of N and C flow from ^{13}C, ^{15}N-glycine and glutamate into individual de novo synthesised soil amino acids. Org Geochem 41:1259–1268. https://doi.org/10.1016/j.orggeochem.2010.09.003

7. BADC. Accessed 21 Jan 2015

8. Defra (2013) Guidance on complying with the rules for Nitrate Vulnerable Zones in England for 2013 to 2016. Report available from http://www.gov.uk/nitrate-vulnerable-zones. Accessed 13 Jan 2016

9. MAFF and ADAS (1986) The analysis of agricultural materials: a manual of the analytical methods used by the Agricultural Development and Advisory Service (ADAS), Reference book 427, 3rd edn., HMSO, London, UK

10. Černohlávková J, Jarkovský J, Nešporová M, Hofman J (2009) Variability of soil microbial properties: effects of sampling, handling and storage. Ecotoxicol Environ Saf 72:2102–2108. https://doi.org/10.1016/j.ecoenv.2009.04.023

11. Ross DJ, Tate KR, Cairns A, Meyrick KF (1980) Influence of storage on soil microbial biomass estimated by three biochemical procedures. Soil Biol Biochem 12:369–374

12. Jones DL, Shannon D (1999) Mineralization of amino acids applied to soils: impact of soil sieving, storage, and inorganic nitrogen additions. Soil Sci Soc Am J 63:1199–1206

13. Cox CS (1993) Roles of water molecules in bacteria and viruses. Orig Life Evol Biosph 23:29–36

14. Potts M (1994) Desiccation tolerance of prokaryotes. Microbiol Rev 58:755–805

15. Burger M, Jackson LE (2003) Microbial immobilization of ammonium and nitrate in relation to ammonification and nitrification rates in organic and conventional cropping systems. Soil Biol Biochem 35:29–36

16. Ross DJ, Speir TW, Tate KR, Orchard VA (1985) Effects of sieving on estimations of microbial biomass, and carbon and nitrogen mineralization, in soil under pasture. Aust J Soil Res 23:319–324

17. Murphy DV, Bhogal A, Shepherd M, Goulding KWT, Jarvis SC, Barraclough D, Gaunt JL (1999) Comparison of ^{15}N labelling methods to measure gross nitrogen mineralisation. Soil Biol Biochem 31:2015–2024

18. Mathieu O, Lévêque J, Hénault C, Ambus P, Milloux M, Andreux F (2007) Influence of ^{15}N enrichment on the net isotopic fractionation factor during the reduction of nitrate to nitrous oxide in soil. Rapid Commun Mass Spectrom 21:1447–1451. https://doi.org/10.1002/rcm.2979

19. Tang FHM, Maggi F (2012) The effect of ^{15}N to ^{14}N ratio on nitrification, denitrification and dissimilatory nitrate reduction. Rapid Commun Mass Spectrom 26:430–442. https://doi.org/10.1002/rcm.6119

20. Roberts P, Jones DL (2008) Critical evaluation of methods for determining total protein in soil solution. Soil Biol Biochem 40:1485–1495. https://doi.org/10.1016/j.soilbio.2008.01.001

21. Fountoulakis M, Lahm HW (1998) Hydrolysis and amino acid composition analysis of proteins. J Chromatogr A 826:109–134

22. Corr LT, Berstan R, Evershed RP (2007) Optimisation of derivatisation procedures for the determination of δ^{13}C values of amino acids by gas chromatography/combustion/isotope ratio mass spectrometry. Rapid Commun Mass Spectrom 21:3759–3771. https://doi.org/10.1002/rcm.3252

23. Corr LT, Berstan R, Evershed RP (2007) Development of N-acetyl methyl ester derivatives for the determination of δ^{13}C values of amino acids using gas chromatography-combustion-isotope ratio mass spectrometry. Anal Chem 79:9082–9090

24. Mariotti A (1983) Atmospheric nitrogen is a reliable standard for natural ^{15}N abundance measurements. Nature 303:685–687

25. Brand WA, Coplen TB, Vogl J, Rosner M, Prohaska T (2014) Assessment of international reference materials for isotope-ratio analysis (IUPAC Technical Report). Pure Appl Chem 86:425–467. https://doi.org/10.1515/pac-2013-1023

26. Coplen TB, Krouse HR, Böhlke JR (1992) Reporting of nitrogen-isotope abundances (IUPAC Technical Report). Pure Appl Chem 64:907–908
27. Meija J, Coplen TB, Berglund M, Brand WA, De Bièvre P, Gröning M, Holden NE, Irrgeher J, Loss RD, Walczyk T, Prohaska T (2016) Isotopic composition of the elements 2013 (IUPAC Technical Report). Pure Appl Chem 88:293–306. https://doi.org/10.1515/pac-2015-0503
28. Werner RA, Brand WA (2001) Referencing strategies and techniques in stable isotope ratio analysis. Rapid Commun Mass Spectrom 15:501–519
29. Coplen TB (1988) Normalization of oxygen and hydrogen isotope data. Chem Geol (Isotope Geosci Sect) 72:293–297
30. Gonfiantini R (1978) Standards for stable isotope measurements in natural compounds. Nature 271:534–536
31. Schimmelmann A, Qi H, Coplen TB, Brand WA, Fong J, Meier-Augustein W, Kemp HF, Toman B, Ackermann A, Assonov S, Aerts-Bijma AT, Brejcha R, Chikaraishi Y, Darwish T, Elsner M, Gehre M, Geilmann H, Gröning M, Hélie J-F, Herrero-Martín S, Meijer HAJ, Sauer PE, Sessions AL, Werner RA (2016) Organic reference materials for hydrogen, carbon, and nitrogen stable isotope-ratio measurements: caffeines, n-alkanes, fatty acid methyl esters, glycines, L-valines, polyethylenes, and oils. Anal Chem 8:4294–4302. https://doi.org/10.1021/acs.analchem.5b04392
32. Brand WA, Coplen TB (2012) Stable isotope deltas: tiny, yet robust signatures in nature. Isot Environ Health Stud 48:393–409
33. Brenna JT, Corso TN, Tobias HJ, Caimi RJ (1997) High-precision continuous-flow isotope ratio mass spectrometry. Mass Spectrom Rev 16:227–258
34. Corso TN, Brenna JT (1997) High-precision position-specific isotope analysis. Proc Natl Acad Sci USA 94:1049–1053
35. Coplen TB (2011) Guidelines and recommended terms for expression of stable-isotope-ratio and gas-ratio measurement results. Rapid Commun Mass Spectrom 25:2538–2560. https://doi.org/10.1002/rcm.5129

Chapter 3
Compound-Specific Amino Acid [15]N Stable Isotope Probing of Nitrogen Assimilation by the Soil Microbial Biomass Using Gas Chromatography-Combustion-Isotope Ratio Mass Spectrometry

3.1 Introduction

As already introduced in Chap. 1 (Sect. 1.4.2), the complexity and heterogeneity of soil organic N is partly responsible for limiting understanding of soil N cycling. Organic N concentrations far exceed those of inorganic N in most soils [2–7] and organic N plays important roles in soil N cycling and balancing the supply of N to microorganisms, plants and loss pathways (in some cases supplying 30–50% of the inorganic N for crop uptake; [8–10]; Fig. 3.1). It is particularly difficult, however, to resolve the soil organic N actively cycling through the soil system from more recalcitrant soil organic N [11], which plays a lesser role in soil nutrient N supply.

3.1.1 Approaches to Investigate Soil Nitrogen Cycling

Many of the most direct approaches to assess soil N supplying capacity have focused on establishing methods to determine PMN. One of the most reliable of these remains, however, an early and time-consuming incubation and leaching method reported by Stanford and Smith in [12, 13]. In addition, these assessments may not accurately reflect the situation in the field, or the true availability of N to plants (which can also assimilate organic N and whose N assimilation depends on complex rhizospheric interactions, for example) and only reflect the N status of the soil at the time of sampling, rather than improving understanding of the dynamic soil processes responsible for nutrient N supply.

Studies monitoring the concentrations of added substrates and potential products (e.g. [5]) are useful because the concentrations of actively cycled components will fluctuate, providing indications of 'reactivity' in the soil. Such approaches, however,

Parts of this chapter have been adapted from Charteris et al. [1] (published by Rapid Communications in Mass Spectrometry under a CC BY 4.0 licence).

© Springer Nature Switzerland AG 2019
A. F. Charteris, [15]N Tracing of Microbial Assimilation, Partitioning and Transport of Fertilisers in Grassland Soils, Springer Theses, https://doi.org/10.1007/978-3-030-31057-8_3

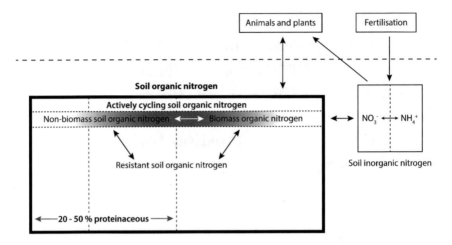

Fig. 3.1 Conceptual diagram of the soil N cycle emphasising the relative contributions of organic and inorganic N to soil N (*ca.* 90:10%) and the actively cycling soil organic N pool which is currently difficult to assess, but which may be estimated via the approach discussed in this chapter. The substrates selected to demonstrate the approach in this chapter represent both input types shown in the figure—'Fertilisation' and 'Animals and plants'. Proteinaceous matter; proteins, peptides and AAs

do not allow elucidation of the pathways of N transformation and results may be confounded by additional processes operating in the soil. The use of ^{15}N-labelled substrates ensures that the applied N can be traced (physically or biochemically) directly into product pools (e.g. in lysimeter leaching studies [14] or to trace the fate of AA N in soil [15]). Or, alternatively, as in ^{15}N pool dilution experiments, that any newly produced N (usually as NH_4^+ or NO_3^-) entering the ^{15}N-labelled pool is iden-tifiable (e.g. to determine gross N mineralisation during RuBisCO decomposition; [16]).

^{15}N stable isotope tracing has contributed much to our understanding of N cycling in the soil system (e.g. [17]), however, the complexity and heterogeneity of soil organic N have prevented interrogation of the *biomolecular* fate of applied ^{15}N in any detail. As a result, a considerable proportion of previous work has either assumed that since the majority of soil N is organic, all of the ^{15}N retained in the soil is organic N (e.g. [18]), or has derived estimates for the N isotopic composition of organic N by extracting/subtracting ^{15}N-enriched inorganic compounds from bulk soils/values (e.g. [19, 20]). A shortcoming of both of these methods is that they only provide an estimate of the bulk N isotopic composition of an extremely complex and non-uniformly ^{15}N-enriched organic N pool. Critically, in these methods the ^{15}N substrates/amendments mostly serve as physical tracers rather than true *biochemical* tracers. A more refined approach has been to use microbial biomass N extraction [21] and subsequent isotopic analysis to determine the N isotopic composition of biomass N, representing the fraction of ^{15}N assimilated by microorganisms or the ^{15}N cycling through the 'living', 'active' or 'available' portion of soil organic N (e.g. [22–24]).

This extraction method can only generate estimates of bulk soil microbial biomass N, however [25–28], and only a small proportion of the soil microbial community are actively growing and assimilating N at a given time [29].

Nucleic acid (DNA/RNA) ^{15}N-SIP represents a more recently developed technique that interrogates the active microbial community in more detail [30–33]. This culture-independent method employs isopycnic centrifugation to separate out ^{15}N-enriched nucleic acids, which, following fingerprinting, may enable the taxonomic identity of actively assimilating microorganisms to be established [30–32]. The technique is reviewed in more detail in Sect. 1.4.3 and although potentially valuable in identifying active microorganisms, nucleic acid ^{15}N-SIP, does not afford insights into the N cycling of the soil community as a whole or into the fate of applied [15]N.

A technique which can offer this capability is compound-specific ^{15}N-SIP of soil organic N using mass spectrometry. As described in Sect. 1.4.3, many studies have continued to use conventional GC-MS in compound-specific ^{15}N-SIP studies, despite the disadvantage that it can generally only be used precisely (± 0.01 atom %) with highly ^{15}N-enriched compounds where the ^{15}N enrichment is easily detectable above natural background values [34–36]. Moreover, few workers have exploited the far higher potential precision (0.5–2.0‰; 0.0002–0.0008 atom %; [35–37]) of GC-C-IRMS in ^{15}N-SIP studies. Indeed, in an extensive search of the literature only three studies exploiting the high sensitivity and selectivity of GC-C-IRMS to trace the fate of applied ^{15}N in soils were found: Knowles et al. [15] traced the fate of dual [13]C- and ^{15}N-labelled Glu and Gly into newly synthesised soil AAs in work that forms the basis of the method presented in this chapter; Dörr et al. [38] investigated the fate of [15]NH_4^+ in a Norwegian spruce forest in relation to changing atmospheric N deposition rates; and Redmile-Gordon et al. [39] compared the response of hydrolysable soil AAs and extracellular polymeric substance AAs to [15]$NH_4^{15}NO_3$ additions with and without glycerol.

The work presented in these studies provides some indication of the potential of ^{15}N-SIP using GC-C-IRMS to be an extremely powerful tool for tracing the fate of ^{15}N in soils. As implied in Sect. 1.4.3, a crucial advantage of the technique is that it offers the selectivity and sensitivity [36, 40–42] required to enable the biomolecular tracing of ^{15}N substrates (applied at environmentally relevant concentrations and appropriately low ^{15}N enrichments to minimise perturbations to native ecosystem conditions) into highly 'diluted' N-containing metabolic products. It is therefore somewhat surprising that ^{15}N-SIP using GC-C-IRMS remains largely unexploited. While this may be partly due to the challenges associated with compound-specific N isotopic analyses via GC-C-IRMS (Sect. 1.4.3), it may also be due to a lack of awareness regarding the potential of compound-specific ^{15}N-SIP using GC-C-IRMS to investigate soil N cycling.

3.1.2   Compound-Specific Amino Acid ^{15}N Stable Isotope Probing to Investigate the Fate of Nitrogen in Soils, Microbial Nitrogen Assimilation and New Protein Biosynthesis

The work of Knowles et al. [15] in particular aimed to "open the 'black box' of organic N cycling in soil" and in relation to this, contributed important knowledge regarding AA degradation/turnover and the disparate fates of AA C and N in soils. It was the realisation that this molecular approach for investigating microbial N processing in soils could be developed into a powerful proxy for estimating soil microbial N assimilation and new soil protein biosynthesis that led to the method presented in this chapter. The power of the approach lies in the analysis of AAs as these are major 'building blocks' of all life, forming the proteins which regulate essential biochemical reactions. Proteinaceous matter generally comprises 20–50% of total soil N and is ubiquitous in living organisms, so is a major 'organic product' of microbial activity/assimilation [2–4, 43]. Since AAs represent major organic nitrogenous products in soil they provide a highly sensitive integrating tool across the many thousands of proteins present in soil, revealing important general features of the dynamics and pathways of assimilation of N-containing substrates into the organic N pool. Critically, the percentage of applied ^{15}N detectable in the total hydrolysable AA pool offers a measure of (or 'proxy' for) the assimilation of an applied ^{15}N-labelled substrate by the soil microbial biomass and an estimate of newly synthesized soil protein.

3.1.3   Characteristics and Occurrence of Amino Acids

Amino acids are bifunctional organic molecules which contain both a carboxyl (–COOH) and an amino (–NH$_2$) group. Asparagine was the first AA to be isolated (from asparagus in 1806; [44]) and by the early 1980s over 500 naturally occurring AAs had been identified [45]. More are discovered each year. Amino acid side chains may be attached to an α-, β-, γ- or δ-C, but those with the side chain attached to the C atom adjacent to the carboxyl group (α- or 2-AAs) are biochemically most important. This includes the 20 canonical AAs (Ala, Arginine [Arg], Asn, Asp, Cys, Gln, Glu, Gly, histidine [His], Ile, Leu, Lys, Met, Phe, Pro, Ser, Thr, Trp, Tyr and Val) encoded by DNA or 22 proteinogenic AAs (if selenocysteine [Sec] and pyrrolysine [Pyl] are included) required to form life's proteins and enzymes [46]. All other AAs are known as 'non-proteinogenic AAs' and may be post-translationally incorporated into proteins and/or act as non-protein AAs and play other important physiological or ecological roles (e.g. citrulline; [47]).

All α-AAs except Gly, have a chiral centre at the α-C and the enantiomers are commonly differentiated as L- and D-AAs according to the arrangement of groups around the α-C (relative to glyceraldehyde; [48]). L-AAs predominate in nature, but some D-AAs play important roles, for example: in bacterial peptidoglycan (e.g. D-Ala and D-Glu), helping to protect cell walls from proteases [48–50]; in antibiotics and antimicrobials (e.g. D-Glu and D-Phe; [48, 51]); to regulate bacterial spore germination, cell wall remodelling and biofilm dispersal [48]; and also in eukaryotes (e.g. D-Asp and D-Ser are important in human/mammalian brain function; [48]), amongst other things. D-Ala and D-Glu have therefore been used as biomarkers for bacteria [50], despite the fact that AA enantiomers that are no longer part of functioning cells racemise over time. For this reason, the accumulation of D-AAs has also been used as a marker of protein aging, although due to microbial activities, this may only be reliable for D-Lys [50, 52].

The distributions of canonical AAs differ considerably between the major kingdoms of life (archaea, bacteria and eukaryotes), but differences between different proteomes considered individually are larger [53]. The relative abundances of AAs in organisms are related to their guanine-cytosine content and phylogeny, but the extent to which the guanine-cytosine content and thus AA composition of organisms is influenced by environmental factors appears to remain a matter of debate [54, 55]. Moura et al. [55] did find some indications, however, for the existence of broad environment-specific AA signatures (grouping environments as ocean waters, sandy soils and sediments, grasslands, forest soils and host-associated environments), presumably resulting from the complex and dynamic feedback relationships between environmental conditions and community compositions. This is in agreement, to an extent, with the common finding that the AA distributions of soils tend to be similar over a wide range of soil properties and under different management/land uses [3, 4, 56, 57]. Aspartate, Gly, Glx and Ala are commonly present in higher abundance, while Met and Tyr are present at low abundance [4, 56, 57]. Clear differences in AA composition are known, however, in: tropical soils rich in amorphous aluminium silicates, which have higher concentrations of in acidic AAs; arctic soils, which have low concentrations of acidic AAs; and tropical soils, which have lower concentrations of basic AAs [3]. It has also been observed that soils from cooler temperate regions have higher concentrations of acidic AAs [56].

The hydrolysable AA compositions of soils result from the disaggregation of the many thousands of proteins present in soils, either generated in situ by soil microorganisms and fauna, or input to soil in excreta, plant matter and rainwater. The mean AA compositions of some soils have been found to be most similar to those of bacteria (as opposed to algae, fungi or yeasts) suggesting that these microorganism exert the strongest influence on the turnover of proteinaceous matter in these soils [3]. In several New Zealand soils, on the other hand, particularly high concentrations of Asx and Glx were suggested to result from the presence of undecomposed plant litter and root tissues, rich in these AAs in the soils [56]. The autotrophic carboxylase enzyme that facilitates oxygenic photosynthesis, ribulose-1,5-bisphosphate carboxylase-oxygenase (RuBisCO) is considered the most abundant protein in terrestrial ecosystems and may therefore be expected to influence the AA composition of soils [58, 59]. RuBisCO is rich in Gly, Leu and Ala and has a low abundance of Cys, Met, Gln and Trp [58], but the occurrence of a RuBisCO AA signature in soils has not been reported.

3.2 Objectives

This chapter aims to demonstrate the advantages of compound-specific AA ^{15}N-SIP using GC-C-IRMS to: investigate the fate of N in soils; obtain a measure of the assimilation of an applied ^{15}N-labelled substrate by the soil microbial biomass; and derive quantitative estimates of newly synthesized soil protein, which is representative of the functioning of the soil microbial biomass and biomass protein production using the applied substrate. This will be done by presenting and discussing the results of three laboratory incubation experiments, two with inorganic N substrates (^{15}NO$_3^-$ and ^{15}NH$_4^+$) and one with an organic N substrate (^{15}N-glutamate; ^{15}N-Glu). This range of contrasting substrates was selected in order to illustrate the utility of the approach for the study of any N-containing soil amendment. Finally, this chapter also aims to highlight the range of potential insights and the wider conceptual applicability of compound-specific AA ^{15}N-SIP using GC-C-IRMS in the investigation of complex N cycling ecosystems.

The specific objectives of this work were to:

(i)   Monitor bulk soil % TN contents at $t = 0$, following N addition and over the course of the experiments.
(ii) Monitor bulk soil δ^{15}N values following application of ^{15}NO$_3^-$, ^{15}NH$_4^+$ or ^{15}N-Glu to soil as indicators of application efficiency and bulk soil processes.
(iii) Monitor hydrolysable soil AA concentrations for response to the N addition over the course of the experiments.
(iv) Determine and examine the patterns in individual hydrolysable soil AA δ^{15}N values in response to the addition of ^{15}NO$_3^-$, ^{15}NH$_4^+$ or ^{15}N-Glu as indicators of the primary biochemical pathways operating in the assimilation of these substrates.
(v)   Determine the percentages of applied ^{15}N (from ^{15}NO$_3^-$, ^{15}NH$_4^+$ and ^{15}N-Glu) incorporated into the total hydrolysable AA or soil protein pool.

Potential findings include:

(i)   Increased understanding of the utility and value of compound-specific ^{15}N-SIP using GC-C-IRMS for investigating N cycling in soils and other complex ecosystems.
(ii) Increased understanding of the ability of the approach to reveal the biomolecular fate and partitioning of the applied ^{15}N and provide valuable insights into the biochemical pathways of microbial N assimilation from chemically different forms of N applied at environmentally relevant concentrations.
(iii) Increased understanding of the value of the approach for revealing differences in the overall rates and fluxes of ^{15}N from contrasting ^{15}N-labelled substrates into the total hydrolysable AA pool or soil protein pool.
(iv) Increased understanding of the ability of the approach to provide estimates of ^{15}N assimilation by the soil microbial biomass and newly synthesized soil protein, enabling comparisons of the relative 'availabilities' of different N sources

in the same soil and of the processing of the same N source by different soils (with different or differently functioning soil microbial populations).

The hypotheses tested in this work are:

(i)   Bulk soil % TN contents will not change as a result of a small, but agriculturally relevant N addition, or over the course of the experiments.
(ii) Bulk soil $\delta^{15}N$ values will rise following application of a ^{15}N-labelled substrate and will remain high over the course of the incubation experiments due to the experimental design which minimises N losses.
(iii) Hydrolysable soil AA concentrations will not change as a result of a small, but agriculturally relevant N addition, or over the course of the experiments.
(iv) As a ubiquitous primary metabolic process, the assimilation of $^{15}NH_4^+$ into Glu via the glutamate dehydrogenase (GDH) or glutamine synthetase (GS), glutamate synthase (glutamine oxoglutarate aminotransferase; GOGAT) pathways, and then into other AAs by transamination reactions, will be reflected in the $\delta^{15}N$ values of individual AAs with time.
(v)   There will be differences in the rates and fluxes of ^{15}N from contrasting ^{15}N-labelled substrates into the total hydrolysable AA pool or soil protein pool.

3.3   Sites, Sampling, Materials and Methods

Soil for the $^{15}NO_3^-$ and $^{15}NH_4^+$ incubation experiments described in this chapter was sampled from the RM site and prepared as described in Sects. 2.2.1 and 2.3. These incubation experiments were carried out as described in Sect. 2.4. The ^{15}N-Glu incubation experiment, on the other hand, was carried out previously by Knowles et al. [15] and results have been represented in this chapter in order to demonstrate the use of compound-specific ^{15}N-SIP using GC-C-IRMS with an organic ^{15}N-labelled substrate (which in this case is also a hydrolysable AA). The two sets of incubation experiments are directly comparable as they were carried out in the same soil and by following the same experiment protocol. The substrates applied and incubation periods at which samples were taken for all experiments discussed in this chapter are displayed in Table 3.1.

As described in Sect. 2.4, the ^{15}N-labelling of 10 atom % for $^{15}NH_4^+$ and $^{15}NO_3^-$ was chosen based on recent research highlighting changes in ^{15}N discrimination and isotopic fractionation in biological mechanisms at very high enrichments [61, 62]. Enrichments of 10 atom % were considered low enough for these effects to be negligible. This was not the case for the ^{15}N-Glu incubation experiment as it was carried out previously by Knowles et al. [15].

For the $^{15}NO_3^-$ and $^{15}NH_4^+$ incubation experiments, hydrolysable soil AAs were extracted, isolated and derivatised as described in Sect. 2.5 and soils analysed (bulk % TN contents and N isotopic compositions and AA concentrations and N isotopic compositions) as described in Sect. 2.6. Experimental protocols and analyses were

Table 3.1 Table summarising the laboratory incubation experiments discussed in this chapter

Substrate	Labelling	Mass substrate applied/10 g soil	Mass N/10 g soil (μg)	Equivalent[a] fertilisation rate/kg N ha^{-1} year^{-1}	Incubation periods
K^{15}NO$_3$	10 atom % ^{15}N	400 μg in 200 μl DDW	55	100	1.5, 3, 6 & 12 h & 1, 2, 4, 8, 16 & 32 days
^{15}NH$_4$Cl	10 atom % ^{15}N	400 μg in 200 μl DDW	105	190	1.5, 3, 6 & 12 h & 1, 2, 4, 8, 16 & 32 days
^{15}N-Glu	98 atom % ^{15}N	2 mg in 200 μl 0.1 M HCl	200	370	3, 6 & 12 h & 1, 2, 4, 8, 16 & 32 days

[a]Equivalent fertilisation rate calculated based on a 0.3 m soil depth and an average of 5–6 applications between February and October. The rates are within or not far from the range recommended for grasslands for dairy grazing (140–340 kg N ha^{-1} year^{-1}; [60])

conducted almost identically for the ^{15}N-Glu incubation experiment [15]. Data processing was carried out as described in Sect. 2.7.

3.4   Results and Discussion

Compound-specific AA ^{15}N-SIP using GC-C-IRMS has the potential to provide hitherto unattainable insights into soil N cycling from any (inorganic or organic) N-containing substrate. The utility of the method is discussed in terms of: (i) limitations of bulk N and AA concentrations to detect appropriate N additions/cycling in soils; (ii) pathways of assimilation of different N-containing substrates; (iii) revealing differences in rates and fluxes of N between applied substrates; and (iv) interpretations of ^{15}N-SIP determinations in relation to complex N dynamics in soils.

3.4.1   Limitations of Bulk Nitrogen and Amino Acid Concentrations to Detect Appropriate Nitrogen Additions/Nitrogen Cycling in Soils

The addition of an agriculturally relevant, but sufficiently low N concentration to prevent alteration of the soil's N status (and thereby limit perturbation), almost by definition results in no notable changes in the % TN content of the soil over the course of the experiment. Tables 3.2, 3.3 and 3.4 confirm this—there is no observable trend

Table 3.2 Soil % TN contents and composition and concentrations of soil hydrolysable AAs for the $^{15}NO_3^-$ -SIP experiment

		Time/days											Mean	SE
		0	0.0625	0.125	0.25	0.5	1	2	4	8	16	32		
Mean concentration/mg g^{-1}	% TN	0.63	0.67	0.65	0.67	0.66	0.65	0.67	0.66	0.66	0.68	0.69	0.66	0.0039
	Ala	2.4	1.9	1.9	2.0	1.9	2.1	2.0	2.2	1.8	1.9	2.3	2.1	0.056
	Asp	1.6	1.9	2.0	2.0	2.0	1.8	2.2	1.8	1.6	1.6	1.2	1.8	0.066
	Glx	1.6	1.8	1.9	1.7	1.8	1.8	1.9	1.8	1.5	1.6	1.1	1.7	0.055
	Gly	1.8	1.3	1.4	1.4	1.4	1.5	1.4	1.4	1.2	1.5	1.8	1.5	0.050
	Hyp	0.12	0.12	0.12	0.12	0.12	0.14	0.12	0.13	0.11	0.10	0.12	0.12	0.0028
	Ile	0.38	0.49	0.49	0.39	0.35	0.25	0.34	0.39	0.34	0.26	0.37	0.37	0.015
	Leu	1.0	1.1	1.1	1.0	0.97	0.95	0.95	1.1	0.85	0.91	0.87	0.98	0.017
	Lys	0.48	0.34	0.41	0.57	0.46	0.39	0.40	0.46	0.53	0.39	0.60	0.44	0.025
	Met	0.065	0.13	0.12	0.12	0.10	0.093	0.10	0.073	0.057	0.081	0.064	0.089	0.0044
	Phe	0.48	0.58	0.59	0.59	0.54	0.56	0.43	0.51	0.37	0.49	0.41	0.50	0.015
	Pro	1.2	1.1	1.1	1.1	1.1	1.2	1.1	1.2	1.0	0.97	1.2	1.1	0.026
	Ser	0.89	0.89	0.95	1.1	1.1	0.90	1.0	0.98	0.87	0.81	0.85	0.93	0.026
	Thr	0.73	0.82	0.90	0.97	0.88	0.68	0.87	0.86	0.77	0.67	0.65	0.80	0.027
	Tyr	0.22	0.34	0.31	0.37	0.34	0.34	0.23	0.26	0.19	0.25	0.21	0.27	0.012
	Val	0.75	0.72	0.77	0.73	0.63	0.44	0.63	0.76	0.67	0.52	0.62	0.67	0.026
	THAA N	1.8	1.7	1.8	1.8	1.7	1.7	1.7	1.8	1.5	1.5	1.7	1.7	0.0053
	% THAA N of TN	29	25	27	27	26	26	26	27	23	23	24	26	0.52

THAA N; total hydrolysable amino acid nitrogen

Table 3.3 Soil % TN contents and composition and concentrations of soil hydrolysable AAs for the $^{15}NH_4^+$-SIP experiment

		Time/days											Mean	SE
		0	0.0625	0.125	0.25	0.5	1	2	4	8	16	32		
	% TN	0.63	0.60	0.62	0.62	0.62	0.56	0.55	0.67	0.69	0.64	0.69	0.63	0.0081
Mean concentration/mg g^{-1}	Ala	2.4	1.9	1.8	1.8	1.9	1.9	1.9	1.9	2.0	2.1	2.1	2.0	0.048
	Asp	1.6	2.2	1.8	1.8	2.0	1.9	2.1	2.0	2.1	2.1	2.0	1.9	0.052
	Glx	1.6	2.0	1.8	1.8	1.9	1.8	1.9	1.8	1.9	1.9	1.9	1.8	0.043
	Gly	1.8	1.3	1.1	1.4	1.3	1.3	1.5	1.4	1.5	1.5	1.4	1.5	0.043
	Hyp	0.12	0.12	0.11	0.12	0.12	0.11	0.12	0.11	0.13	0.13	0.13	0.12	0.0024
	Ile	0.38	0.61	0.54	0.43	0.46	0.52	0.55	0.45	0.48	0.48	0.47	0.48	0.015
	Leu	1.0	1.2	1.2	1.0	1.1	1.0	1.1	1.0	1.1	1.1	1.1	1.1	0.012
	Lys	0.48	0.41	0.27	0.27	0.25	0.34	0.48	0.40	0.51	0.59	0.55	0.42	0.028
	Met	0.065	0.12	0.12	0.085	0.13	0.10	0.081	0.13	0.12	0.11	0.13	0.10	0.0051
	Phe	0.48	0.55	0.58	0.45	0.46	0.46	0.48	0.54	0.60	0.59	0.64	0.53	0.014
	Pro	1.2	1.1	1.1	1.2	1.1	1.1	1.1	1.1	1.1	1.2	1.2	1.2	0.020
	Ser	0.89	1.0	0.85	0.84	0.94	0.92	1.0	0.95	1.1	1.1	1.0	0.96	0.024
	Thr	0.73	1.1	0.89	0.87	0.90	0.86	0.96	0.92	0.99	1.0	0.93	0.91	0.025
	Tyr	0.22	0.29	0.29	0.23	0.24	0.24	0.26	0.30	0.36	0.35	0.35	0.28	0.0098
	Val	0.75	0.95	0.80	0.76	0.72	0.83	0.90	0.78	0.82	0.83	0.78	0.81	0.021
	THAA N	1.8	1.8	1.6	1.7	1.7	1.7	1.8	1.7	1.9	1.9	1.9	1.8	0.027
	% THAA N of TN	29	31	26	27	28	30	34	26	27	30	27	29	0.56

Table 3.4 Soil % TN contents and composition and concentrations of soil hydrolysable AAs for the ^{15}N-Glu-SIP experiment

Mean concentration/mg g^{-1}		Time/days										Mean	SE
		0	0.125	0.25	0.5	1	2	4	8	16	32		
	% TN	0.76	0.76	0.78	0.77	0.79	0.77	0.76	0.77	0.78	0.77	0.77	0.0033
	Ala	5.4	7.9	7.3	7.4	5.7	4.1	4.4	5.5	3.5	4.5	5.6	0.39
	Asp	3.6	4.2	3.7	3.2	3.6	3.1	3.6	3.3	2.2	2.4	3.3	0.19
	Glx	2.9	3.9	3.6	3.1	2.8	2.5	3.0	3.0	1.9	2.0	2.9	0.17
	Gly	4.7	6.4	6.0	6.2	4.8	3.7	3.5	5.2	3.2	3.8	4.7	0.30
	Ile	1.2	1.4	1.4	1.3	1.1	0.94	1.0	2.3	1.0	0.72	1.3	0.12
	Leu	0.69	0.97	0.90	0.78	0.70	0.53	0.84	0.60	0.33	0.48	0.68	0.047
	Lys	0.64	0.21	0.14	0.32	0.15	0.92	0.46	1.3	0.98	0.28	0.53	0.078
	Met	0.17	0.23	0.23	0.20	0.17	0.17	0.24	0.14	0.069	0.10	0.17	0.012
	Phe	0.41	0.40	0.41	0.40	0.36	0.35	0.36	0.49	0.28	0.26	0.37	0.020
	Pro	2.5	3.6	3.8	3.7	2.7	1.8	2.2	3.0	1.6	2.3	2.7	0.20
	Ser	2.7	3.7	3.3	3.0	2.5	2.2	2.5	2.6	1.8	1.8	2.6	0.16
	Thr	2.3	2.2	1.9	1.7	1.9	1.4	1.9	1.8	1.2	1.1	1.7	0.11
	Val	1.3	1.4	1.4	1.1	1.1	0.5	1.0	0.75	0.46	0.48	0.91	0.079
	THAA N	3.9	5.0	4.6	4.5	3.7	3.0	3.3	4.1	2.5	2.8	3.7	0.21
	% THAA N of TN	51	66	60	58	47	39	44	53	33	36	48	2.7

Table 3.5 Bulk soil δ^{15}N values for the ^{15}NO$_3^-$, ^{15}NH$_4^+$ and ^{15}N-Glu incubation experiments

	$t = 0$		$t = 3$ h		Overall incubation mean			
	Mean δ^{15}N value	SE	Mean δ^{15}N value	SE	Mean δ^{15}N value	SE	% Retention ^{15}N	SE
^{15}NH$_4^+$	4.47	0.0432	87.3	4.03	85.4	1.25	109	2.07
^{15}NO$_3^-$	4.47	0.0432	35.8	1.69	36.8	1.09	88	3.1
^{15}N-Glu	7.16	0.773	1050	67.0	1070	14.0	100	1.32

in the % TN content of the incubation mesocosms and the standard errors (SEs) of the means of the % TN contents for all incubation mesocosms are small. Thus, the application of ^{15}N-labelled amendments is clearly valuable in allowing added N to be differentiated from native soil N. Following addition of all three substrates, however, bulk soil δ^{15}N values, after the initial rise, remained relatively constant throughout the rest of the incubation experiment (i.e. overall percentage retentions of ^{15}N in the system were high and close to 100%; Table 3.5). The elevated δ^{15}N values compared to $t = 0$ values confirm the continued presence of the ^{15}N tracer in the soil, but no insights can be gained about the form or internal processing of the amendments within the soil, i.e. is the ^{15}N still present as ^{15}NO$_3^-$, ^{15}NH$_4^+$ or ^{15}N-Glu or has it been assimilated by the soil microbial biomass?

As essential biomolecules, proteinaceous AAs are likely products of amendment assimilation, however, the concentrations of individual AAs show little change over the course of the incubation and there is no observable increase in concentration with incubation duration, as might be expected from the synthesis of new AAs using the applied NO$_3^-$, NH$_4^+$ or Glu (Tables 3.2, 3.3 and 3.4).

This could imply that the supplied NO$_3^-$, NH$_4^+$ or Glu has not been used in the synthesis of AAs, but it could also be that the concentration of NO$_3^-$, NH$_4^+$ or Glu added has not stimulated protein biosynthesis above that present prior to the additions. Accordingly, the total hydrolysable AA N content of the soil is on average $26 \pm 0.5\%$, $29 \pm 0.6\%$ and $48 \pm 3\%$ of total soil N throughout the ^{15}NO$_3^-$, ^{15}NH$_4^+$ and ^{15}N-Glu incubation experiments, respectively (Tables 3.2, 3.3 and 3.4). These concentrations fall within the range (20–50%) generally reported for total hydrolysable soil AAs [2–4, 43] and are equated to the concentration of the soil protein pool [43]. It is noted that acid hydrolysis does not extract all proteinaceous AAs and extracts some non-proteinaceous AAs [4, 43], however, the fraction of soil AAs recovered is constant. These results emphasise the need to undertake compound-specific N isotopic analysis (using GC-C-IRMS) of the newly biosynthesised AAs (mostly new protein) to gain detailed insights into the dynamics of the assimilation of an N-containing amendment into the soil organic N pool to be traced.

3.4.2 Pathways of Assimilation of Different Nitrogen-Containing Substrates

The use of GC-C-IRMS allows precise (0.5–2.0‰; 0.0002–0.0008 atom %) determination of the $\delta^{15}N$ values of individual hydrolysable soil AAs [35–37]. Figure 3.2 shows the trends in hydrolysable AA $\delta^{15}N$ values over the course of the incubation experiments applying $^{15}NO_3^-$, $^{15}NH_4^+$ or ^{15}N-Glu. The advantages of the compound-specific approach are immediately apparent, with readily detectable changes being seen in the $\delta^{15}N$ values of all AAs in all experiments (Fig. 3.2). This is the first time the assimilation of NO_3^- and NH_4^+ by the soil microbial biomass has been measured in this way and the results clearly emphasise the importance of investigating N cycling from different N-containing soil amendments.

Incorporation of $^{15}NH_4^+$ (Fig. 3.2a) occurs in two phases for all AAs except Glx—fast over the first 2 to 4 days, then more slowly for the remainder of the experiment. Assimilation into Glx occurs more quickly over the first 2 days than into any other AA and the degree of ^{15}N enrichment is two- to five-fold greater, before declining during the rest of the experiment. These differences in the patterns of ^{15}N incorporation relate to the fundamental biosynthetic pathways most microorganisms use for the assimilation of NH_4^+, i.e. via the reductive amination of α-ketoglutarate to $_L$-Glu, catalysed by GDH or via the GS-GOGAT pathway [63–65]. Glutamate is of central importance to the biosynthesis of new proteins as other AAs are synthesised from Glu, using it as a substrate for the amination of appropriate α-ketoacid C skeletons. The fast rise in the $\delta^{15}N$ values of Glx reflects the initial incorporation of $^{15}NH_4^+$, with the subsequent decline after 2 days reflecting redistribution of the ^{15}N into newly synthesised AAs, hence their $\delta^{15}N$ values rise.

The results for the $^{15}NO_3^-$ experiment are extremely interesting, further emphasising the importance of this approach (Fig. 3.2b). Broadly, most AAs initially, and somewhat surprisingly, show lower $\delta^{15}N$ values before rising slightly over the rest of the incubation. The initial dip indicates that at the start of the incubation AAs are ^{15}N-depleted compared with at $t = 0$. The reason for this is unknown, but one possibility is that contact with initial high NO_3^- concentrations causes cell lysis providing non-proteinaceous substrates with low $\delta^{15}N$ values for AA biosynthesis. The subsequent smaller and more irregular rise in $\delta^{15}N$ values for all AAs compared to the $^{15}NH_4^+$ experiment is likely because NO_3^- requires reduction prior to incorporation into AAs [66].

Comparison with the ^{15}N-Glu incubation carried out by Knowles et al. [15]; data represented for this discussion in Fig. 3.2c, d exposes a more complex situation as the applied substrate is itself an AA in the total hydrolysable AA pool. As might be expected, the $\delta^{15}N$ value of Glx falls with time, whilst those of the other AAs rise (Fig. 3.2c, d). Interestingly, a comparable pattern to that of $^{15}NH_4^+$ incorporation emerges for the transfer of ^{15}N-Glu into other hydrolysable AAs, but in this case all non-substrate AAs, except Asx, exhibit two-phase incorporation. As Knowles et al. [15] concluded, this is again likely due to the fundamental biosynthetic pathways that operate in most microorganisms; Asp is produced by the transamination of

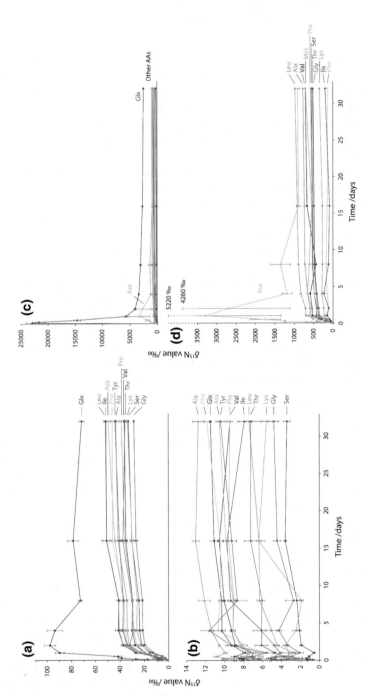

Fig. 3.2 δ^{15}N values of individual hydrolysable soil AAs over the course of 32-day incubation experiments: **a** ^{15}NH$_4^+$ incubation, **b** ^{15}NO$_3^-$ incubation, **c** ^{15}N-Glu incubation, including the applied ^{15}N-Glu and **d** ^{15}N-Glu incubation, excluding the applied ^{15}N-Glu. Error bars are ± SE (n = 3)

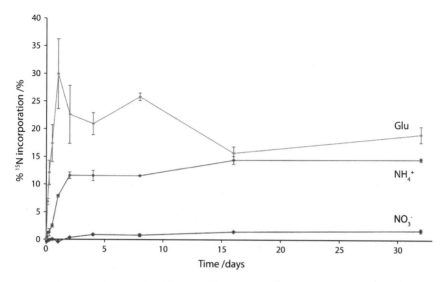

Fig. 3.3 Percentage of applied $^{15}NO_3^-$, $^{15}NH_4^+$ and ^{15}N-Glu incorporated into the total hydrolysable AA pool or soil protein pool. Error bars are $\pm$ SE (n = 3). Calculations for $^{15}NO_3^-$ and $^{15}NH_4^+$ are straightforward summations of the percentage of the applied ^{15}N incorporated into each AA, while results for ^{15}N-Glu incubation were, in this case, calculated excluding the ^{15}N residing in Glx

oxaloacetate using an amino group from Glu [67], the remaining C-skeleton of which is α-ketoglutarate, which is used in the tricarboxylic acid (TCA) cycle, an essential metabolic process that generates energy in aerobic respiration. Decarboxylation of α-ketoglutarate as part of the cycle then generates another molecule of oxaloacetate. Interpreting the rate data alongside this known biochemistry, Knowles et al. [15] concluded that the patterns of isotope incorporation are consistent with Asp being the AA closest in biosynthetic proximity to Glu.

3.4.3 Revealing Differences in the Rates and Fluxes of Nitrogen Between Applied Substrates

Quantifying the fate of N-containing substrates (inorganic or organic) in different soils is essential to understanding the N cycle in natural or semi-natural ecosystems but is especially important in agricultural systems where managing fertiliser applications has ecological and economic relevance. The new insights gained into N cycling through this novel approach offer the potential to enhance fundamental understanding in this area. Using Eqs. (2.16) and (2.18)–(2.20), increases in AA $\delta^{15}N$ values can be used to determine the percentage of the applied ^{15}N incorporated into each AA and by summation, the percentage incorporated into the total hydrolysable AA or soil protein pool and cycling through the 'living', 'active' or 'available' portion of soil organic N at that time (Fig. 3.3).

These calculations are straightforward where the applied substrate is not a hydrolysable AA (e.g. ^{15}NO$_3^-$ and ^{15}NH$_4^+$) as any ^{15}N enrichment in the hydrolysable AA pool must be derived from the applied substrate via microbial processing during the experiment. The assessment is more complicated however when the applied substrate is a hydrolysable AA (e.g. ^{15}N-Glu) as this must be accounted for in the analytical approach and calculations. Knowles et al. [15] briefly describe a 'backflush' GC-C-IRMS method to obtain accurate δ^{15}N values for both the highly enriched substrate hydrolysable AA and subsequently eluting AAs, which is a first step in tackling such analyses. Somewhat more difficult is the question of whether or not the ^{15}N retained in the initially ^{15}N-labelled substrate should be included in the summation of ^{15}N in the newly synthesised total hydrolysable AA or soil protein pool. At very short incubation periods, it is likely that a high proportion of ^{15}N-labelled substrate remains unaltered and it therefore would not be reasonable to include these data. Once the δ^{15}N values of the substrate hydrolysable AA have reached an apparent equilibrium, however, some of the ^{15}N in the substrate hydrolysable AA will have been used by soil microorganisms in protein synthesis (either intact or via some cycling of the supplied ^{15}N and regeneration of the substrate AA). The best solution to this dilemma depends on the specific purpose of the experiment being undertaken.

For the ^{15}N-Glu experiment in this case, a relatively high level of enrichment remains in Glx at the apparent equilibrium compared with the enrichment of the other AAs (Fig. 3.2c), indicating considerable intact use of the applied ^{15}N in preference of de novo AA biosynthesis. This suggests that Glx should be excluded from the calculation of newly synthesised soil protein, but noting that this will result in an underestimation of activity relative to an inorganic substrate where all hydrolysable AA contributions to newly synthesised soil protein are included. It is also worth considering, however, that the addition of Glu (even at low concentrations) may inhibit its de novo biosynthesis to some degree during the incubation, as microorganisms are extremely efficient and tend not to produce compounds to which they already have access. On one hand, this could represent an unwanted perturbation of the system under study, but on the other, will result in a lower underestimation of ^{15}N incorporated into newly synthesised soil protein if Glx is excluded.

The use of several different treatments applied separately to the same soil allows comparison of their relative 'availabilities' to the soil microbial biomass—in the case of NO$_3^-$, NH$_4^+$ and Glu here, clear differences in the assimilation of these substrates into newly synthesised hydrolysable soil AAs are revealed. Note that these differences are not due to differences in percentage retention of ^{15}N between incubation experiments as all were close to 100% (Table 3.5). Alternatively, the technique can also be used to compare the fate of particular N amendments in different soils to provide hitherto unattainable estimates of the relative 'activity' of the microbial biomass of the soils under selected incubation conditions. In both cases, a measure of newly synthesised protein can be obtained by summing the ^{15}N enrichments of all the AAs (Eq. 2.18) at each time point for each treatment to give the moles of

[15]N in the soil protein pool at that time. Note, however, that new protein will also be biosynthesised from non-labelled sources during the experiment, e.g. following cell lysis or concomitant organic matter mineralisation.

3.4.4 Interpretations of [15]N Stable Isotope Probing Determinations in Relation to Complex Nitrogen Dynamics

Due to the dynamic nature of the soil system any estimates of ^{15}N in the soil protein pool represent the balance of assimilation into/loss from the pool at a given point in time. [15]N incorporated into the soil protein pool does not simply accumulate with time, but is turned over as native soil N turns over, e.g. via catabolic mineralisation. Insights into the dynamics of this aspect of the N cycle in soil can now be gained at the AA level. In these experiments, applied labile substrates (NH_4^+ and Glu) are initially assimilated rapidly, with the amount assimilated increasing considerably between each time point until a transient equilibrium with slower soil N turnover/loss develops (Figs. 3.2a, d and 3.3). For NO_3^- (an energy demanding substrate), on the other hand, the dynamics are more complex and the rate of assimilation is always closer to that of turnover (Figs. 3.2b and 3.3). In natural systems these assimilation-turnover dynamics would be subject to external forcings (e.g. rainfall events, soil type, etc.). Time course incubations of this type allow the overall assimilation-turnover dynamics of the substrate with time, and other environmental variables, to be investigated and provide a measure of substrate availability/lability and value (via rate of incorporation and flux). Although this approach cannot currently generate absolute values for the assimilation of an applied ^{15}N-labelled substrate by the soil microbial biomass or the amount of newly synthesized soil protein, it does provide enhanced insights compared to other currently available methods. It is reassuring that the percentages of applied ^{15}N recovered in soil microbial biomass N studies (e.g. [22, 24]) are comparable (0.8–15.3% across these two studies) to those obtained herein.

3.5 Conclusions

The compound-specific AA ^{15}N-SIP technique described in this chapter to investigate the fate of N in soils offers a number of advantages that existing techniques cannot to reveal a range of new insights, in particular:

(i) The method provides a sensitive and relatively selective means of assessing the microbial assimilation of ^{15}N-labelled substrates applied at environmentally relevant concentrations and appropriately low ^{15}N enrichments to minimise perturbations and ^{15}N discrimination, respectively. Substrate-product concentration monitoring and bulk N isotopic analysis cannot provide such insights, while GC-MS ^{15}N-SIP studies require highly ^{15}N-labelled substrates for sufficient product AA ^{15}N enrichment, making such studies more susceptible to ^{15}N discrimination/isotopic fractionation effects and much more expensive.

(ii)   Valuable insights into microbial biochemical assimilation pathways can be gained and differences are readily revealed in the microbial processing of N-containing substrates of differing chemical/biochemical natures, e.g. inorganic versus organic or different types of inorganic or organic amendment.

(iii) Estimates are provided for newly synthesized soil protein, which are inaccessible based on currently available methods.

(iv)   Detailed quantitative insights can be gained into the dynamics of N cycling from an applied substrate through the soil protein pool.

(v)   Scope exists for using this new approach to probe microbial N cycling behaviour in relation to a wide range of soil biota, ecosystem variables and anthropogenic management regimes. Opportunities for further refinement of the method are exemplified by Redmile-Gordon et al. [39], wherein additional insights were gained by considering different soil protein fractions.

(vi)   The method is potentially adaptable to investigate N cycling into other N-containing biochemical pools, e.g. amino sugars.

Specific findings, with reference to the hypotheses presented in Sect. 3.2, include:

(i)   There were no observable changes in bulk soil % TN contents as a result of a small, but agriculturally relevant N addition, or over the course of the experiments.

(ii)   Bulk soil δ^{15}N values rose following the application of a ^{15}N-labelled substrate and then remained high over the course of the incubation experiments, ^{15}N losses were low.

(iii) There were no observable changes in hydrolysable soil AA concentrations as a result of a small, but agriculturally relevant N addition, or over the course of the experiments.

(iv)   The pattern of ^{15}NH$_4^+$ assimilation into Glx was distinctly different from that into other AAs and occurred more quickly and to a greater extent, reflecting the known, ubiquitous biochemical assimilation pathway of NH$_4^+$, via the reductive amination of α-ketoglutarate to $_L$-Glu catalysed by GDH or via the GS-GOGAT pathway [63–65]. The δ^{15}N values of Glx then decline earlier than those of other AAs as Glu-^{15}N is transaminated into other AAs.

(v)   There were differences in the rates and fluxes of ^{15}N from contrasting ^{15}N-labelled substrates into the total hydrolysable AA pool or soil protein pool.

References

1. Charteris AF, Knowles TDJ, Michaelides K, Evershed RP (2016) Compound-specific amino acid ^{15}N stable isotope probing of nitrogen assimilation by the soil microbial biomass using gas chromatography/combustion/isotope ratio mass spectrometry. Rapid Commun Mass Spectrom 30:1846–1856. https://doi.org/10.1002/rcm.7612
2. Stevenson FJ (1982) Organic forms of soil nitrogen. In: Stevenson FJ (ed) Nitrogen in agricultural soils. American Society of Agronomy, Madison, Wisconsin, USA, pp 67–74
3. Schulten H-R, Schnitzer M (1998) The chemistry of soil organic nitrogen: a review. Biol Fertil Soils 26:1–15
4. Friedel JK, Scheller E (2002) Composition of hydrolysable amino acids in soil organic matter and soil microbial biomass. Soil Biol Biochem 34:315–325
5. Jones DL, Kielland K (2012) Amino acid, peptide and protein mineralization dynamics in a taiga forest soil. Soil Biol Biochem 55:60–69. https://doi.org/10.2136/sssaj2011.0252
6. Michaelides K, Lister D, Wainwright J, Parsons AJ (2012) Linking runoff and erosion dynamics to nutrient fluxes in a degrading dryland landscape. J Geophys Res 117:G00N15. https://doi.org/10.1029/2012jg002071
7. van Groenigen JW, Huygens D, Boeckx P, Kuyper TW, Lubbers IM, Rütting T, Groffman PM (2015) The soil N cycle: new insights and key challenges. Soil 1:235–256. https://doi.org/10.5194/soil-1-235-2015
8. Macdonald AJ, Poulton PR, Powlson DS, Jenkinson DS (1997) Effects of season, soil type and cropping on recoveries, residues and losses of ^{15}N-labelled fertilizer applied to arable crops in spring. J Agric Sci 129:125–154
9. Murphy DV, Macdonald AJ, Stockdale EA, Goulding KWT, Fortune S, Gaunt JL, Poulton PR, Wakefield JA, Webster CP, Wilmer WS (2000) Soluble organic nitrogen in agricultural soils. Biol Fertil Soils 30:374–387
10. Dungait JAJ, Cardenas LM, Blackwell MSA, Wu L, Withers PJA, Chadwick DR, Bol R, Murray PJ, Macdonald AJ, Whitmore AP, Goulding KWT (2012) Advances in the understanding of nutrient dynamics and management in UK agriculture. Sci Total Environ 434:39–50
11. Nannipieri P, Eldor P (2009) The chemical and functional characterization of soil N and its biotic components. Soil Biol Biochem 41:2357–2369
12. Stanford G, Smith SJ (1972) Nitrogen mineralization potentials of soils. Soil Sci Soc Am J 36:465–472
13. Sharifi M, Zebarth BJ, Burton DL, Grant CA, Cooper JM (2007) Evaluation of some indices of potentially mineralisable nitrogen in soil. Soil Sci Soc Am J 71:1233–1239. https://doi.org/10.2136/sssaj2006.0265
14. Barraclough D, Geens EL, Maggs JM (1984) Fate of fertiliser nitrogen applied to grassland. II. Nitrogen-15 leaching results. J Soil Sci 35:191–199
15. Knowles TDJ, Chadwick DR, Bol R, Evershed RP (2010) Tracing the rate and extent of N and C flow from ^{13}C,[15]N-glycine and glutamate into individual de novo synthesised soil amino acids. Org Geochem 41:1259–1268. https://doi.org/10.1016/j.orggeochem.2010.09.003
16. Gibbs P, Barraclough D (1998) Gross mineralisation of nitrogen during the decomposition of leaf protein I (Ribulose 1,5-diphosphate carboxylase) in the presence or absence of sucrose. Soil Biol Biochem 30:1821–1827
17. Barraclough D (1995) ^{15}N isotope dilution techniques to study soil nitrogen transformations and plant uptake. Fertiliser Res 42:185–192
18. Sebilo M, Mayer B, Nicolardot B, Pinay G, Mariotti A (2013) Long-term fate of nitrate fertilizer in agricultural soils. Proc Natl Acad Sci USA 110:18185–18189. https://doi.org/10.1073/pnas.1305372110
19. Ladd JN, Amato M (1986) The fate of nitrogen from legume and fertiliser sources in soils successively cropped with wheat under field conditions. Soil Biol Biochem 18:417–425
20. Pilbeam CJ, Hutchison D (1998) Fate of nitrogen applied in different fertilisers to the surface of a calcareous soil in Syria. Nutr Cycl Agroecosyst 52:55–60

21. Brookes PC, Landman A, Pruden G, Jenkinson DS (1985) Chloroform fumigation and the release of soil nitrogen: a rapid direct extraction method to measure microbial biomass nitrogen in soil. Soil Biol Biochem 17:837–842
22. Nannipieri P, Falchini L, Landi L, Benedetti A, Canali S, Tittarelli F, Ferri D, Convertini G, Badalucco L, Grego S, Vittori-Antisari L, Raglione M, Barraclough D (1999) Nitrogen uptake by crops, soil distribution and recovery of urea-N in a sorghum-wheat rotation in different soils under Mediterranean conditions. Plant Soil 208:43–56
23. Zhou S, Sugawara S, Riya S, Sagehashi M, Toyota K, Terada A, Hosomi M (2011) Effect of infiltration rate in nitrogen dynamics in paddy soil after high-load nitrogen application containing ^{15}N tracer. Ecol Eng 37:685–692. https://doi.org/10.1016/j.ecoleng.2010.04.032
24. Liang B, Yang XY, Murphy DV, He XH, Zhou JB (2013) Fate of ^{15}N-labeled fertilizer in soils under dryland agriculture after 19 years of different fertilizations. Biol Fertil Soils 49:977–986. https://doi.org/10.1007/s00374-013-0789-3
25. Dalal RC (1998) Soil microbial biomass—what do the numbers really mean? Aust J Exp Agric 38:649–665
26. Haubensak KA, Hart SC, Stark JM (2002) Influences of chloroform exposure time and soil water content on C and N release in forest soils. Soil Biol Biochem 34:1549–1562
27. Müller C, Stevens RJ, Laughlin RJ, Ottow JCG, Jäger H-J (2003) Ammonium immobilisation during chloroform fumigation. Soil Biol Biochem 35:651–665. https://doi.org/10.1016/S0038-0717(03)00014-2
28. Jenkinson DS, Brookes PC, Powlson DS (2004) Measuring soil microbial biomass. Soil Biol Biochem 36:5–7. https://doi.org/10.1016/j.soilbio.2003.10.002
29. Jenkinson DS, Fox RH, Rayner JH (1985) Interactions between fertiliser nitrogen and soil nitrogen—the so-called 'priming' effect. J Soil Sci 36:425–444
30. Cadisch G, Espana M, Causey R, Richter M, Shaw E, Morgan JAW, Rahn C, Bending GD (2005) Technical considerations for the use of ^{15}N-DNA stable-isotope probing for functional microbial activity in soils. Rapid Commun Mass Spectrom 19:1424–1428. https://doi.org/10.1002/rcm.1908
31. Buckley DH, Huangyutitham V, Hsu S-F, Nelson TA (2007) Stable isotope probing with ^{15}N achieved by disentangling the effects of genome G + C content and isotope enrichment on DNA density. Appl Environ Microbiol 73:3189–3195. https://doi.org/10.1128/AEM.02609-06
32. Cupples AM, Shaffer EA, Chee-Sanford JC, Sims GK (2007) DNA buoyant density shifts during ^{15}N-DNA stable isotope probing. Microbiol Res 162:328–334. https://doi.org/10.1016/j.micres.2006.01.016
33. Addison SL, McDonald IR, Lloyd-Jones G (2010) Stable isotope probing: technical considerations when resolving ^{15}N-labeled RNA in gradients. J Microbiol Methods 80:70–75. https://doi.org/10.1016/j.mimet.2009.11.002
34. Lee TA, Forrest TM, Wilson GE, Hardy JK (1990) The use of multiple mass spectral line pairs for enhanced precision in isotope enrichment studies of ^{15}N-labeled amino acids. Anal Biochem 185:24–28
35. Metges CC, Petzke K-J, Hennig U (1996) Gas chromatography/combustion/isotope ratio mass spectrometric comparison of N-acetyl- and N-pivaloyl amino acid esters to measure ^{15}N isotopic abundances in physiological samples: a pilot study on amino acid synthesis in the upper gastrointestinal tract of minipigs. J Mass Spectrom 31:367–376
36. Meier-Augenstein W (1999) Applied gas chromatography coupled to isotope ratio mass spectrometry. J Chromatogr A 842:351–371
37. Merritt DA, Hayes JM (1994) Nitrogen isotopic analyses by isotope-ratio-monitoring gas chromatography/mass spectrometry. J Am Soc Mass Spectrom 5:387–397
38. Dörr N, Kaiser K, Sauheitl L, Lamersdorf N, Stange CF, Guggenberger G (2012) Fate of ammonium ^{15}N in a Norway spruce forest under long-term reduction in atmospheric N deposition. Biogeochemistry 107:409–422. https://doi.org/10.1007/s10533-010-9561-z
39. Redmile-Gordon MA, Evershed RP, Hirsch PR, White RP, Goulding KWT (2015) Soil organic matter and the extracellular microbial matrix show contrasting responses to C and N availability. Soil Biol Biochem 88:257–267. https://doi.org/10.1016/j.soilbio.2015.05.025

40. Evershed RP, Crossman ZM, Bull ID, Mottram H, Dungait JAJ, Maxfield PJ, Brennand EL (2006) ^{13}C-Labelling of lipids to investigate microbial communities in the environment. Curr Opin Biotechnol 17:72–82. https://doi.org/10.1016/j.copbio.2006.01.003

41. Maxfield PJ, Dildar N, Hornibrook ERC, Stott AW, Evershed RP (2012) Stable isotope switching (SIS): a new stable isotope probing (SIP) approach to determine carbon flow in the soil food web and dynamics in organic matter pools. Rapid Commun Mass Spectrom 26:997–1004. https://doi.org/10.1002/rcm.6172

42. Petzke KJ, Metges CC (2012) Practical recommendations for the reduction of memory effects in compound-specific ^{15}N/^{14}N-ratio analysis of enriched amino acids by gas chromatography/combustion/isotope ratio mass spectrometry. Rapid Commun Mass Spectrom 26:195–204. https://doi.org/10.1002/rcm.5319

43. Roberts P, Jones DL (2008) Critical evaluation of methods for determining total protein in soil solution. Soil Biol Biochem 40:1485–1495. https://doi.org/10.1016/j.soilbio.2008.01.001

44. Vauquelin LN, Robiquet PJ (1806) The discovery of a new plant principle in *Asparagus sativus*. Ann Chim 57:88–93

45. Wagner I, Musso H (1983) New naturally occurring amino acids. Angew Chem, Int Ed Engl 22:816–828

46. Hertweck C (2011) Biosynthesis and charging of pyrrolysine, the 22nd genetically encoded amino acid. Angew Chem, Int Ed 50:9540–9541. https://doi.org/10.1002/anie.201103769

47. Vranova V, Rejsek K, Skene KR, Formanek P (2011) Non-protein amino acids: plant, soil and ecosystem interactions. Plant Soil 342:31–48

48. Cava F, Lam H, de Pedro MA, Waldor MK (2011) Emerging knowledge of regulatory roles of d-amino acids in bacteria. Cell Mol Life Sci 68:817–831. https://doi.org/10.1007/s00018-010-0571-8

49. Schleifer KH, Kandler O (1972) Peptidoglycan types of bacterial cell walls and their taxonomic implications. Bacteriological Rev 36:407–477

50. Amelung W (2003) Nitrogen biomarkers and their fate in soil. J Plant Nutr Soil Sci 166:677–686. https://doi.org/10.1002/jpln.200321274

51. Nagata Y, Fujiwara T, Kawaguchi-Nagata K, Fukumori Y, Yamanaka T (1998) Occurrence of peptidyl D-amino acids in soluble fractions of several eubacteria, archaea and eukaryotes. Biochem Biophys Acta 1379:76–82

52. O'Dowd RW, Barraclough D, Hopkins DW (1999) Nitrogen and carbon mineralization in soil amended with D- and L-leucine. Soil Biol Biochem 31:1573–1578

53. Bogatyreva NS, Finkelstein AV, Galzitskaya OX (2006) Trend of amino acid composition of proteins of different taxa. J Bioinform Comput Biol 4:597–608

54. Foerstner KU, von Mering C, Hooper SD, Bork P (2005) Environments shape the nucleotide composition of genomes. Eur Mol Biol Organ 6:1208–1213. https://doi.org/10.1038/sj.embor.7400538

55. Moura A, Savageau MA, Alves R (2013) Relative amino acid composition signatures of organisms and environments. PLoS ONE 8(10):e77319. https://doi.org/10.1371/journal.pone.0077319

56. Goh KM, Edmeades DC (1979) Distribution and partial characterisation of acid hydrolysable organic nitrogen in six New Zealand soils. Soil Biol Biochem 11:127–132

57. Senwo ZN, Tabatabai MA (1998) Amino acid composition of soil organic matter. Biol Fertil Soils 26:235–242

58. Wang M, Kapralov MV, Anisimova M (2011) Coevolution of amino acid residues in the key photosynthetic enzyme Rubisco. Evol Biol 11(1):266

59. Raven JA (2013) Rubisco: still the most abundant protein of Earth? New Phytol 198:1–3

60. Defra (2010) Fertiliser manual (RB209), pp 60, 65. Report available from http://www.ahdb.org.uk/projects/CropNutrition.aspx. Accessed 26 Mar 2014

61. Mathieu O, Lévêque J, Hénault C, Ambus P, Milloux M, Andreux F (2007) Influence of ^{15}N enrichment on the net isotopic fractionation factor during the reduction of nitrate to nitrous oxide in soil. Rapid Commun Mass Spectrom 21:1447–1451. https://doi.org/10.1002/rcm.2979

62. Tang FHM, Maggi F (2012) The effect of ^{15}N to ^{14}N ratio on nitrification, denitrification and dissimilatory nitrate reduction. Rapid Commun Mass Spectrom 26:430–442. https://doi.org/10.1002/rcm.6119

63. Meers JL, Tempest DW, Brown CM (1970) 'Glutamine(amide): 2-Oxoglutarate amino transferase oxido-reductase (NADP)', an enzyme involved in the synthesis of glutamate by some bacteria. J Gen Microbiol 64:178–194

64. Tempest DW, Meers JL, Brown CM (1970) Synthesis of glutamate in *Aerobacter aerogenes* by a hitherto unknown route. Biochem J 117:405–407

65. Santero E, Hervás A, Canosa I, Govantes F (2012) Glutamate dehydrogenases: enzymology, physiological role and biotechnological relevance. In: Canuto RA (ed) Dehydrogenases, InTech, pp 289–291. Published online Nov 2014

66. Puri G, Ashman MR (1999) Microbial immobilization of ^{15}N-labelled ammonium and nitrate in a temperate woodland soil. Soil Biol Biochem 31:929–931

67. Gelfand DH, Steinberg RA (1977) Escherichia coli mutants deficient in the aspartate and aromatic amino acid in aminotransferases. J Bacteriol 130:429–440

Chapter 4
Biosynthetic Routing, Rates and Extents of Microbial Fertiliser Nitrogen Assimilation in Two Grazed Grassland Soils

4.1 Introduction

4.1.1 Context

Freshwater resources worldwide are affected by high and rising NO_3^- concentrations which threaten human health and the natural environment, reducing diversity, resilience and value [1–7]. On a global scale, agriculture represents the main source of this pollution [2, 6, 7], primarily due to the run-off and leaching of mobile NO_3^- ions derived from manures and synthetic fertilisers [1, 3, 6, 7]. In fact, on average, only about 50% of the fertiliser N applied to crops is taken up and, under certain conditions, (soil type, season/climate, crop and management practices, especially fertiliser application rates) leaching losses can be substantial [1, 8].

Improving crop NUE has become increasingly important, primarily for environmental reasons, but also as the economics of agricultural production tighten and the wider costs of the subsequent pollution have been realised [2, 9, 10]. Accordingly, this has been the focus of much research (e.g. [11–14]), as have N balance investigations [10] and studies tracing the fate of applied N under various conditions (often using ^{15}N-labelled fertilisers; e.g. [15–20]) all with the aim of identifying and then reducing losses. Yet despite over 100 years of research, the N balance, even in 'simpler' arable systems, has proven elusive [10, 21] and the complexity and heterogeneity of fertiliser-derived organic N products prevents thorough interrogation of the biomolecular fate of applied N in detail (Sect. 3.1).

Historically, the rates of N mineralisation, immobilisation and nitrification in soils (commonly measured by isotope pool dilution of $^{15}NH_4^+$ or $^{15}NO_3^-$) and the availability of inorganic N species for crop uptake/loss have been the focus for improving crop yields and reducing nutrient losses [22–24]. Even in agricultural soils, however, N stored in organic forms dominates inorganic N [8] and although direct plant uptake of organic N [24–26] may not play a substantial role in most

© Springer Nature Switzerland AG 2019
A. F. Charteris, ^{15}N Tracing of Microbial Assimilation, Partitioning
and Transport of Fertilisers in Grassland Soils, Springer Theses,
https://doi.org/10.1007/978-3-030-31057-8_4

agricultural settings (up to 20% of plant N uptake; [27], this large and heterogeneous soil N pool is still important in soil N cycling and balancing the supply of N to microorganisms, plants and loss pathways (in some cases supplying 30–50% of the inorganic N for crop uptake; [8, 28, 29]).

The wide range of possible organic N compounds in soil complicates their investigation and subsequently, the microbially-mediated transformations between inorganic and organic N species and the cycling of organic N in the soil are poorly understood [24–26, 30]. Accordingly, a better understanding of the soil N cycle as a whole, (and not simply focused on inorganic N species) and then of the interactions between the soil microbial biomass (as the major mediators of the soil N cycle and thus regulators of N available for plant uptake/leaching) and plants (whose rhizosphere zones and roots also play an active role in N cycling) with respect to N uptake and cycling [8, 31] could contribute to the development of more sustainable agricultural soil management practices.

The work described in this chapter investigating the soil microbial assimilation of a range of commonly applied fertiliser N compounds in two soils using the ^{15}N-SIP method described in Chap. 3 [32] constitutes a small step toward greater understanding of the microbially-mediated transformations of inorganic fertiliser N to organic N and could ultimately contribute to the development of a more complete picture of soil N cycling in response to fertiliser N applications. The next section (Sect. 4.1.2) is a short review of microbial N uptake and assimilation, the biochemistry of which may be expected to play an important role in the microbial processing of N-containing compounds in soils. Following this, in Sect. 4.1.3 the known soil processing of N fertiliser compounds at various (macroscopic) scales is summarised.

4.1.2 Microbial Nitrogen Uptake and Assimilation

Although microbial mineralisation to provide inorganic N for plant uptake is no longer considered a necessity [24, 26, 33], this does not detract from the major role microbial activity plays in soil N cycling and the effects this activity has on nutrient availability and leaching [34–37]. Competition between plants and microorganisms for available N simply occurs over a wider range of N sources and, as microorganisms generally outcompete plants for all forms of N in the short term [38–40], they will also exert a control on N availability in this way (e.g. Jackson et al. [34] found that microbial uptake strongly controlled NO_3^- availability to plants). Improved soil biotic diversity and abundance has been suggested to enhance agricultural sustainability [41] and in terms of applied N, the soil microbial biomass is likely to be a major mediator controlling its fate, both by uptake and transformation. Work by Abaas et al. [37] demonstrated the importance of the microbial biomass in controlling the physical leaching of applied NH_4^+ and DON (represented by Ala and trialanine) and the persistent mobility of NO_3^-, but the mechanisms by which this regulation occurred were not investigated.

Microorganisms can assimilate N as NO_3^-, ammonia $(NH_3)/NH_4^+$ (either applied as, or extracellularly mineralised to, commonly known as the MIT route), or low molecular weight (LMW) organic compounds, such as AAs and even short peptides (commonly known as the direct route; [42–47]). The relative importance of these pathways is thought to be controlled primarily by the form of N available and the availability and form of C [44]. Nitrogen uptake and transformation processes that are facilitated by enzymes require C, N and energy for the synthesis of these enzymes and their subsequent expression [44]. Many soil microorganisms are able to adapt their N assimilation strategies to expend minimal energy, based on the availability of N and C (both form/bioavailability/energy required to assimilate and actual availability in terms of concentration), in order to fulfil their nutritional requirements and this is clearly a competitive advantage [44, 48]. In theory, therefore, it should be possible to manage the soil N available for plant uptake/leaching by manipulating soil N and C availabilities (sources and concentrations), but microbial responses to this and its effects on the interactions of the soil N and C cycles and associated feedbacks are complex and require much further investigation before such a strategy could be successfully adopted [8, 49–51].

4.1.2.1 Microbial Uptake and Assimilation of Nitrate

Assimilatory NO_3^- reduction can be performed by bacteria, fungi, algae and higher plants and possibly, genetic and functional evidence suggests, some archaea [52, 53]. Nitrate is thought to be actively transported into bacterial cells for assimilation by an ABC-type transporter (adenosine triphosphate [ATP]-dependent) that requires a periplasmic binding protein or a NarK-type transporter (proton-motive force dependent; [52–54]). Sequence data and some functional studies suggest that appropriate ABC-type transporters are also present in some archaea [52]. Once taken up, NO_3^- has to be reduced, first to NO_2^- (by an assimilatory NO_3^- reductase; Eq. (4.1); [55] and then to NH_4^+ (by an assimilatory NO_2^- reductase; Eq. (4.2); [56]) which is assimilated, mainly via the GS-GOGAT pathway [44, 52].

$$NO_3^- + 2H^+ + 2e^- \rightarrow NO_2^- + H_2O \qquad (4.1)$$

$$NO_2^- + 6H^+ + 6e^- \rightarrow NH_4^+ + 2OH^- \qquad (4.2)$$

Bacterial assimilatory NO_3^- reductases are ferredoxin-/flavodoxin-dependent or NADH-dependent cytoplasmic Mo-containing enzymes [52, 53, 55] and bacterial assimilatory NO_2^- reductases are ferredoxin-dependent or NADH-dependent enzymes which contain a siroheme and a [4Fe-4S] centre and are also located in the cytoplasm [53]. Archaeal NO_3^- assimilation has not been widely demonstrated, but genes encoding putative NO_3^- transporters, NO_3^- reductases and NO_2^- reductases have been found in a wide range of archaea, some functional studies have been carried out and assimilatory NO_3^- and NO_2^- reductases have been purified from an extreme halophilic archaeon, *Haloferax mediterranei* [52, 57, 58].

Nitrate assimilation is stimulated by NO_3^- availability under conditions of NH_4^+ limitation (either due simply to low NH_4^+ concentrations, or in soils, perhaps as a result of NH_4^+ depletion in microsites or competition with nitrifiers; [44, 49, 59]). Nitrate assimilation is strongly and quickly (but not completely) inhibited by NH_4^+, even at low concentrations (possibly due either to direct or indirect repression of NO_3^- transport or NO_3^- reductase synthesis; [44, 59]). These effects, however, do vary for different microorganisms and under different soil conditions [59].

4.1.2.2 Microbial Uptake and Assimilation of Ammonia and Ammonium

Gaseous NH_3 can diffuse through biological membranes (rates depend on concentration gradients and pH) and microorganisms can also take up NH_3 and NH_4^+ via a range membrane-bound proteins (possibly passively as well as actively; [44]). Synthesis of these proteins is inhibited if external NH_4^+ concentrations exceed 10–20 mM when presumably, NH_3 diffusional inflow is sufficient [44]. Following uptake, NH_4^+ can be assimilated via the reductive amination of α-ketoglutarate to $_L$-Glu catalysed by GDH or via the GS-GOGAT pathway (Fig. 4.1; [60–63]).

The GDH pathway requires relatively high NH_4^+ concentrations to drive the reaction equilibrium to produce $_L$-Glu while the GS-GOGAT pathway is operative at much lower NH_4^+ concentrations, but consumes 18–30% more energy [44, 60, 62–64]. It is thought that bacteria primarily use the GS-GOGAT pathway, while fungi mainly use the GDH pathway [65, 66]. Ammonium is generally the preferred source of inorganic N for assimilation by soil microorganisms due to the lower energy costs associated with its uptake and incorporation into cell material [44, 48, 59, 67, 68], but some bacterial NO_3^- assimilation has been observed at high NH_4^+ concentrations [69].

4.1.2.3 Microbial Uptake and Assimilation of Urea

Urea is a LMW organic N molecule that is agriculturally significant as it is the most extensively applied synthetic N fertiliser in the world and an important constituent of urine [70]. Its popularity results from its high stability, lower cost and high N content (46% by weight; [70]). Following deposition or application to soil, urea is hydrolysed by the ubiquitous enzyme urease, yielding NH_3 and carbamate (NH_2CO_2H) which decomposes to another molecule of NH_3 and carbonic acid (H_2CO_3; [71, 72]):

$$(NH_2)_2CO + H_2O \; \rightarrow \; NH_3 + NH_2CO_2H \overset{H_2O}{\rightarrow} 2NH_3 + H_2CO_3 \qquad (4.3)$$

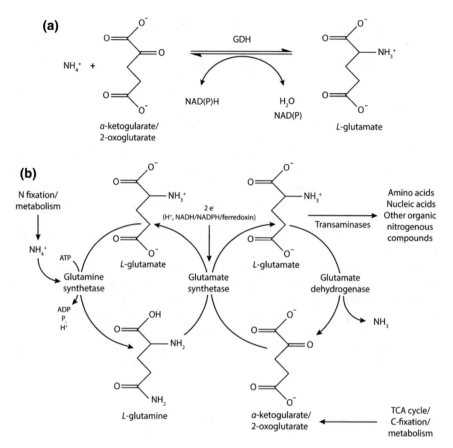

Fig. 4.1 Ammonium assimilation pathways: **a** the GDH pathway, **b** the GS-GOGAT pathway. ADP; Adenosine diphosphate

Soil pH is increased by urea hydrolysis as in soil solution, NH_3 and carbonic acid are in equilibrium [71]:

$$H_2CO_3 \rightleftharpoons H^+ + HCO_3^- \tag{4.4}$$

$$2NH_3 + 2H_2O \rightleftharpoons 2NH_4^+ + 2OH^- \tag{4.5}$$

Ureases are Ni-containing enzymes produced by bacteria, yeasts, fungi, algae and plants and their synthesis is usually regulated by N availability [44, 71]. Although most urease activity in soils is extracellular, urease is also a cytoplasmic enzyme in bacteria and yeasts [44, 71]. The NH_3 produced by urease hydrolysis of urea may be volatilised or taken up as NH_3 or NH_4^+ as described above. As urease activity raises soil pH, the equilibrium in Eq. (4.5) favours the left hand side and as much as 50% of the N applied as urea may be lost by volatilisation of NH_3 [73, 74].

4.1.2.4   Microbial Uptake and Assimilation of Organic Nitrogen

Microbial uptake of other LMW organic N-containing molecules is facilitated by a wide range of transport enzymes [44]. Bacteria are known to possess up to twelve kinetically-defined cytoplasmic transport systems for groups of structurally related AAs, while fungi have fewer, less specific systems able to transport groups of AAs with similar properties [44]. Amino acid uptake is an active process and for most transport enzymes, regulated by N availability; transport is inhibited by NH_4^+ and high intracellular AA concentrations and stimulated by N, C or S limitation [44]. Following uptake, AAs may be directly incorporated intact into new proteins or enzymatically deaminated or transaminated and used in the synthesis of other compounds [75–78]. Different AAs are taken up and assimilated at different rates and to different extents, partly due to the differing strengths of their sorptive interactions with soil and their differing positions and roles in central biochemical pathways [59, 77, 79–81].

Peptide uptake is similarly active, but favoured over AA uptake, presumably because less energy is generally required to transport a peptide than a mixture of the same AAs and peptides are thought to be nutritionally superior [43, 44, 46, 82, 83]. There is an upper size limit, however, for peptide transport across the cytoplasmic membrane of apparently approximately 600 Da (penta- or hexapeptides; [44]). Amino sugar transport systems have likewise been found in both bacteria and fungi and are stimulated by amino sugar availability and inhibited by glucose [44].

4.1.3   Fertiliser Nitrogen Immobilisation in Soils

Given the widespread use and importance of N-containing amendments, many studies on a variety of scales (size and time) have assessed the immobilisation of fertiliser/organic waste N following soil application. In work where the immobilisation of different N fertilisers/organic wastes (predominantly N as NO_3^-, NH_4^+, urea or manure/slurry N as NH_4^+) was compared, NH_4^+-N or urea-N was generally immobilised into soil 'organic N' to a much greater extent than NO_3^--N, often with little difference between different soils where this was also tested (e.g. [19, 34, 67, 68, 84–88]). This is unsurprising given the common preference of soil microorganisms for NH_4^+ [44, 48, 59, 67, 68].

A few investigations have, however, recorded higher soil microbial immobilisations of NO_3^- than NH_4^+ (e.g. [39, 49]) and these were accounted for by evoking low soil NH_4^+ concentrations, the lower diffusional mobility of NH_4^+ and potentially strong competition for NH_4^+ with nitrifiers. It has also been suggested that the microbial populations of soils regularly receiving relatively large quantities of synthetic fertilisers could adapt and shift their N demand to NO_3^- [39]. Evidence for such a shift was not found by Burger and Jackson [49] in their comparison of the microbial immobilisation of NO_3^- and NH_4^+ in an organic versus a conventionally managed system—in both soils more NO_3^- than NH_4^+ was immobilised and microbial NO_3^-

assimilation was actually higher in the organic system (this was ascribed to a greater availability of C supporting a more active microbial biomass with a larger demand for N). More recent work by Bunch and Bernot [89], however, did demonstrate the ability of stream sediment microbial communities to adapt to some extent to higher NO_3^- and NH_4^+ inputs.

The majority of studies investigating the immobilisation of N-containing amendments in soils have estimated the applied N immobilised as 'organic N' using ^{15}N-labelled fertiliser compounds/organic wastes by either: assuming that all of the ^{15}N retained in the soil is in an organic form (e.g. [20, 85]); extracting/subtracting ^{15}N-labelled inorganic compounds from bulk soils/values (e.g. [19, 34, 68, 84, 87, 88, 90, 91]); or by extraction and isotopic analysis of microbial biomass N (e.g. [92–95]). As discussed in Chap. 3, none of these approaches are able to reveal the biomolecular fate of applied ^{15}N and only very few studies have investigated this in any more detail.

A study of fertiliser N immobilisation and remobilisation in cultivated soils by Sulçe et al. [96] attained some more detail by fractionating the soils using acid hydrolysis in order to access a 'predominantly AA' fraction which acted as the most 'active' or 'labile' pool of soil organic N, containing most of the immobilised ^{15}N and supplying most of the remobilised ^{15}N. Works such as those of He et al. [97] and Geisseler and Horwath [98] go further and do provide insights into the biomolecular fate of applied ^{15}N (from inorganic ^{15}N into amino sugars, and from ^{15}N-AAs into microbial biomass [intact] and keto acids, respectively) using GC-MS and while valuable, the disadvantages of this approach are highlighted in Chap. 3. In an extensive search of the literature, only three studies exploiting the high sensitivity and selectivity of GC-C-IRMS to trace the fate of applied ^{15}N in soils were found: Knowles et al. [77] traced the fate of dual ^{13}C- and ^{15}N-labelled Glu and Gly into newly synthesised soil AAs in work not related to fertiliser N inputs, but complimentary to that discussed in this chapter; Dörr et al. [99] investigated the fate of ^{15}NH$_4^+$ in a Norwegian spruce forest in relation to changing atmospheric N deposition rates; and Redmile-Gordon et al. [100] compared the response of hydrolysable soil AAs and extracellular polymeric substance AAs to ^{15}NH$_4$^{15}NO$_3$ additions with and without glycerol.

4.2 Objectives

The work presented in this chapter is the first to investigate the biomolecular fate of commonly applied fertiliser compounds via microbially-mediated transformations into the total hydrolysable AA or soil protein pool. It will provide more detailed insights into the processing of fertiliser compounds in soils and contribute to a more complete picture of the soil N cycle (particularly in response to fertiliser N additions). For simplicity and in order to gain detailed understanding of this part of the system without other influences, bare soils free of plants have been investigated. Although this is in some ways quite unrealistic, so are the small-scale laboratory incubation experiments employed and both compromises are considered necessary first steps in slowly unpicking such a complex system. The experiments could also be thought of as representing bare, fallow or 'just sown' soils, which have been shown to be particularly susceptible to leaching losses [1, 10, 101].

Nitrate, NH_4^+ and urea were chosen for study as common fertiliser N sources: NH_4NO_3 is the most used fertiliser in Great Britain [102], followed by urea, which is the most used worldwide [70]. Although much of the decline in N fertiliser use in Great Britain over the past 20–30 years (Fig. 1.4) results from decreased inputs to grasslands, overall application rates of 60 kg N ha^{-1} were recorded in 2014 [103]. In this work, $^{15}NO_3^-$ and $^{15}NH_4^+$ were applied separately because they are chemically quite different—bearing opposite charges, disparate oxidation states (+V and −III, respectively) and occupying different positions in the N cycle—and because this would obviate the added complexity of their interactions in soils as different N sources for microbial uptake.

Some mixed treatments (i.e. inorganic N plus cattle slurry or glucose) were also included in the investigation. Manures/slurries are commonly applied to crops as supplementary N (and organic matter) sources around the world; in Great Britain in 2014, for example, 66% of farms used manures/slurries (mostly cattle-derived) on at least one field (most commonly on grassland; [102]). Glucose is a simple, highly labile C source, which may be released in soils as a degradation product of plant litter polysaccharides and which is also the most commonly rhizodeposited sugar [104, 105]. These treatments were applied as mixtures partly due to the often lower and less predictable N content of slurries [10], but also because the supply of available N and C can improve microbial N immobilisation [22, 68, 88] and thereby reduce N losses [8, 51].

The specific objectives of this work were to:

(i)   Monitor bulk soil $\delta^{15}N$ values following application of $^{15}NO_3^-$, $^{15}NH_4^+$ or ^{15}N-urea to two soils and $^{15}NO_3^-$ or $^{15}NH_4^+$ applied with slurry and glucose (separately) in one soil as indicators of application efficiency and bulk soil processes.

(ii) Reveal and compare the patterns in individual AA $\delta^{15}N$ values and the partitioning of ^{15}N between AAs in response to the addition of $^{15}NO_3^-$, $^{15}NH_4^+$ or ^{15}N-urea in two soils and $^{15}NO_3^-$ or $^{15}NH_4^+$ applied with slurry and glucose (separately) in one soil as indicators of the primary biochemical pathways operating in the assimilation of these substrates.

(iii) Determine and compare the rates and fluxes of the conversion of $^{15}NO_3^-$, $^{15}NH_4^+$ and ^{15}N-urea in two soils and $^{15}NO_3^-$ and $^{15}NH_4^+$ applied with slurry and glucose (separately) in one soil into individual AAs.

(iv) Determine and compare the percentages of applied (fate) and retained (partitioning) ^{15}N (as $^{15}NO_3^-$, $^{15}NH_4^+$ or ^{15}N-urea in two soils and $^{15}NO_3^-$ or $^{15}NH_4^+$ applied with slurry or glucose in one of the soils) incorporated into the total hydrolysable AA or soil protein pool as a whole (and the rates of this incorporation).

Potential findings include:

(i)   Valuable insights into the biochemical pathways, rates and extents of microbial assimilation of different fertiliser N forms applied at environmentally relevant concentrations.

(ii) A greater understanding of the biomolecular fate and partitioning of three fertiliser N forms in two soils.

(iii) Estimates of newly synthesised soil protein using the applied ^{15}N, enabling comparisons of the relative 'availabilities' of different N sources in the same soil and of the processing of the same N source by different soils (different or differently functioning soil microbial populations).

The hypotheses tested in this work are:

(i)   Losses of ^{15}N over the course of the incubation experiments in both soils will be very small as the experiments have been designed to exclude them as far as possible.

(ii) As a ubiquitous primary metabolic process, the assimilation of ^{15}NH$_4$$^+$ into Glu via GDH or the GS-GOGAT pathway, and then into other AAs by transamination reactions, will be reflected in the δ^{15}N values of individual AAs with time.

(iii) ^{15}N-labelled NH$_4$$^+$ will be incorporated into the total hydrolysable AA pool to a greater extent than ^{15}N-urea and ^{15}NO$_3$$^-$ will be least incorporated for both soils.

(iv) Co-additions with slurry or glucose will increase microbial N assimilation, with potentially an inhibitory effect being observed for ^{15}NO$_3$$^-$ with slurry due to the addition (and preferential assimilation) of slurry-NH$_4$$^+$.

(v)   There will be little difference between the two soils as the extracted AAs will integrate the thousands of proteins present in soils, providing an 'averaged' signal that would be expected to be similar across many soils and also due to the ubiquity of primary metabolic processes that operate in microorganisms. The soil that regularly received fertiliser applications may process the applied N fertilisers more efficiently.

4.3 Sites, Sampling, Materials and Methods

The soils used for the laboratory incubation experiments were sampled from the RM and WA sites and prepared for the experiments as described in Sects. 2.2 and 2.3. Incubation experiments were carried out as described in Sect. 2.4 and the substrates applied and incubation periods at which samples were taken are displayed in Table 4.1. All incubation experiments with the RM soil were carried out using the soil sampled in February 2013. The incubation experiments with the WA soil were carried out using soil sampled in January 2013, except the ^{15}NH$_4$$^+$ plus glucose (^{15}NH$_4$$^+$ + G) and urea (^{15}N-U) experiments for which soil sampled in October 2014 was used.

Table 4.1 Table summarising the laboratory incubation experiments discussed in this chapter

Soil	Substrate	Key	Labelling	Mass substrate applied/10 g soil	Mass N/10 g soil (µg)	Equivalent[a] fertilisation rate (kg N ha⁻¹ year⁻¹)	Incubation periods
Rowden Moor	$K^{15}NO_3$	RM $^{15}NO_3{}^-$	10 atom % ^{15}N	400 µg in 200 µl DDW	55	100	1.5, 3, 6 & 12 h and 1, 2, 4, 8, 16 & 32 days
	$^{15}NH_4Cl$	RM $^{15}NH_4{}^+$	10 atom % ^{15}N	400 µg in 200 µl DDW	105	190	1.5, 3, 6 & 12 h and 1, 2, 4, 8, 16 & 32 days
	$K^{15}NO_3$ + slurry	RM $^{15}NO_3{}^-$ + 2	<10 atom % ^{15}N	400 µg in 200 µl DDW + 130 µl slurry	195	355	3, 6 & 12 h and 1, 2, 4, 8, 16, 32 & 64 days
	$CO(^{15}NH_2)_2$	RM ^{15}N-U	10 atom % ^{15}N	400 µg in 200 µl DDW	185	340	3 h and 2, 16 & 32 days
Winterbourne Abbas	$K^{15}NO_3$	WA $^{15}NO_3{}^-$	10 atom % ^{15}N	400 µg in 200 µl DDW	55	100	1.5, 3, 6 & 12 h and 1, 2, 4, 8, 16 & 32 days[b]
	$^{15}NH_4Cl$	WA $^{15}NH_4{}^+$	10 atom % ^{15}N	400 µg in 200 µl DDW	105	190	1.5, 3, 6 & 12 h and 1, 2, 4, 8, 16 & 32 days[c]
	$K^{15}NO_3$ + slurry	WA $^{15}NO_3{}^-$ + 2	<10 atom % ^{15}N	400 µg in 200 µl DDW + 130 µl slurry	195	355	2, 32 & 64 days
	$^{15}NH_4Cl$ + slurry	WA $^{15}NH_4{}^+$ + 2	<10 atom % ^{15}N	400 µg in 200 µl DDW + 130 µl slurry	245	445	2, 32 & 64 days

(continued)

Table 4.1 (continued)

Soil	Substrate	Key	Labelling	Mass substrate applied/10 g soil	Mass N/10 g soil (µg)	Equivalent[a] fertilisation rate (kg N ha^{-1} year^{-1})	Incubation periods
	K^{15}NO$_3$ + glucose	WA ^{15}NO$_3^-$ + G	10 atom % ^{15}N	400 µg + 1390 µg glucose in 400 µl DDW	55	100	32 days
	^{15}NH$_4$Cl + glucose	WA ^{15}NH$_4^+$ + G	10 atom % ^{15}N	400 µg + 1390 µg glucose in 400 µl DDW	105	190	32 days
	CO(^{15}NH$_2$)$_2$	WA ^{15}N-U	10 atom % ^{15}N	400 µg in 200 µl DDW	185	340	3 h and 2, 16 & 32 days

[a] Equivalent fertilisation rate calculated based on a 0.3 m soil depth and an average of 5–6 applications between February and October. The rates are generally within the range recommended for grasslands for dairy grazing (140–340 kg N ha^{-1} year^{-1}; [106]), with the exception of the slurry treatments, in which only approximately half of the N is actually 'available' reducing calculated application rates to 225 and 330 kg N ha^{-1} year^{-1} for the ^{15}NO$_3^-$ + 2 and ^{15}NH$_4^+$ + 2 treatments

[b] Not all time-points analysed for AAs, only 3 and 6 hours and 2, 4, 16 and 32 days

[c] Not all time-points analysed for AAs, only 1.5, 3 and 12 hours and 2, 8 and 32 days

Table 4.2 Basic characteristics of the slurry used in the X + 2 incubation experiments

	% DMC	% TC	% TN
Mean	2.86	45	3.8
SE (n = 3)	0.0996	0.024	0.0088

The $^{15}NO_3{}^-$ and $^{15}NH_4{}^+$ incubation experiments were conducted first, followed by the slurry (X + 2) and $^{15}NO_3{}^-$ plus glucose ($^{15}NO_3{}^- + G$) incubation experiments and finally the $^{15}NH_4{}^+ + G$ and ^{15}N-U incubation experiments.

Initially, time points from 1.5 h to 32 days were selected based on the ^{15}N-AA incubation experiments of Knowles [80] which were carried out over 3 h to 32 days and captured ^{15}N transfer dynamics well. As rates of ^{15}N transfer between different AAs were high only over the first 1–2 days for Knowles [80], a sampling point at 1.5 h was added. In later incubation experiments, this time point was deemed unnecessary and a time point at 64 days was added to make sure that the system had reached a (temporary) equilibrium or steady state by $t = 32$ days.

The slurry applied in the X + 2 incubation experiments was obtained from fresh dairy cow pats collected from Little Broadheath, the field also sampled for soil at the WA site. The cow pats were homogenised and DDW added over a 2 mm sieve to produce a low DMC slurry [106] suitable for application by injection. Basic characteristics of the slurry are shown in Table 4.2. The additional N added in the form of slurry in the X + 2 incubation experiments will have diluted the ^{15}N-labelling of the treatment based on the $NO_3{}^-$ or $NH_4{}^+$ content of the slurry and its N isotopic composition. This dilution may be calculated using Eq. (2.13). Based on Defra [106] guidance, it is estimated that approximately half of slurry TN is 'available N' (urea/$NH_4{}^+$/$NO_3{}^-$) and most of this available N is likely to be in the form of $NH_4{}^+$ with almost no $NO_3{}^-$ [8, 107]. In addition, slurry-$NH_4{}^+$ generally has an N isotopic composition in the region of 10‰ and slurry-$NO_3{}^-$, an N isotopic composition of *ca.* 15‰. Using these approximations in Eq. (2.13) it is possible that the N isotopic composition of the applied $^{15}NO_3{}^-$ was reduced only slightly in the $^{15}NO_3{}^- + 2$ incubation experiments and that of the applied $^{15}NH_4{}^+$ in the $^{15}NH_4{}^+ + 2$ incubation experiments was reduced to *ca.* 6.5–7.3 atom %.

The addition of 1390 μg glucose was selected to provide 560 μg C, about ten times the amount of N supplied as $^{15}NO_3{}^-$ in order to ensure that available C could not be limiting in these incubation experiments.

Hydrolysable soil AAs were extracted, isolated and derivatised as described in Sect. 2.5 and soils analysed (bulk % TC and % TN contents and N isotopic compositions and AA concentrations and N isotopic compositions) as described in Sect. 2.6. Data processing was carried out as described in Sect. 2.7.

4.4 Results

4.4.1 Bulk Soil Percentage Total Carbon and Percentage Total Nitrogen Contents

Bulk soil % TC and % TN contents varied in individual incubation samples, but there are no observable trends in the bulk soil content of these components over the course of the incubation experiments (Table 4.3). Bulk soil % TC contents are summarised as average values for all replicates at every time point (including $t = 0$ pre-incubation values) for each treatment as associated SEs are small and these data simply provide background soil information for each incubation experiment. For the RM soils, the later incubation experiments ($^{15}NO_3^- + $ 2 and ^{15}N-U) have slightly lower average bulk soil % TC contents, whilst for the later WA experiments average bulk soil % TC contents are higher. The average bulk soil % TC content of the WA soil sampled in January 2013 was slightly higher (7.33–7.84%) than that of the RM soil (6.94–7.23%), while the average bulk soil % TC content of the WA soil sampled in October 2014 (incubation experiments WA $^{15}NH_4^+ + $ G and WA ^{15}N-U) is lower (5.99 and 5.66%, respectively). The addition of C as slurry (1700 μg C per 10 g soil) and as glucose (560 μg C per 10 g soil) represented 0.23% (slurry) of the C present in 10 g of the RM soil and 0.22% and 0.07–0.10% (respectively) of the C present in 10 g of the WA soil.

Bulk soil % TN contents are displayed for every time point sampled in each incubation experiment alongside average bulk soil % TN contents (for all replicates at every time point, including $t = 0$ pre-incubation values) and associated SEs. These SEs are small as a result of the narrow range of bulk soil % TN contents recorded throughout each incubation experiment. The RM soil had the highest TN content throughout the incubation experiments, while the WA soil in January 2013 had the lowest. N additions represented between 0.087% and 0.54% of the N present in 10 g of soil.

4.4.2 Hydrolysable Soil Amino Acid Concentrations

The distributions of hydrolysable AAs in the RM and WA (at both sampling dates) soils at $t = 0$ prior to incubation are shown in Fig. 4.2. The RM soil has the highest concentrations of all hydrolysable AAs (except Hyp and Thr) and considerably higher concentrations of Ala and Gly than the WA soils. The WA soil sampled in January 2013 has the lowest concentrations of all hydrolysable AAs. There is a significant difference between the AA distributions of the RM and (averaged) WA soils (Wilcoxon Signed Ranks test, $Z = -3.408$, $p = 0.001$, $p < 0.01$).

Amino acid concentrations and the contribution of total hydrolysable AAs to the TN pool at each time point analysed are shown for each incubation experiment in Appendix A1 (Tables A1.1–A1.11). Considering the patterns in each incubation

Table 4.3 Bulk soil % TC and % TN contents for each incubation experiment

	% TC			% TN														
	Mean % TC	SE	n =	Time/days													Mean % TN	SE
				0	0.0625	0.125	0.25	0.5	1	2	4	8	16	32	64			
RM $^{15}NO_3^-$	7.23	0.0578	36	0.63	0.67	0.65	0.67	0.66	0.65	0.67	0.66	0.66	0.68	0.69	2013	0.66	0.0039
RM $^{15}NH_4^+$	7.21	0.0431	36	0.63	0.60	0.62	0.62	0.62	0.56	0.55	0.67	0.69	0.64	0.69	–	0.63	0.0081
RM $^{15}NO_3^-$ + S	7.12	0.0322	33	0.64	–	0.69	0.69	0.68	0.67	0.68	0.70	0.67	0.68	0.67	0.68	0.68	0.0040
RM ^{15}N-U	6.94	0.0488	15	0.64	–	0.65	–	–	–	0.64	–	–	0.67	0.67	–	0.65	0.0050
WA $^{15}NO_3^-$	7.33	0.0346	36	0.45	0.48	0.48	0.47	0.48	0.47	0.48	0.47	0.47	0.48	0.46	–	0.47	0.0026
WA $^{15}NH_4^+$	7.57	0.0216	36	0.45	0.49	0.49	0.48	0.47	0.48	0.49	0.49	0.47	0.48	0.49	–	0.48	0.0040
WA $^{15}NO_3^-$ + S	7.84	0.0441	12	0.45	–	–	–	–	–	0.46	–	–	–	0.45	0.45	0.45	0.0040
WA $^{15}NH_4^+$ + S	7.83	0.0352	12	0.45	–	–	–	–	–	0.45	–	–	–	0.45	0.45	0.45	0.0032
WA $^{15}NO_3^-$ + G	7.78	0.0882	6	0.45	–	–	–	–	–	–	–	–	–	0.44	–	0.44	0.0075

(continued)

Table 4.3 (continued)

	% TC			% TN															
	Mean % TC	SE	n =	Time/days														Mean % TN	SE
				0	0.0625	0.125	0.25	0.5	1	2	4	8	16	32	64			
WA $^{15}NH_4^+$ +G	5.99	0.239	6	0.56	–	–	–	–	–	–	–	–	–	0.58	–	0.57	0.0162
WA ^{15}N-U	5.66	0.0689	15	0.56	–	0.58	–	–	–	0.57	–	–	0.57	0.56	–	0.57	0.0051

Bulk soil % TC contents are summarised as average values for all replicates at every time point (including $t = 0$ pre-incubation values) for each treatment alongside associated SEs. Bulk soil % TN contents are displayed for every time point sampled in each incubation experiment ($n = 3$ or 6 for some experiments at $t = 0$; bars, '–' indicate that no samples were incubated for this time period) alongside mean values (for all replicates at every time point, including $t = 0$ pre-incubation values) and their associated SEs

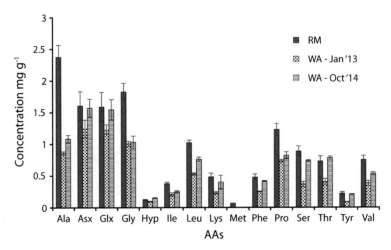

Fig. 4.2 Distributions of hydrolysable AAs in the RM and WA (at both sampling dates) soils at $t = 0$ prior to incubation

experiment separately, the concentrations of individual AAs vary from time point-to-time point and in general the same variation is reflected in all AAs at that time point (i.e. the concentrations of all AAs are slightly lower/higher than at the previous time point). There are no clear trends in the variation of individual hydrolysable AA concentrations, total hydrolysable AA concentrations or percentage contributions of total hydrolysable AA concentrations to TN (considering the incubation experiments with data at several time points) over time in all incubation experiments. There is, however, an apparent decrease over time in the concentration of individual hydrolysable AAs, total hydrolysable AA concentrations (linear regression gives $m = -0.0202$, $R^2 = 0.895$) and their percentage contribution to TN (linear regression gives $m = -0.0453$, $R^2 = 0.905$) for the WA $^{15}NO_3^- + 2$ incubation experiment.

Across different incubation experiments carried out in the same soil, there are also small variations in the mean concentrations of each AA, but no clear differences between treatments. The mean total hydrolysable AA contents of the earlier RM incubation experiments ($^{15}NO_3^-$ and $^{15}NH_4^+$), however, are higher (13.3 and 14.1, respectively) than that of the later performed incubation experiments ($^{15}NO_3^- + 2$ and ^{15}N-U; 12.9 and 11.5, respectively). Comparing the two soils, at $t = 0$ (Fig. 4.2), the concentrations of hydrolysable AAs are higher in the RM soil than in the WA soils, and as a result of little overall change over the course of the incubation experiments, this disparity remains throughout (Tables A1.1–A1.11). This is reflected in larger total hydrolysable AA concentrations for all RM incubation experiments compared with the WA experiments (Tables A1.1–A1.11). This does not, however, translate into a larger percentage contribution of hydrolysable AA N to TN for all RM incubation experiments compared with WA incubation experiments due to the larger % TN content of the RM soil. Nevertheless, the percentage contribution of total hydrolysable AA N to TN on average overall is a little higher across all of the

RM incubation experiments, $25.6 \pm 0.4\%$, compared with $23.8 \pm 0.3\%$ for all of the WA incubation experiments. The highest AA concentration (2.44 ± 0.04 mg g^{-1}) was recorded for Glx at $t = 3$ h in the RM $^{15}NO_3^- + 2$ incubation experiment, the highest total hydrolysable AA concentration (15.6 ± 0.20 mg g^{-1}) was recorded at $t = 3$ h in the RM $^{15}NO_3^- + 2$ incubation experiment and the highest percentage contribution of total hydrolysable AA N to TN ($33.5 \pm 1.90\%$) was recorded in the RM $^{15}NH_4^+$ incubation experiment.

Plots of the concentrations of individual AAs with time for the NH_4^+ incubation experiment in the RM and WA soils are shown in Fig. 4.3, as examples, in order to highlight the variation, but lack of overall trends in individual hydrolysable AA concentrations and for comparison of the same incubation experiment in different soils. As evidenced in Tables A1.2 and A1.6, the distribution of hydrolysable AAs

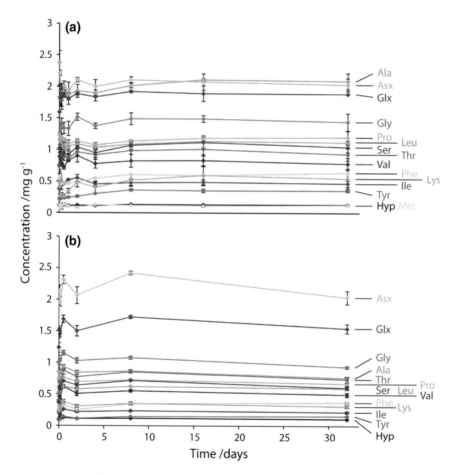

Fig. 4.3 Concentrations of individual AAs with time for: **a** the RM $^{15}NH_4^+$ incubation experiment and, **b** the WA $^{15}NH_4^+$ incubation experiment. Error bars are $\pm$ SE (n = 3)

remains fairly stable throughout the incubation experiment in both soils, with the AA hierarchy echoing the $t = 0$ pre-incubation distribution presented in Fig. 4.2. In addition, as at $t = 0$ (Fig. 4.2), the concentrations of individual AAs are generally higher in the RM $^{15}NH_4^+$ than in the WA $^{15}NH_4^+$ incubation experiment.

4.4.3 Bulk Soil Nitrogen Isotopic Compositions

The N isotopic composition of bulk soils prior to treatment application are shown in Table 4.4. At $t = 0$, the RM soil had the lowest bulk N isotopic composition and the WA soil sampled in October 2014 had the highest bulk N isotopic composition, although $\delta^{15}N$ values are not widely different. Application of a ^{15}N-labelled substrate to soil immediately raised the bulk $\delta^{15}N$ value of the soil from $t = 0$ pre-incubation values. The efficiency of the application and subsequent processing of the applied ^{15}N in the soil with time will dictate the $\delta^{15}N$ values each soil attains. The mean bulk soil $\delta^{15}N$ value for each incubation experiment is shown in Table 4.4 alongside the mean percentage retention of ^{15}N, based on the excess moles of ^{15}N applied, calculated using Eq. (2.17). The incubation experiments in which $^{15}NO_3^-$ was applied have the lowest mean bulk soil $\delta^{15}N$ values (26.5–41.3‰), followed by those in which $^{15}NH_4^+$ was applied (85.4–96.7‰) and mean bulk soil $\delta^{15}N$ values are highest in incubation experiments where ^{15}N-urea was applied (124 and 131‰). Mean percentage retention of ^{15}N ranges from 38 to 115% in the incubation experiments, but is low (38–55%) only in three experiments (WA $^{15}NO_3^- + G$, WA $^{15}NO_3^- + 2$ and WA $^{15}NO_3^-$).

4.4.4 Individual Hydrolysable Amino Acid Nitrogen Isotopic Compositions and the Partitioning of ^{15}N Between Them

The $t = 0$ $\delta^{15}N$ values of AAs in the two different soils were comparable for individual AAs, with $\delta^{15}N$ values in the WA soil generally slightly higher. Values ranged from 1.3‰ for Ser in the RM soil to 12.2‰ for Ala in the WA soil. Following the application of a ^{15}N-labelled substrate to soil, microbial uptake and incorporation of this applied ^{15}N into cell matter over time alters the N isotopic composition of N-containing microbial products. As highlighted in Chap. 3, AAs are the most important of these microbial products and time course plots of the $\delta^{15}N$ values (or $\Delta^{15}N$ values; sample $\delta^{15}N$ values minus control $\delta^{15}N$ values) of individual hydrolysable AAs can be used to: (i) compare the patterns in AA N isotopic compositions/proportions of N derived from the ^{15}N-labelled substrate (between different AAs in the same incubation experiment and between AAs in different treatments or soils); and (ii) assess the rate(s) at which ^{15}N is incorporated into each hydrolysable AA. Amino acid $\delta^{15}N$ values do not, however, expose the amount of extra ^{15}N in each hydrolysable AA pool

Table 4.4 Bulk soil $\delta^{15}N$ values at $t = 0$ and mean bulk soil $\delta^{15}N$ values and percentage retentions for all incubation samples in each incubation experiment

	$t = 0$		Overall incubation mean			
	Mean $\delta^{15}N$ value	SE	Mean $\delta^{15}N$ value	SE	% Retention ^{15}N	SE
RM $^{15}NO_3^-$	4.47	0.0432 (n = 6)	36.8	1.09 (n = 30)	88	3.1
RM $^{15}NH_4^+$	4.47	0.0432 (n = 6)	85.4	1.25 (n = 30)	109	2.07
RM $^{15}NO_3^-$ + S	4.99	0.0167 (n = 3)	41.3	1.15 (n = 30)	97	3.2
RM ^{15}N-U	4.99	0.0167 (n = 3)	131	1.31 (n = 12)	97	1.0
WA $^{15}NO_3^-$	5.83	0.0571 (n = 6)	31.3	2.48 (n = 30)	55	5.3
WA $^{15}NH_4^+$	5.83	0.0571 (n = 6)	96.7	1.54 (n = 30)	107	1.95
WA $^{15}NO_3^-$ + S	6.08	0.162 (n = 3)	27.3	5.11 (n = 9)	44	10
WA $^{15}NH_4^+$ + S	6.08	0.162 (n = 3)	91.0	5.15 (n = 9)	93	5.6
WA $^{15}NO_3^-$ + G	6.08	0.162 (n = 3)	26.5	10.1 (n = 3)	38	17
WA $^{15}NH_4^+$ + G	6.38	0.109 (n = 3)	94.4	3.92 (n = 3)	115	8.06
WA ^{15}N-U	6.38	0.109 (n = 3)	124	3.24 (n = 12)	81	2.5

Note Percentage retentions for all incubation experiments were calculated based on substrate ^{15}N-labelling of 10 atom % although this may have been diluted by natural abundance slurry-NO_3^- or NH_4^+ in the X + 2 incubation experiments

or the distribution of the additional ^{15}N amongst the AAs, as those present in larger concentrations will require larger amounts of ^{15}N to raise their overall N isotopic compositions. It can therefore also be useful to consider the excess moles of ^{15}N in each AA and, in order to provide some context, the excess moles of ^{15}N in comparison with what was applied, or alternatively, what is present (or *retained*) in the soil at that time (calculated from AA $\delta^{15}N$ values using Eqs. 2.18–2.21). Percentage applied and retained ^{15}N incorporations can be used to assess: (i) the partitioning (and patterns thereof) of the applied/retained ^{15}N in each hydrolysable AA; and (ii) the rate(s) at which applied/retained ^{15}N is partitioned into each hydrolysable AA.

Calculations of the percentage of applied ^{15}N incorporated into each AA (Eq. 2.20) will be affected by poor treatment applications and/or any losses of the applied ^{15}N from the system (as less ^{15}N than assumed will be available for assimilation into AAs). Those based on the excess moles ^{15}N retained (Eq. 2.21) could instead be affected by volatile losses of lighter ^{14}N over time, (giving rise to greater ^{15}N excesses in the system), but more accurately reflect the partitioning of the excess ^{15}N present in the soil. The data generated by these percentage ^{15}N incorporation calculations will be very similar where bulk percentage retentions of the applied ^{15}N are close to 100% (Table 4.4). For brevity, and since the partitioning of retained ^{15}N between the individual AAs is more relevant than the overall fate of the ^{15}N when considering individual hydrolysable AAs, only results relating to the latter calculation are presented here (Sect. 4.4.4.3).

4.4.4.1 Time Course Plots of Amino Acid Nitrogen Isotopic Compositions

The $\delta^{15}N$ values of individual hydrolysable AAs over the course of 32-day incubation experiments with $^{15}NO_3^-$ and $^{15}NH_4^+$ in the RM and WA soils are shown in Fig. 4.4. AA $\delta^{15}N$ values, rather than $\Delta^{15}N$ values, have been presented in order to highlight the changes in AA $\delta^{15}N$ values from their (different) $t = 0$ values in each soil. For the RM $^{15}NO_3^-$ incubation experiment (Fig. 4.4a), the $\delta^{15}N$ values of all AAs initially either fluctuate around $t = 0$ pre-incubation values or fall below these values for the first 24 h of incubation. Thereafter, the $\delta^{15}N$ values of almost all AAs increase slightly until 16 or 32 days of incubation, in general by similar degrees (e.g. Ala is the most enriched AA both at $t = 0$ and at $t = 32$ days) and not much above $t = 0$ pre-incubation values. Similar trends are exhibited for the WA $^{15}NO_3^-$ incubation experiment (Fig. 4.4c), except that AA $\delta^{15}N$ values are initially more stable and in the absence of a sampling point at $t = 24$ h, an increase is observed from $t = 12$ h. In addition, AA $\delta^{15}N$ values in the WA $^{15}NO_3^-$ incubation experiment rise further above $t = 0$ pre-incubation values than in the RM $^{15}NO_3^-$ incubation experiment, although the error bars for AA $\delta^{15}N$ values in the former experiment are generally larger.

In general, the increases in AA $\delta^{15}N$ values over the course of the $^{15}NH_4^+$ incubation experiment in each soil (Fig. 4.4b and d) are much larger than for $^{15}NO_3^-$. The two soils show similar trends in AA $\delta^{15}N$ values, but higher ^{15}N enrichments are observed for all AAs in the RM soil (with the exception of Hyp, which does not appear to incorporate ^{15}N in either soil). In general, for all AAs except Glx, $\delta^{15}N$ values rise in two phases—fast over the first 2–4 days of incubation and then more slowly for the remainder of the experiment. Incorporation of the applied ^{15}N into Glx occurs more quickly over the first 2 days than into any other AA and at their peak, Glx $\delta^{15}N$ values are two and a half- to five-fold greater than those of other AAs in the RM soil and one and a half- to four-fold greater than those of other AAs

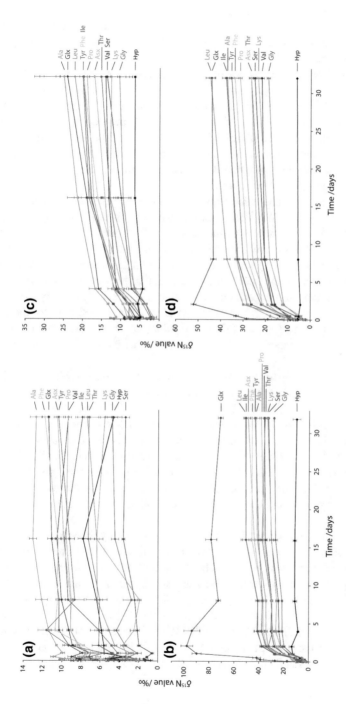

Fig. 4.4 Individual AA $\delta^{15}N$ values over the course of a 32-day incubation experiment: **a** RM $^{15}NO_3^-$, **b** RM $^{15}NH_4^+$, **c** WA $^{15}NO_3^-$ and, **d** WA $^{15}NH_4^+$. Error bars are $\pm$ SE (n = 3)

in the WA soil (excluding Hyp). After 2 days, Glx $\delta^{15}N$ values decline in both soils (although it is possible that higher peak values were reached after 2 days, but missed due to sampling frequency) and then appear to stabilise at elevated levels during the rest of the experiment.

Figure 4.5 shows the $\delta^{15}N$ values of individual AAs over the course of 64-day incubation experiments involving slurry in the RM and WA soils. Trends in AA $\delta^{15}N$ values for the RM $^{15}NO_3^- + 2$ incubation experiment (Fig. 4.5a) are comparable to, but less pronounced than, those of the RM $^{15}NO_3^-$ incubation (Fig. 4.4a) and AA $\delta^{15}N$ values fluctuate only slightly above $t = 0$ values. The same is true for the WA $^{15}NO_3^- + 2$ incubation experiment (Fig. 4.5b) compared with the WA $^{15}NO_3^-$ incubation experiment (Fig. 4.4c) and AA $\delta^{15}N$ values in the former rise to a lesser extent and in a slightly different pattern.

In the WA $^{15}NH_4^+ + 2$ incubation experiment (Fig. 4.5c), on the other hand, $\delta^{15}N$ values rise more than in the WA $^{15}NH_4^+$ incubation experiment (Fig. 4.4d), although the trend for each AA in the two experiments is very similar despite a lower sampling frequency. The WA X + G incubation experiments are not shown as line graphs as samples were only taken at $t = 32$ days. AA $\delta^{15}N$ values at $t = 32$ days for the WA $^{15}NO_3^- + G$ incubation experiment show a similar pattern of enrichment as in the WA $^{15}NO_3^-$ incubation experiment, but values are generally slightly lower by 1.3–6‰, depending on the AA. Amino acid $\delta^{15}N$ values at $t = 32$ days for the WA $^{15}NH_4^+ + G$ incubation experiment show a similar pattern of enrichment as in the WA $^{15}NH_4^+$ incubation experiment, but values are higher by 7–43‰, depending on the AA.

Amino acid $\delta^{15}N$ values for the urea incubations in the two soils (Fig. 4.6a and b) appear to display more different behaviour than the other treatments when comparing the soils. As with the NH_4^+ incubation experiments, urea-^{15}N is initially incorporated into Glx more quickly and to a greater extent than into other AAs for both soils. Peak Glx $\delta^{15}N$ values, however, do not occur at $t = 2$ days in both soils, but apparently only for the WA ^{15}N-U incubation experiment and at $t = 16$ days for the RM ^{15}N-U incubation experiment. As no other samples were taken between $t = 2$ and 16 days, it is possible that the 'true' peak in Glx $\delta^{15}N$ values in both soils occurred at a more similar time somewhere between these sampling points, but was missed due to sampling frequency.

Although as Glx $\delta^{15}N$ values initially rise more gradually than in the WA ^{15}N-U incubation experiment, it is also possible that peak Glx $\delta^{15}N$ values in the RM ^{15}N-U incubation experiment were reached at a later time than in the WA ^{15}N-U incubation, as a preliminary comparison suggests. The $\delta^{15}N$ values of other AAs generally increase until $t = 16$ days to comparable values in both soils, but again, this initially occurs more slowly in the RM than in the WA soil (gradient from $t = 0$ to 3 h steeper in the latter). In addition, for the RM soil, three AAs (Gly, Ile and Val) have higher $\delta^{15}N$ values than the other AAs at $t = 3$ h, but more comparable values at $t = 2$ days, after which their $\delta^{15}N$ values follow the same trend as other AAs.

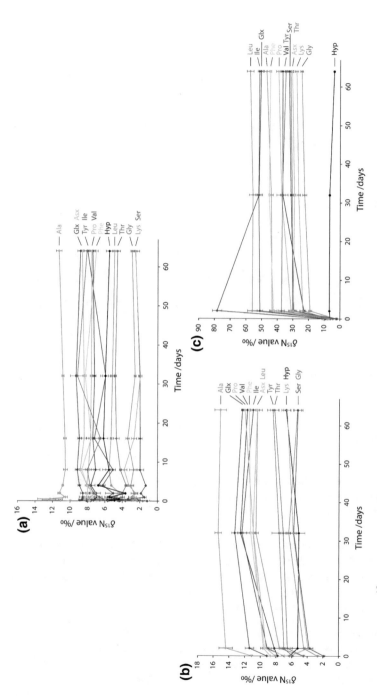

Fig. 4.5 Individual AA δ^{15}N values over the course of a 64-day incubation experiment: **a** RM ^{15}NO$_3^-$ + 2, **b** WA ^{15}NO$_3^-$ + 2 and, **c** WA ^{15}NH$_4^+$ + 2. Error bars are ± SE (n = 3)

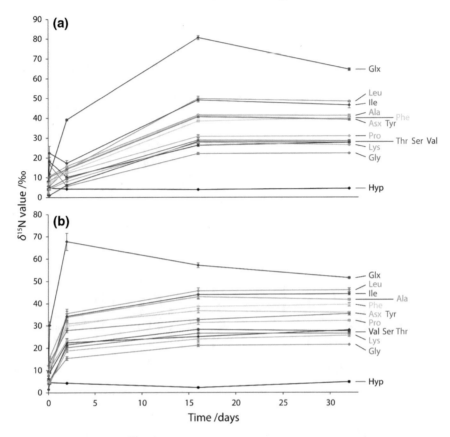

Fig. 4.6 Individual AA δ^{15}N values over the course of a 32-day incubation experiment: **a** RM ^{15}N-U, **b** WA ^{15}N-U. Error bars are ± SE (n = 3)

4.4.4.2   Rates of ^{15}N Incorporation and Proportions of Nitrogen Derived from the ^{15}N-Labelled Substrate in Each Hydrolysable Amino Acid

In order to obtain quantitative information about the rates at which AA isotopic compositions change and the overall change in AA isotopic compositions at 'steady state', time course plots of AA Δ^{15}N values (selected in order to exclude the difference of $t = 0$ starting values between AAs and between treatments) were fitted (where sufficient time points are available) to rate equations as described in Sect. 2.7.6:

(i)   Linear regressions (Eq. 2.22) were fitted to the first 0.5, 2 and 4 days of the incubation experiments in order to determine the initial rates at which Δ^{15}N values change. (Initial periods of 0–0.5 days were selected for the ^{15}NH$_4^+$ incubation experiments as a linear trend was present over this period in the AA N isotopic composition data; initial periods of 0–2 days were selected for the

^{15}N-U incubation experiments as a result of sampling frequency; and initial periods of 0–4 days were selected for the ^{15}NO$_3^-$ incubation experiments due to the initial dip and lag in AA δ/Δ^{15}N values).

(ii) Non-linear single first-order exponential assimilation regressions (Eq. 2.23) were fitted to data generated over 32 days for all incubations.

The results of these regressions are presented in Tables 4.5 and 4.6. The linear and non-linear regressions selected do not fit the time course AA Δ^{15}N data for most AAs in the RM ^{15}NO$_3^-$ and ^{15}NO$_3^- + 2$ incubation experiments, but acceptable and good fits are generally present for the other experiments (Tables 4.5 and 4.6). Considering each incubation experiment individually, the hierarchy of plateau δ^{15}N values is easily observed in Figs. 4.4, 4.5 and 4.6 and the quantitative data generated by the non-linear regressions is more valuable for comparison of Δ^{15}N values between experiments. For the RM incubation experiments, the plateau Δ^{15}N values of the ^{15}NH$_4^+$ and ^{15}N-urea treatments are comparable and much higher than those of the ^{15}NO$_3^-$ and ^{15}NO$_3^- + 2$ treatments, whereas for the WA incubation experiments, the plateau Δ^{15}N values of ^{15}N-urea $> ^{15}$NH$_4^+ > ^{15}$NO$_3^-$. Comparing the same treatments between the two different soils, plateau Δ^{15}N values are greater for the RM soil for both the ^{15}NH$_4^+$ and ^{15}N-urea treatments, but for ^{15}NO$_3^-$, WA $>$ RM and RM ^{15}NO$_3^- >$ RM ^{15}NO$_3^- + 2$.

It is more difficult to qualitatively compare the rates at which AA δ^{15}N values change using Figs. 4.4, 4.5 and 4.6 and the quantitative data presented in Tables 4.5 and 4.6 enables comparison both within a particular incubation experiment and between experiments (where the selected regressions fit the data). These rates, how-ever, reflect how the proportion of N derived from the ^{15}N-labelled substrate changes rather than a rate of ^{15}N incorporation (because the Δ^{15}N values of AAs present in higher concentrations will require larger amounts of ^{15}N to affect changes and there-fore may be expected to change more slowly). Thus, they are most valuable for comparing the same AAs between incubation experiments (particularly where they are present at the same/similar concentrations, i.e. usually in the same soil).

When comparing the zero- and first-order rate constants between substrates in the same soil: ^{15}NH$_4^+ > ^{15}$N-U $> ^{15}$NO$_3^-$ (although it should be noted that the zero-order rate constants for different substrates are calculated over different lengths of time). The fit of the linear regressions to determine the zero-order rate constants for AAs in the RM ^{15}NO$_3^-$ and ^{15}NO$_3^- + 2$ incubation experiments is poor or very poor for most AAs (Table 4.5), but values are of the same order of magnitude and comparable for over half of the AAs. There is insufficient first-order rate constant data for these incubation experiments to make the analogous comparison.

Tables 4.5 and 4.6 also facilitate comparisons of the zero- and first-order rate constants for each of the three treatments in the two soils (noting that differences in AA concentrations between the soils will affect the rates):

(i) For the ^{15}NO$_3^-$ treatment, initial rates are higher for all AAs in the WA soil, but first-order rate constants are more comparable (where data for both are available and except Hyp).

Table 4.5 Results of linear (zero-order rate constants) and non-linear single first-order exponential assimilation (first-order rate constants and plateau levels) regressions (alongside measures of the equation fits, R^2 and S, which is the standard error of the regression) of the $\Delta^{15}N$ values of individual hydrolysable AAs against time for the RM incubation experiments with sufficient data points

| | RM $^{15}NO_3^-$ | | | | | RM $^{15}NH_4^+$ | | | | | RM $^{15}NO_3^- + S$ | | | | | RM ^{15}N-U | | | | |
| | 0–4 days | | 0–32 days | | | 0–0.5 days | | 0–32 days | | | 0–4 days | | 0–32 days | | | 0–2 days | | 0–32 days | | |
	Zero-order	R^2	First-order	Plateau	S	Zero-order	R^2	First-order	Plateau	S	Zero-order	R^2	First-order	Plateau	S	Zero-order	R^2	First-order	Plateau	S
Ala	0.33	0.64	0.15	3.00	0.34	8.60	0.91	0.55	29.7	2.75	–	–	32.9	1.17	0.73	2.90	1.00	0.14	33.1	1.95
Asx	0.21	0.29	0.07	2.93	0.52	10.7	0.92	0.57	35.6	3.46	–	–	16.2	0.58	0.26	3.46	1.00	0.15	33.7	2.14
Glx	0.27	0.32	0.10	2.82	0.61	80.4	0.71	2.03	75.1	11.1	–	–	0.84	0.95	0.32	15.6	0.99	0.34	64.5	6.70
Gly	0.51	0.74	0.19	2.25	0.64	7.07	0.36	0.53	23.4	2.11	–	–	–	–	–	–	–	36.3	15.0	7.83
Hyp	0.59	0.17	0.64	2.28	1.01	–	–	2.39	6.88	1.76	–	–	–	–	–	–0.73	0.30	8.10	–1.46	0.21
Ile	0.17	0.04	0.16	3.77	1.11	11.0	0.94	0.56	42.5	3.69	0.52	0.50	0.61	1.84	0.68	–	–	0.22	44.0	10.0
Leu	–	–	–	–	–	11.4	0.96	0.56	43.4	3.70	–	–	0.32	0.93	0.46	4.86	0.99	0.15	46.5	2.59
Lys	–	–	–	–	–	7.01	0.71	0.52	28.2	2.29	0.23	0.00	26.9	0.48	0.51	2.92	0.83	0.16	25.5	1.72
Phe	–	–	–	–	–	8.88	0.95	0.51	34.4	3.13	0.24	0.58	0.28	1.25	0.47	2.90	1.00	0.12	34.7	1.50
Pro	–	–	–	–	–	4.49	0.57	0.41	27.1	2.56	–	–	1.45	–0.18	0.31	2.10	0.98	0.13	24.3	1.35
Ser	0.18	0.35	–	–	–	8.41	0.93	0.55	27.8	2.59	–	–	29.8	1.19	1.11	2.52	0.99	0.14	28.1	1.72
Thr	–	–	–	–	–	10.3	0.83	0.77	29.0	3.30	–	–	29.5	0.85	0.61	2.95	0.97	0.16	26.0	1.24
Tyr	–	–	–	–	–	3.82	0.50	0.59	31.9	3.34	–	–	1.12	2.13	0.86	4.42	0.57	0.17	35.2	2.78
Val	–	–	–	–	–	3.78	0.59	0.50	25.7	2.45	–	–	2.22	–0.40	0.33	–	–	11.6	13.7	7.95

Units of the zero-order rate constant are ‰ day^{-1} and units of the first-order rate constant are day^{-1}. Poor and very poor fits are highlighted in orange and red, respectively and a dash (–) indicates either that a horizontal line would have produced a better fit than the linear regression or the non-linear regression selected did not fit the data

Table 4.6 Results of linear (zero-order rate constants) and non-linear single first-order exponential assimilation (first-order rate constants and plateau levels) regressions (alongside measures of the equation fits, R^2 and S, which is the standard error of the regression) of the $\Delta^{15}N$ values of individual hydrolysable AAs against time for the WA incubation experiments with sufficient data points

	WA $^{15}NO_3^-$					WA $^{15}NH_4^+$					WA ^{15}N-U				
	0–4 days			0–32 days		0–0.5 days			0–32 days		0–2 days			0–32 days	
	Zero-order	R^2	First-order	Plateau	S	Zero-order	R^2	First-order	Plateau	S	Zero-order	R^2	First-order	Plateau	S
Ala	1.15	0.90	0.08	14.0	0.79	12.9	0.93	0.60	23.5	1.44	10.9	0.98	0.65	30.2	0.91
Asx	0.57	0.72	0.05	9.4	0.74	7.00	0.91	0.63	16.2	1.20	11.0	0.99	0.81	27.4	0.34
Glx	1.66	0.99	0.11	14.5	1.29	56.3	0.38	4.04	36.3	5.17	29.7	0.83	4.37	49.4	6.76
Gly	0.78	0.88	0.12	7.45	0.55	4.31	0.79	0.39	14.5	1.22	5.87	0.98	0.56	17.6	0.46
Hyp	–	–	–	–	–	−3.25	0.06	23.5	−1.19	0.57	−0.56	0.12	0.52	−1.83	1.11
Ile	1.42	0.85	0.11	14.6	0.99	17.2	0.82	0.62	29.6	1.91	14.0	1.00	0.67	37.9	0.19
Leu	0.98	0.78	0.06	19.3	1.04	13.2	0.98	0.49	35.9	2.38	15.0	0.99	0.68	40.2	0.31
Lys	1.16	0.85	0.13	11.8	0.87	6.34	0.19	0.31	20.6	1.56	6.75	0.98	0.56	19.7	1.35
Phe	0.60	0.94	0.04	17.2	0.37	9.13	0.99	0.38	24.9	1.62	10.8	1.00	0.60	30.8	0.36
Pro	0.67	0.74	0.08	11.5	0.83	5.96	0.87	0.26	21.8	1.36	7.25	0.99	0.50	22.9	0.48
Ser	1.14	0.93	0.09	11.6	0.93	8.42	0.96	0.55	20.9	1.55	10.0	0.99	0.71	26.5	0.36
Thr	0.74	0.65	0.07	10.5	1.07	6.76	0.87	0.40	19.2	1.44	7.98	1.00	0.62	22.3	0.32
Tyr	1.16	0.70	0.10	14.8	0.94	10.0	0.37	0.39	27.5	1.78	10.4	0.89	0.86	26.2	2.41
Val	0.78	0.74	0.14	6.56	0.93	5.24	0.93	0.43	14.0	0.44	6.77	1.00	0.69	17.7	1.42

Units of the zero-order rate constant are ‰ day^{-1} and units of the first-order rate constant are day^{-1}. Poor and very poor fits are highlighted in orange and red, respectively and a dash (−) indicates either that a horizontal line would have produced a better fit than the linear regression or the non-linear regression selected did not fit the data

(ii)   For the NH_4^+ treatment, zero- and first-order rate constants are similar for all
       AAs except Glx (and Hyp), which has a higher zero-order rate constant in the
       RM soil (ca. 80‰ day^{-1} compared with ca. 56‰ day^{-1}) and a higher first-order
       rate constant in the WA soil (ca. 4 day^{-1} compared with ca. 2 day^{-1})
(iii) Both rate constants are larger for all AAs in the WA soil for the ^{15}N-U incubation
       experiments.

   (Note that data from non-linear regressions with very poor fits have been excluded
from these comparisons.)

4.4.4.3   Percentages of Retained ^{15}N Incorporated into Individual Hydrolysable Amino Acids

At first glance, the graphs of percentage retained ^{15}N incorporated into each AA
for all the incubation experiments (Figs. 4.7, 4.8 and 4.9) appear similar to those of
AA δ^{15}N values (Figs. 4.4, 4.5 and 4.6). The differences in the ordering/hierarchy
of AAs results from the effects of AA concentration and further differences in the
general trend result from a combination of this and the moles ^{15}N present/retained
in the soil for that incubation tube (based on bulk soil δ^{15}N values). In addition, the
effect of different $t = 0$ AA δ^{15}N values has been removed in Figs. 4.7, 4.8 and
4.9 as at $t = 0$, all AAs contain 0% excess ^{15}N. This effect is noticeable only for
the ^{15}NO$_3^-$ treatments, as in these treatments differences in $t = 0$ AA δ^{15}N values
remain clear throughout the incubation experiments due to only small increases in
AA δ^{15}N values. Thus, the data for the percentage retained ^{15}N incorporated into
each AA lie more closely together, generally only a little above $t = 0$.

   For the RM ^{15}NO$_3^-$ incubation experiment, the general trend of a lag (and some
cases a dip) in AA δ^{15}N values (Fig. 4.4a) over the first 24 h, followed by a small rise
is also apparent in Fig. 4.7a. The AAs Ala and Gly, however, appear to behave a little
differently and this is due a combination of 'large' increases in δ^{15}N values (compared
with most of the other AAs in the experiment) and more importantly, high concentrations (compared with most of the other AAs in the experiment). Close inspection
of the trends for each of these AAs reveals little difference between Figs. 4.4a and
4.7a (other than scale). The AAs which incorporate the next highest amounts of
^{15}N toward the end of the incubation experiment in Fig. 4.7a are Lys, Glx and Asp.
Although Lys is not present in particularly high concentrations, as it contains two
N atoms, its concentration is double weighted and, like Ala and Gly, Glx and Asp
are present in high concentrations (compared with most of the other AAs in the
experiment).

   The picture for the percentage of retained ^{15}N incorporated into individual AAs
in the WA ^{15}NO$_3^-$ incubation experiment is very similar. The AAs present at higher
concentrations are generally higher in position, but the larger and more consistent
rise in ^{15}N enrichment with time also has more of an effect (compared with the
RM ^{15}NO$_3^-$ incubation experiment) and percentage retained ^{15}N incorporations are
much larger (maximums of ca. 1% compared with 0.35%). In addition, the more

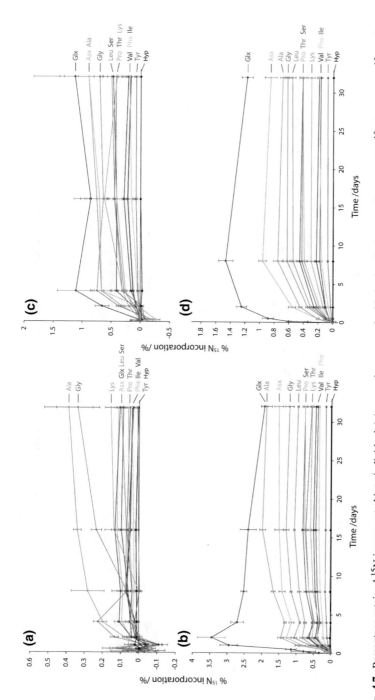

Fig. 4.7 Percentage retained ¹⁵N incorporated into individual AAs over the course of a 32-day incubation experiment: **a** RM ¹⁵NO₃⁻, **b** RM ¹⁵NH₄⁺, **c** WA ¹⁵NO₃⁻ and, **d** WA ¹⁵NH₄⁺. Error bars are ± SE (n = 3)

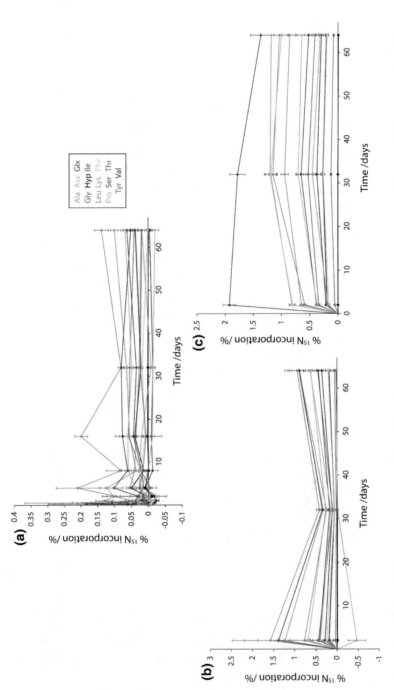

Fig. 4.8 Percentage retained ^{15}N incorporated into individual AAs over the course of a 64-day incubation experiment: **a** RM ^{15}NO$_3^-$ + ℇ, **b** WA ^{15}NO$_3^-$ + ℇ and, **c** WA ^{15}NH$_4^+$ + ℇ. Error bars are ± SE (n = 3)

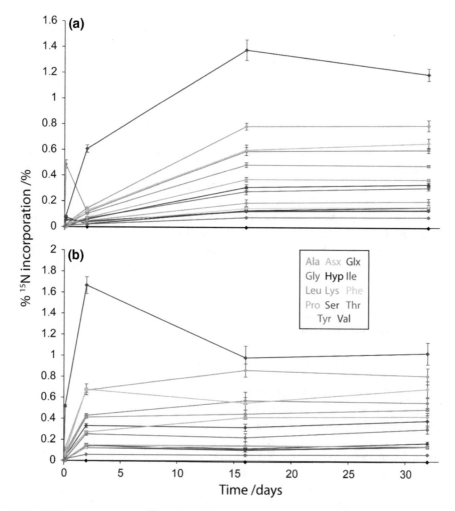

Fig. 4.9 Percentage retained ^{15}N incorporated into individual AAs over the course of a 32-day incubation experiment: **a** RM ^{15}N-U and, **b** WA ^{15}N-U. Error bars are ± SE (n = 3)

inconsistent bulk soil ^{15}N percentage retentions at earlier time points have led to higher observed percentage ^{15}N incorporations at $t = 2$ and 4 days than expected from the δ^{15}N values at these times.

Percentage retained ^{15}N incorporations for AAs in the RM ^{15}NO$_3^-$ + 2 incubation experiment (Fig. 4.8a) are even lower than those of the RM ^{15}NO$_3^-$ incubation experiment (Fig. 4.7a) and fluctuate just above 0%. As enrichments are very small, calculated percentage incorporations are strongly affected by AA concentrations. Percentage retained ^{15}N incorporations for AAs in the WA ^{15}NO$_3^-$ + 2 incubation experiment (Fig. 4.8b) are higher than those of the WA ^{15}NO$_3^-$ incubation experiment (Fig. 4.7c) at $t = 2$ days (maximum of *ca.* 1.5%), but lower at $t = 32$ days (*ca.* 0.5%), before rising again at $t = 64$ days (*ca.* 1%). The AA hierarchy in this experiment is affected mainly by AA concentration and the trend observed (based on only a few data points) is controlled mainly by inconsistent bulk soil percentage ^{15}N retentions (Fig. 4.8b).

The higher δ^{15}N values generally attained in the ^{15}NH$_4^+$ and ^{15}N-U incubation experiments leads to a dominance of enrichment effects over AA concentration in Figs. 4.7b, d, 4.8c, 4.9a and b and, since bulk soil percentage retentions are also generally close to 100%, trends in the percentage retained ^{15}N incorporated into AAs are very similar to those in AA δ^{15}N values. With the exception of Glx, which is most highly enriched by a considerable amount in all of these incubation experiments, AA concentration still plays a role in the vertical ordering of AAs by percentage retained ^{15}N incorporated. Thus, AAs such as Ala, Asp, Glx and Leu, which are present in higher concentrations, incorporated more of the retained ^{15}N. Considering peak recorded Glx percentage ^{15}N incorporations in each incubation experiment, RM ^{15}NH$_4^+$ was highest (3.5%) and RM ^{15}N-U was lowest (1.4%), although it is possible in all experiments that 'true' peak values were missed due to sampling frequency.

4.4.4.4   Rates and Percentages of ^{15}N Partitioning into Each Hydrolysable Amino Acid Pool

In order to obtain quantitative information about the rates and plateau percentages of ^{15}N partitioning, plots of percentage retained ^{15}N incorporation for individual hydrolysable AAs were fitted (where sufficient time points are available) to linear and non-linear first-order exponential assimilation equations as in Sect. 4.4.4.2. Since percentage incorporations take into account the moles ^{15}N in each AA pool, they can provide experimentally determined quantitative estimates of the rates of ^{15}N partitioning/incorporation into that AA pool and plateau/steady state percentage retained ^{15}N incorporations. As in Sect. 4.4.4.2, however, the linear and non-linear regressions selected do not fit the percentage retained ^{15}N incorporation data for many AAs in the RM ^{15}NO$_3^-$ and ^{15}NO$_3^-$ + 2 incubation experiments, but acceptable and good fits are generally present for the other incubation experiments (Tables 4.7 and 4.8).

First, considering each incubation experiment individually, the rate constants presented in Tables 4.7 and 4.8 can be used to compare the different rates at which the AA pools in each experiment incorporated the ^{15}N present in the soil. For the RM ^{15}NH$_4^+$ incubation experiment (Table 4.7), Glx has the highest zero-order rate constant (2.61% day^{-1}) and Hyp, Tyr and Lys have the lowest (0% day^{-1} [very poor fit], 0.01% day^{-1} [poor fit] and 0.06% day^{-1} [poor fit], respectively); Phe and Val have the lowest zero-order rate constants with acceptable fits (both 0.07% day^{-1}). First-order rate constants for the RM ^{15}NH$_4^+$ incubation experiment are highest for Hyp and Glx (2.08 day^{-1} and 1.97 day^{-1}, respectively) with poor fits and for Thr with an acceptable fit (0.81 day^{-1}). Lys has the lowest first-order rate constant (0.35 day^{-1}). For the RM ^{15}N-U incubation experiment (Table 4.7), the highest zero-order rate constant is also found for Glx (0.30% day^{-1}) and several AAs have zero-order rate constants of 0% day^{-1} (Gly, Hyp, Ile and Val), but the fit of the linear regressions for

Table 4.7 Results of linear (zero-order rate constants) and non-linear single first-order exponential assimilation (first-order rate constants and plateau levels) regressions (alongside measures of the equation fits, R^2 and S, which is the standard error of the regression) of the percentage retained ^{15}N incorporated into individual hydrolysable AAs against time for the RM incubation experiments with sufficient data points

| | RM $^{15}NO_3^-$ | | | | | RM $^{15}NH_4^+$ | | | | | RM $^{15}NO_3^-$ + S | | | | | RM ^{15}N-U | | | | |
| | 0–4 days | | 0–32 days | | | 0–0.5 days | | 0–32 days | | | 0–4 days | | 0–32 days | | | 0–2 days | | 0–32 days | | |
	Zero-order	R^2	First-order	Plateau	S	Zero-order	R^2	First-order	Plateau	S	Zero-order	R^2	First-order	Plateau	S	Zero-order	R^2	First-order	Plateau	S
Ala	0.05	0.52	0.16	0.39	0.05	0.48	0.95	0.55	1.82	0.17	–	–	26.9	0.11	0.07	0.07	1.00	0.13	0.84	0.04
Asx	0.02	0.17	0.14	0.13	0.05	0.41	0.98	0.65	1.46	0.14	–	–	25.7	0.05	0.01	0.06	1.00	0.12	0.68	0.02
Glx	0.02	0.14	0.16	0.13	0.06	2.61	0.79	1.97	2.65	0.48	–	–	–	–	–	0.30	0.99	0.33	1.28	0.08
Gly	0.06	0.59	0.10	0.33	0.08	0.33	0.47	0.59	1.20	0.12	–	–	–	–	–	–	–	–	–	–
Hyp	–	–	1.85	0.01	0.01	–	–	2.08	0.02	0.01	–	–	–	–	–	0.00	0.10	10.2	0.00	0.00
Ile	–	–	0.18	0.05	0.02	0.10	0.84	0.77	0.40	0.05	0.01	0.74	0.65	0.03	0.01	–	–	0.20	0.13	0.03
Leu	–	–	–	–	–	0.24	0.97	0.61	0.97	0.10	0.01	0.15	0.45	0.04	0.02	0.05	0.99	0.15	0.50	0.02
Lys	–	–	0.04	0.24	0.04	0.06	0.44	0.35	0.56	0.05	–	–	–	–	–	0.02	0.80	0.14	0.21	0.01
Phe	–	–	–	–	–	0.07	0.78	0.39	0.34	0.02	0.00	0.62	0.34	0.02	0.01	0.01	1.00	0.11	0.17	0.01
Pro	–	–	0.04	0.09	0.05	0.12	0.60	0.44	0.72	0.07	–	–	1.49	–0.01	0.02	0.03	0.97	0.14	0.39	0.02
Ser	0.01	0.25	0.12	0.10	0.03	0.19	0.86	0.55	0.75	0.08	–	–	–	–	–	0.03	0.99	0.12	0.35	0.01
Thr	–	–	–	–	–	0.20	0.77	0.81	0.63	0.09	–	–	–	–	–	0.03	0.96	0.13	0.32	0.01
Tyr	–	–	–	–	–	0.01	0.36	0.42	0.16	0.01	–	–	12.4	0.02	0.01	0.01	0.57	0.15	0.08	0.01
Val	0.01	0.02	0.20	0.05	0.03	0.07	0.51	0.64	0.47	0.06	–	–	2.98	–0.01	0.01	0.01	–	0.09	0.17	0.04

Units of the zero-order rate constant are % day^{-1} and units of the first-order rate constant are day^{-1}. Poor and very poor fits are highlighted in orange and red, respectively and a dash (–) indicates either that a horizontal line would have produced a better fit than the linear regression or the non-linear regression selected did not fit the data

Table 4.8 Results of linear (zero-order rate constants) and non-linear single first-order exponential assimilation (first-order rate constants and plateau levels) regressions (alongside measures of the equation fits, R^2 and S, which is the standard error of the regression) of the percentage retained ^{15}N incorporated into individual hydrolysable AAs against time for the WA incubation experiments with sufficient data points

| | WA $^{15}NO_3^-$ | | | | | WA $^{15}NH_4^+$ | | | | | WA ^{15}N-U | | | | |
| | 0–4 days | | 0–32 days | | | 0–0.5 days | | 0–32 days | | | 0–2 days | | 0–32 days | | |
	Zero-order	R^2	First-order	Plateau	S	Zero-order	R^2	First-order	Plateau	S	Zero-order	R^2	First-order	Plateau	S
Ala	0.16	0.91	0.29	0.84	0.11	0.42	0.95	0.54	0.72	0.04	0.34	0.98	0.83	0.83	0.03
Asx	0.13	0.66	0.18	0.85	0.21	0.38	0.91	0.48	0.91	0.05	0.34	1.00	1.59	0.64	0.07
Glx	0.29	0.99	0.58	1.04	0.14	1.99	0.57	3.64	1.27	0.15	0.85	0.88	4.42	1.22	0.31
Gly	0.18	0.88	0.45	0.71	0.15	0.21	0.80	0.39	0.64	0.02	0.22	0.98	0.74	0.56	0.02
Hyp	–	–	–	–	–	-0.01	0.19	–	–	–	–	–	0.55	-0.01	0.00
Ile	0.03	0.84	0.25	0.19	0.03	0.10	0.95	0.49	0.17	0.01	0.06	1.00	1.35	0.13	0.02
Leu	0.06	0.78	0.15	0.50	0.07	0.19	0.99	0.39	0.56	0.01	0.21	0.99	1.01	0.47	0.02
Lys	0.08	0.84	0.25	0.51	0.08	0.10	0.65	0.31	0.31	0.01	0.07	0.98	1.15	0.15	0.02
Phe	0.01	0.88	0.08	0.21	0.01	0.06	0.98	0.22	0.19	0.01	0.06	1.00	0.95	0.14	0.00
Pro	0.06	0.67	0.18	0.43	0.08	0.13	0.87	0.29	0.42	0.01	0.14	0.99	0.53	0.42	0.01
Ser	0.09	0.95	0.30	0.46	0.06	0.17	0.97	0.45	0.44	0.02	0.17	1.00	1.20	0.36	0.03
Thr	0.06	0.60	0.15	0.46	0.10	0.15	0.90	0.37	0.43	0.02	0.13	1.00	1.14	0.27	0.04
Tyr	0.01	0.50	0.17	0.08	0.01	0.02	0.83	0.25	0.08	0.00	0.03	0.92	2.30	0.06	0.00
Val	0.03	0.66	0.19	0.25	0.06	0.09	0.92	0.39	0.21	0.01	0.07	1.00	1.14	0.15	0.03

Units of the zero-order rate constant are % day^{-1} and units of the first-order rate constant are day^{-1}. Poor and very poor fits are highlighted in orange and red, respectively and a dash (–) indicates either that a horizontal line would have produced a better fit than the linear regression or the non-linear regression selected did not fit the data

these AAs is very poor and Phe and Tyr have zero-order rates of 0.01% day^{-1} with very good and acceptable fits, respectively. The highest first-order rate constant for the RM ^{15}N-U incubation experiment is for Hyp (10.2 day^{-1}), but with a very poor fit and Gly has the highest first order rate constant with an acceptable fit (0.33 day^{-1}). Val has the lowest first-order rate constant (0.09 day^{-1}).

For the WA incubation experiments (Table 4.8), Glx has the highest zero-order rate constant in all three incubation experiments with sufficient data to carry out regressions (0.29% day^{-1}, 1.99% day^{-1} and 0.85% day^{-1} for the ^{15}NO$_3$$^-$, ^{15}NH$_4$$^+$ and ^{15}N-U incubation experiments, respectively). Similarly, Hyp has the lowest zero-order rate constant in all three incubation experiments (0% day^{-1}, -0.01% day^{-1} and 0% day^{-1} for the ^{15}NO$_3$$^-$, ^{15}NH$_4$$^+$ and ^{15}N-U incubation experiments, respectively), but with very poor regression fits. Phe and Tyr in the WA ^{15}NO$_3$$^-$ and Tyr in the WA ^{15}NH$_4$$^+$ and ^{15}N-U incubation experiments have the lowest zero-order rate constants (0.01–0.03% day^{-1}) with acceptable/good fits (Table 4.8). Glx also has the highest first-order rate constants for all three WA incubation experiments (0.58 day^{-1}, 3.64 day^{-1} and 4.42 day^{-1} for the ^{15}NO$_3$$^-$, ^{15}NH$_4$$^+$ and ^{15}N-U incubation experiments, respectively), but the fits of the regressions are poor and very poor for the WA ^{15}NH$_4$$^+$ and ^{15}N-U incubation experiments, respectively. For these incubation experiments, Ala and Tyr (respectively) have the highest first-order rate constants with acceptable fits (0.54 day^{-1} and 2.30 day^{-1}, respectively). Phe has the lowest first-order rate constant in the WA ^{15}NO$_3$$^-$ and ^{15}NH$_4$$^+$ incubation experiments (0.08 day^{-1} and 0.22 day^{-1}, respectively) and Pro has the lowest first-order rate constant in the WA ^{15}N-U incubation experiment (0.53 day^{-1}).

The quantitative estimates of plateau percentages presented in Tables 4.7 and 4.8 can be used to assist assessment of the different steady state percentage retained ^{15}N incorporations of each AA in a particular incubation experiment. For the RM ^{15}NO$_3$$^-$ incubation experiments, the Ala pool contains the highest plateau percentages of ^{15}N, while for the RM ^{15}NH$_4$$^+$ and ^{15}N-U incubation experiments, most ^{15}N is present in the Glx pool (2.65 and 1.28%, respectively, but with a poor fit for the RM ^{15}NH$_4$$^+$ experiment), followed by the Ala pool. Plateau percentage retained ^{15}N incorporations are lowest for Hyp in all RM incubation experiments in Table 4.7, followed by Tyr for the RM ^{15}NH$_4$$^+$ incubation experiment, in which the fit for Hyp is poor. This is with the exception of the RM ^{15}NO$_3$$^-$ + 2 incubation experiment, where the non-linear regression applied did not fit the data for Hyp and Pro and Val have negative plateau percentage ^{15}N incorporations (−0.01%) with poor regression fits and Phe and Tyr have the next lowest plateau percentage ^{15}N incorporations (0.02%) with poor and acceptable fits, respectively.

For the WA incubation experiments most ^{15}N was partitioned into the Glx pool (followed by the Asx pool for the WA ^{15}NH$_4$$^+$ incubation experiment in which the regression fit for Glx was poor.) The non-linear regression selected did not fit Hyp in the WA ^{15}NO$_3$$^-$ and ^{15}NH$_4$$^+$ incubation experiments, but gives the lowest plateau percentage (−0.01% [with a poor fit]) in the WA ^{15}N-U incubation experiment for Hyp. Tyr has the lowest plateau percentage with an acceptable fit in all WA incubation experiments (0.06–0.08%). Finally, there is a correlation between the zero-order and first-order rate constants only for the WA ^{15}NO$_3$$^-$ (good) and ^{15}NH$_4$$^+$ (weak) percentage retained ^{15}N incorporation data and the zero-order rate constants and plateau percentages are strongly correlated for the RM ^{15}NH$_4^+$, WA ^{15}NO$_3^-$, WA ^{15}NH$_4^+$ and WA ^{15}N-U incubation experiments and weakly correlated for the RM ^{15}N-U incubation experiment (correlations not shown).

Comparing incubation experiments between different treatments in the same soil, both the zero- and first-order rate constants of the RM ^{15}NH$_4^+$ incubation experiment are greater than those of the RM ^{15}N-U incubation experiment (there are insufficient regression parameters with acceptable fits for the ^{15}NO$_3^-$ and ^{15}NO$_3^-$ + ℰ experiments). For the WA soil, the zero-order rate constants of ^{15}NH$_4^+$ > ^{15}N-U > ^{15}NO$_3^-$ and the first-order rate constants of ^{15}N-U > ^{15}NH$_4^+$ > ^{15}NO$_3^-$ (noting however, that zero order rate constants for different substrates were generated over different lengths of time). Due to the poor and very poor fits of both regressions in the RM ^{15}NO$_3^-$ and ^{15}NO$_3^-$ + ℰ incubation experiments, rate constant comparisons of the same treatment in the two different soils are only possible for the ^{15}NH$_4^+$ and ^{15}N-U treatments. For the former, zero-order rate constants in the two soils are comparable, with the RM soil generally slightly higher, and first-order rate constants are larger (to varying degrees) in the RM soil. For ^{15}N-U, both zero- and first-order rate constants were greater in the WA soil.

Comparing the plateau percentages of retained ^{15}N incorporation between incubation experiments, the RM ^{15}NH$_4^+$ incubation experiment generally has the highest percentages. Plateau percentages in the RM ^{15}N-U, WA ^{15}NO$_3^-$, WA ^{15}NH$_4^+$ and WA ^{15}N-U incubation experiments are relatively comparable and the experiment with the highest/lowest percentage incorporation between these four varies for each AA. Amino acids in the RM ^{15}NO$_3^-$ and ^{15}NO$_3^-$ + ℰ incubation experiments have the lowest plateau percentage retained ^{15}N incorporations (and where data is available for comparison, the RM ^{15}NO$_3^-$ > RM ^{15}NO$_3^-$ + ℰ).

4.4.5 Percentage Incorporation of Applied and Retained ^{15}N into the Total Hydrolysable Amino Acid Pool

A summation of the results from Eqs. (2.20) and (2.21) for each hydrolysable AA gives the percentage of applied and the percentage of retained ^{15}N incorporated into the total hydrolysable AA pool, respectively. As described in Sect. 4.4.4, the results of each of these calculations could be affected by differences from the expected moles ^{14}N and ^{15}N in the soil. Percentage applied ^{15}N incorporations are useful in providing an indication of the overall fate of the applied ^{15}N (affected by poor treatment applications and any losses of the applied ^{15}N from the system, as would be likely to occur in a more 'real-world' situation) and percentage retained ^{15}N incorporations reflect the partitioning of the ^{15}N present (or retained) in the system at that time, but as these data are calculated based on bulk soil δ^{15}N values, could be affected by volatile losses of lighter ^{14}N. In order to investigate both the overall fate of the applied ^{15}N and the partitioning of ^{15}N into the total hydrolysable AA pool, the data from both calculations for all incubation experiments are presented in this section.

Figure 4.10a shows the percentage of applied ^{15}N incorporated into the total hydrolysable AA pool in the RM soil and the results of linear and non-linear regressions applied to these data are presented in Table 4.9. ^{15}N applied as $^{15}NO_3^-$ was poorly incorporated into the total hydrolysable AA pool (*ca.* 0.5% and 1.80% applied ^{15}N), urea-^{15}N was 3.5 to 11 times better incorporated and $^{15}NH_4^+$ was incorporated

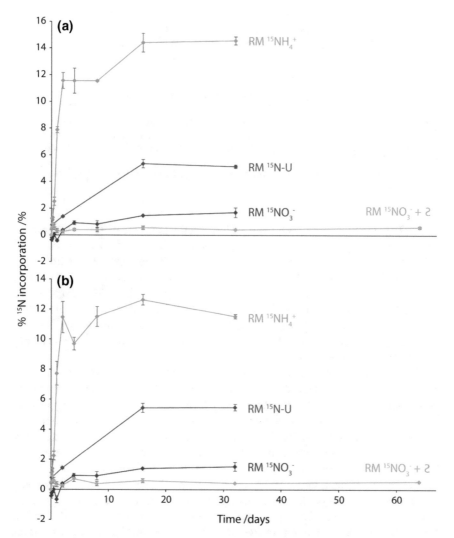

Fig. 4.10 Percentage ^{15}N incorporated into the total hydrolysable AA pool for all treatments in the RM soil: **a** applied and, **b** retained. Error bars are $\pm$ SE (n = 3)

Table 4.9 Results of linear (zero-order rate constants) and non-linear single first-order exponential assimilation (first-order rate constants and plateau levels) regressions (alongside measures of the equation fits, R^2 and S, which is the standard error of the regression) of the percentage applied and retained ^{15}N incorporated into the total hydrolysable AA pool for all treatments in the RM soil with sufficient data points

		Zero-order[a]	R^2	First-order	Plateau	S
% applied ^{15}N incorporated	RM $^{15}NO_3^-$	0.19	0.55	0.10	1.80	0.28
	RM $^{15}NH_4^+$	5.32	0.90	0.73	13.4	1.19
	RM $^{15}NO_3^- + 2$	–	–	–	–	–
	RM ^{15}N-U	0.72	0.54	0.18	5.38	0.44
% retained ^{15}N incorporated	RM $^{15}NO_3^-$	0.17	0.38	0.11	1.61	0.36
	RM $^{15}NH_4^+$	4.91	0.89	0.87	11.7	1.16
	RM $^{15}NO_3^- + 2$	–	–	35.1	0.52	0.19
	RM ^{15}N-U	0.74	0.54	0.17	5.63	0.42

Units of the zero-order rate constant are % day^{-1} and units of the first-order rate constant are day^{-1}. Very poor fits are highlighted in red and a dash (–) indicates either that a horizontal line would have produced a better fit than the linear regression or the non-linear regression selected did not fit the data
[a]Zero-order rate constants were determined over 0.5, 2 and 4 days for the $^{15}NH_4^+$, ^{15}N-urea and $^{15}NO_3^-$ treatments, respectively

7.5 to 27 times better than $^{15}NO_3^-$ and 2.5 times better than urea-^{15}N. Results are similar when the percentage of ^{15}N incorporated is calculated based on the moles of ^{15}N retained or still available in the RM soil for partitioning into the total hydrolysable AA pool (Fig. 4.10b and Table 4.9). The only notable difference is for the RM NH_4^+ incubation experiment, for which calculated percentage ^{15}N incorporations are affected by percentage retentions (mostly slightly) greater than 100% at all time points except $t = 1$ day. This leads to smaller calculated plateau percentage ^{15}N incorporations (11.7% compared with 13.4%) due to the larger moles of ^{15}N apparently available in the soil based on bulk soil $\delta^{15}N$ values.

Figure 4.11a shows the percentage of applied ^{15}N incorporated into the total hydrolysable AA pool in the WA soil and the results of linear and non-linear regressions applied to these data are presented in Table 4.10. In general ^{15}N applied as $^{15}NO_3^-$ was incorporated to a lesser extent than urea-^{15}N (at 4.40%) and $^{15}NH_4^+$ (at 6.27 and 7.32% [and 8.65% for WA $^{15}NH_4^+ + G$ at $t = 32$ days]) with the exception of the WA $^{15}NO_3^-$ incubation experiment in which a plateau of 6.39% of the applied ^{15}N is present in the total hydrolysable AA pool. The error bars for the WA $^{15}NO_3^-$ incubation experiment, however, are quite large, particularly at $t = 32$ days when they stretch above the point for the WA $^{15}NH_4^+ + G$ incubation experiment (which at 8.65% has the highest percentage incorporation of ^{15}N into the total hydrolysable AA pool) and below the point for the WA $^{15}NO_3^- + G$ incubation experiment (at 4.08%).

Calculating the percentage of ^{15}N incorporated into the total hydrolysable AA pool based on the moles of ^{15}N retained (or still available) in the WA soil produces

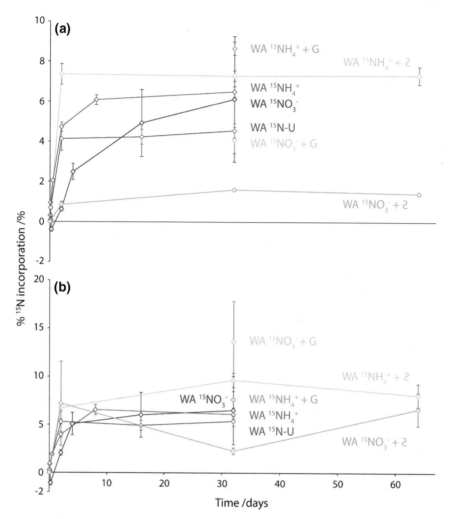

Fig. 4.11 Percentage ^{15}N incorporated into the total hydrolysable AA pool for all treatments in the WA soil: **a** applied and, **b** retained. Error bars are $\pm$ SE (n = 3), the error bars of the WA ^{15}NO$_3^-$ treatment are highlighted in red as the bar at $t = 32$ days is large and otherwise difficult to distinguish

much more of a difference than for the RM soil (Fig. 4.11 and Table 4.10). Percentage retained ^{15}N incorporations in Fig. 4.11b are bunched more closely together between 5 and 8% and the calculation affects WA ^{15}NO$_3^-$ + 2 and ^{15}NO$_3^-$ + G incubation experiments most strongly. The trend for the WA ^{15}NO$_3^-$ + 2 incubation experiment changes considerably due to particularly low percentage ^{15}N retentions at $t = 2$ and 64 days, which cause larger percentage retained ^{15}N incorporations (from 0.9 and 1.4% to 7.2 and 6.6%), which are not matched by the smaller increase at $t = 32$ days (1.6–2.2%). A similar disparity in the differences between applied and

118

4 Biosynthetic Routing, Rates and Extents of Microbial …

Table 4.10 Results of linear (zero-order rate constants) and non-linear single first-order exponential assimilation (first-order rate constants and plateau levels) regressions (alongside measures of the equation fits, R^2 and S, which is the standard error of the regression) of the percentage applied and retained ^{15}N incorporated into the total hydrolysable AA pool for all treatments in the WA soil with sufficient data points

		Zero-order[a]	R^2	First-order	Plateau	S
% applied ^{15}N incorporated	WA $^{15}NO_3^-$	0.55	0.89	0.10	6.39	0.39
	WA $^{15}NH_4^+$	4.30	0.93	0.75	6.27	0.23
	WA $^{15}NO_3^- + 2$	–	–	0.42	1.51	0.10
	WA $^{15}NH_4^+ + 2$	–	–	13.4	7.32	0.04
	WA $^{15}NO_3^- + G$	–	–	–	–	–
	WA $^{15}NH_4^+ + G$	–	–	–	–	–
	WA ^{15}N-U	2.08	0.98	1.39	4.40	0.13
% retained ^{15}N incorporated	WA $^{15}NO_3^-$	1.20	0.91	0.27	6.40	0.83
	WA $^{15}NH_4^+$	4.00	0.93	0.57	6.29	0.33
	WA $^{15}NO_3^- + 2$	–	–	12.0	5.35	2.68
	WA $^{15}NH_4^+ + 2$	–	–	0.71	8.84	0.79
	WA $^{15}NO_3^- + G$	–	–	–	–	–
	WA $^{15}NH_4^+ + G$	–	–	–	–	–
	WA ^{15}N-U	2.66	0.98	1.80	5.23	0.24

Units of the zero-order rate constant are % day^{-1} and units of the first-order rate constant are day^{-1}. Very poor fits are highlighted in red and a dash (–) indicates either that a horizontal line would have produced a better fit than the linear regression or the non-linear regression selected did not fit the data

[a]Zero-order rate constants were determined over 0.5, 2 and 4 days for the $^{15}NH_4^+$, ^{15}N-urea and $^{15}NO_3^-$ treatments, respectively

retained percentage ^{15}N incorporations at each time point also alters the trend for the WA $^{15}NH_4^+ + 2$ incubation experiment (although to a lesser extent), and the plateau percentage of retained ^{15}N incorporated into the total hydrolysable AA pool is 1.52% higher. The percentage retained ^{15}N in the total hydrolysable AA pool is much higher than percentage applied ^{15}N in this pool for the WA $^{15}NO_3^- + G$ experiment (*cf.* 13.64 and 4.08%, respectively) due to the low retention of applied ^{15}N in this experiment (38%) and thus this point has shifted up considerably in Fig. 4.11b. The percentage retained ^{15}N incorporated into the total hydrolysable AA pool for the WA $^{15}NH_4^+ + G$ incubation experiment, on the other hand, is *ca.* 1% lower than that for the applied ^{15}N (Figs. 4.11a and b). Differences in percentage ^{15}N incorporations between applied and retained ^{15}N are smaller for the WA ^{15}N-U, $^{15}NH_4^+$ and $^{15}NO_3^-$ incubation experiments (with plateau percentages changing from 4.40 to 5.23%, 6.27 to 6.29% and 6.39 to 6.40%, respectively; Table 4.10), although for the latter, larger discrepancies are present at earlier time points and the overall trend is more curved in Fig. 4.11b as a result.

4.5 Discussion

4.5.1 Comparison of the Underlying Differences Between Soils Used in the Incubation Experiments

The two soils used in the incubation experiments described in this chapter were sampled from grazed grasslands in south-west England (Sect. 2.2). The soils have contrasting compositions (non-calcareous vs. calcareous) and different management histories, both of which are likely to affect the processing of applied N [108]. While the RM soil has only received cattle slurry (*ca.* 200–250 kg N ha^{-1} year^{-1}) for almost 25 years, the WA soil has received regular additions of $(NH_4)_2SO_4$ or urea (40 kg N ha^{-1} every 40 days from spring until mid-September) as well as cattle depositions since it was converted from spring cropping to a grass ley in 2011.

The soils also have significantly different hydrolysable AA distributions (Sect. 4.4.2). The concentration of free AAs in soils is low [30, 109] and the hydrolysable AA distribution of soils derives from the integration of the AA compositions of the many thousands of proteins present in soil. The major sources of these proteins are those input into the soil (e.g. plant matter, excreta, in rainfall; [79]) and those generated (often from these inputs) in the soil by soil microorganisms and fauna. As a result, the AA distributions of soils tend to be similar over a wide range of soil properties and under different management/land uses [110–112] and large differences are usually only found for soils rich in amorphous aluminium silicates or between soils from very different climatic regions [110, 112]. The micaceous nature of the RM soil's clay minerals may therefore be responsible for the difference in soil AA profiles observed [113].

4.5.2 Limited Insights Gained from Bulk Soil Percentage Total Carbon and Nitrogen Contents and Amino Acids Concentrations

No observable changes or trends were found in the bulk soil % TC content of soil samples in any incubation experiment (Table 4.3). This is not unexpected as no C was added to the $^{15}NO_3^-$, $^{15}NH_4^+$ and ^{15}N-U incubation experiments and only small amounts of C were added in the slurry and glucose incubation experiments (within the normal variability of % TC contents between samples, i.e. the standard deviation [SD] of untreated soil is larger than the amount of C added). The same is true for the bulk soil % TN content of all soil samples (Table 4.3) as N additions were selected at agriculturally relevant (18–80 kg N ha^{-1}; Table 4.1; [106]), but sufficiently low N concentrations to obviate alteration of the soil's N status and limit any perturbation of the system.

The small co-variations and lack of trends in the concentrations of individual hydrolysable AAs and in the total hydrolysable AA N in each incubation experiment (Tables A1.1–A1.11) indicate that substrate addition was insufficient to stimulate AA biosynthesis above native levels and provide a further indication that nutrient cycling within the soil systems was not perturbed substantially by the incubation experiments as the turnover/biosynthesis of AAs was not noticeably altered. Thus, no added nitrogen interaction (ANI) or PE resulting from the ^{15}N addition is observable in this data (note, however, this does not necessarily mean that the addition did not affect soil nutrient cycling at all; [114, 115]). Indeed, excepting the pre-existing differences between the soils prior to incubation (e.g. the higher hydrolysable AA concentration of the RM), there were no differences in the behaviour of the RM and WA soils under the incubation conditions based on the bulk parameters discussed in this section.

4.5.3   Bulk Soil Nitrogen Isotopic Compositions Reveal Differing Substrate Application Efficiencies and Some Gaseous Nitrogen Losses

Disparities in the mean bulk soil δ^{15}N values between treatments (i.e. those of $^{15}NO_3^-$<$^{15}NH_4^+$< ^{15}N-U incubation experiments; Table 4.4) are the result of the differing N contents of the substrates applied at 400 μg of *substrate* at 10 atom % (Table 4.1) giving rise to the application of a different number of moles of ^{15}N between different treatments. Mean percentage retention calculations take the actual moles of ^{15}N applied in each treatment into account, so do not reflect this pattern. There is, however, quite a range in mean percentage retentions, and those of the X + 2 incubation experiments (in which the 10 atom % ^{15}N substrate applied may have been diluted by co-applied natural abundance slurry N) are not consistently the lowest. This is somewhat unexpected as based on the slurry N content and likely NO_3^- and NH_4^+ concentrations and δ^{15}N values, it was estimated that the N isotopic composition of the X + 2 treatments were *ca.* ≤10 and 6.5–7.3 atom % for the $^{15}NO_3^-$ + 2 and $^{15}NH_4^+$ + 2 incubation experiments, respectively. In fact, the RM $^{15}NO_3^-$ + 2 and WA $^{15}NH_4^+$ + 2 incubation experiments had mean percentage retentions (calculated based on 10 atom % treatments) of 97 and 93 atom %, respectively (Table 4.4), suggesting that the ^{15}N dilution by natural abundance slurry N in the X + 2 treatments was minimal.

This premise then indicates that another factor was responsible for the low mean percentage retention (44%) of the WA $^{15}NO_3^-$ + 2 incubation experiment. The SE for this mean percentage retention was large (10%) and closer inspection of the percentage retentions for individual incubations revealed wide variations in values between replicates at a particular time point and no strong trend with time (Fig. 4.12). Based on this, poor (and inconsistent) treatment application, rather than loss of applied ^{15}N with time is the most likely explanation. Indeed, the larger volume

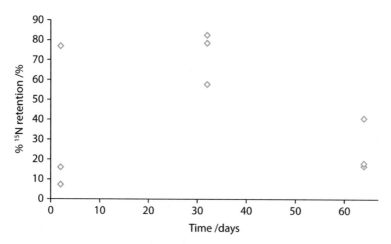

Fig. 4.12 Percentage ^{15}N retentions of individual incubation samples for the WA ^{15}NO$_3$$^-$ + 2 incubation experiment

of + 2 and + G treatments was more difficult to apply evenly and without some leaching of the application into the glass wool plug at the base of the incubation tube (note that at no point in any incubation experiment did any liquid leach further and escape from the base of the tube).

The same explanation may be adopted for the three $t = 32$ days samples for the WA ^{15}NO$_3$$^-$ + G incubation experiment, which had the lowest mean percentage retention (38%) and largest SE (17%). The only other incubation experiment with a low/modest mean percentage retention was the WA ^{15}NO$_3$$^-$ experiment (55%). In this incubation experiment, percentage ^{15}N retentions were generally low in incubation samples from $t = 1.5$ h to 2 days (37 ± 4.1%; mean ± SE) and higher in samples from later time points ($t = 8$ to 32 days; 92 ± 2.7%; Fig. 4.13). These results could be caused by a poorer injection technique in the earlier samples, but the trend of increasing percentage retentions with time suggests that volatilisation loss of lighter ^{14}N from the soil with time (most likely via denitrification with possibly some contribution from DNRA; [116]), leading to an increase in the ^{15}N/^{14}N ratio is also a possibility. It should also be noted that the greater propensity of the WA soil incubation experiments for low/inconsistent treatment applications (Table 4.4) may result from this soil's more clay-like and less structured texture at 50% WHC than the RM soil, resulting in a smaller volume of soil in the incubation tube and greater difficulties in applying treatments evenly and without some leaching into the glass wool plug at the base of the incubation tube.

An accordant positive trend ($m = 2.13$) is present in the mean bulk soil δ^{15}N values of the WA ^{15}NO$_3$$^-$ incubation experiment with time (Fig. 4.14) and interestingly, the mean bulk soil δ^{15}N values of the RM ^{15}NO$_3$$^-$ and RM ^{15}NO$_3$$^-$ + 2 incubation experiments also rise slightly with time (Fig. 4.14). Although the gradients and R^2 values of the latter two incubation experiments are low(er), rising trends in mean bulk soil δ^{15}N values over time for all three of these incubations with ^{15}NO$_3$$^-$ provide

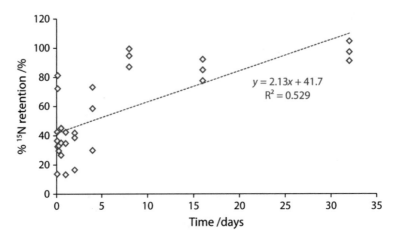

Fig. 4.13 Percentage ^{15}N retentions of individual incubation samples for the WA ^{15}NO$_3^-$ incubation experiment. Note: although at $t = 0$ the percentage ^{15}N retained is 0%, at any $t > 0$ the percentage ^{15}N retained would be expected to be 100% (provided treatment application was perfect), thus the function that governs the relationship between percentage ^{15}N retention and time is discontinuous and the linear regression line does not pass through zero

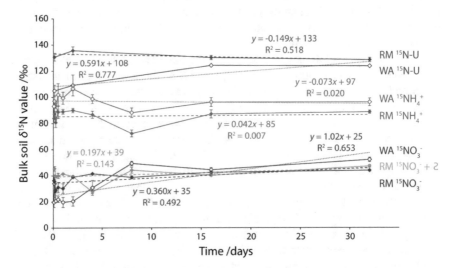

Fig. 4.14 Bulk soil δ^{15}N values from $t > 0$ until $t = 32$ days for incubation experiments with sufficient sampling points within this period. Error bars are $\pm$ SE (n = 3)

evidence that small, most likely denitrification-derived losses of ^{14}N from anaerobic microsites may have occurred [86, 117]. It should be noted, however, that these relatively weak trends could also result from inconsistent treatment applications and/or a combination of these and other processes.

Other incubation experiments that exhibit positive trends with time are WA ^{15}N-U (0.591) and RM ^{15}NH$_4$$^+$ (0.042) and this could, in each case, also result from either inconsistent treatment applications or preferential volatilisation losses of ^{14}N over ^{15}N with time (most likely as ^{14}NH$_3$, but other processes could also have occurred). As some correlations are relatively weak and, with the exception of the matching trend for the ^{15}NO$_3$$^-$ incubation experiments, particular treatments or soils do not show consistent patterns, it is difficult to draw firm conclusions about which causative factor was most important. It is perhaps surprising that volatilisation losses of ^{14}NH$_3$ from treatments with 10 atom % ^{15}NH$_4$$^+$, slurry, and particularly 10 atom % urea-^{15}N were not more consistently evident given the known propensity of these treatments for NH$_3$ losses [73, 74, 118–120]. This is likely due, partly at least, to the method of application utilised—injection, which reduces volatile losses [119, 120].

That no consistent differences between the two soils are apparent is not unusual in the context of the literature. The reported effects of soil type on the retention/volatilisation loss of applied ^{15}N vary; in some studies no clear differences between different soils were found (e.g. [85, 87]), while in others, some influence of soil type was discernible (e.g. [86, 94]). Although not necessarily dependent on soil type, soil moisture content is one factor that is known to affect volatile losses and which may also be influenced by soil type. At higher soil MCs, anaerobic microsites suitable for denitrification and DNRA to take place are more likely to develop, leading to higher gaseous N losses [121, 122]. The RM and WA soils used in these experiments were both maintained at 50% WHC, at which the water content of the RM soil is approximately twice that of the WA soil. Thus, higher denitrification-derived losses might be expected from the RM soil, but no such differences were observed (Fig. 4.14). The lack of a clear difference between the two soils in the context of NH$_3$ volatilisation is also surprising as pH is considered the most important soil characteristic determining these losses and the RM soil is more acidic (pH *ca.* 5 *cf.* 6 for the WA soil) and therefore supposedly less vulnerable to NH$_3$ loss [118]. This is likely due to the general lack of NH$_3$ volatilisation throughout all the incubation experiments.

4.5.4 Amino Acid Nitrogen Isotope Compositions and Percentage Retained ^{15}N Incorporations Provide Insights into the Biochemical Assimilation Pathways of Nitrogen

Time course plots of AA δ^{15}N values and percentage retained ^{15}N incorporations are described in Sect. 4.4.4 alongside (where possible) the results of linear and non-linear regressions. The rises above $t = 0$ $\delta^{15}N$ values and positive percentage retained ^{15}N incorporations for most AAs in all plots in Sect. 4.4.4 (on different scales) demonstrate that ^{15}N from all of the applied substrates has been incorporated into hydrolysable AAs (to different extents). Thus, the ^{15}N additions have resulted in (probably 'apparent') pool substitution ANIs (as additions were small and AA concentrations show no evidence of perturbation, although this does not constitute conclusive evidence that only apparent ANIs/PEs were present; [114, 115, 123]).

As explained in Sect. 4.4.4, AA $\delta^{15}N$ (or $\Delta^{15}N$) values reveal the proportions of N derived from the ^{15}N-labelled substrates (at a particular time) whereas percentage retained ^{15}N incorporations reveal the partitioning of ^{15}N in the system. Time course plots show how the proportions/partitioning of ^{15}N change/s over time and regression analyses enable quantitative estimates of the rates at which these changes occur and plateau proportions/partitioning of ^{15}N into individual hydrolysable AAs during the incubation experiments to be determined. Specifically, the rates at which AA $\Delta^{15}N$ values change reflect how the proportions of N derived from the ^{15}N-labelled substrate change, rather than a rate of ^{15}N incorporation and as a result cannot be used in the same way as the rate constants for percentage ^{15}N incorporations for which:

(i)   Initial rates (from zero-order rate constants) correspond to the time taken for ^{15}N to be transferred from the substrate applied into the AA and can provide a measure of the 'biosynthetic proximity' of the substrate to the product AA.
(ii) First-order rate constants are related to the overall rates at which AAs were biosynthesised from the applied ^{15}N-labelled substrate [77].

Similarly, plateau $\Delta^{15}N$ values are strongly affected by the moles ^{15}N applied as well as by the extent to which ^{15}N is incorporated. As substrates were applied at the same concentrations and therefore with different moles ^{15}N, plateau $\Delta^{15}N$ values can only reliably be used to assess the extent to which a substrate is incorporated in comparisons involving the same substrate (e.g. within incubation experiments, between different soils or if applied with additional substrates). Plateau percentage retained ^{15}N incorporations, on the other hand, provide an estimate of the (steady state) flux of ^{15}N into each AA pool [77]. As AA concentrations (and thus the balance of AA degradation/biosynthesis/turnover) do not noticeably change over the course of the incubation experiments (Sect. 4.5.2), ^{15}N may be expected to be distributed (after the initial uptake phase) to some extent in proportion to AA concentration, or more specifically, to the moles of N in each AA pool. Deviations from this proportional distribution could result from different (active and inactive) pools of the AA within the soil (^{15}N will only be incorporated into actively cycling pools, so an AA with a large, stable pool contributing to the overall soil concentration will incorporate less ^{15}N than expected based on the moles of N in that AA pool) and/or from the different biochemical routing of the ^{15}N. These deviations are reflected in differing plateau AA $\Delta^{15}N$ values (as, if ^{15}N is distributed exactly in proportion with AA concentration, AA $\Delta^{15}N$ values would be approximately equal in a given experiment).

Together, the time course plots and the regression parameters for individual AAs can be used to provide insights into the biochemical pathways through which the applied ^{15}N-labelled substrates are processed. Accordingly, over the next few sections, each substrate is discussed separately and then the contrasting behaviour of the three different substrates is rationalised in terms of incorporation patterns into individual hydrolysable AAs. A more thorough comparison of the incubation experiments in terms of the overall fate and partitioning of the applied ^{15}N-labelled substrates (i.e. different substrates, same soil and same substrate in different soils) is made in Sect. 4.5.5 by considering the total hydrolysable AA pool as a whole and using both percentage applied and retained ^{15}N incorporation data.

4.5.4.1 Assimilation of ^{15}N-Labelled Nitrate into Hydrolysable Amino Acids

^{15}N-labelled NO_3^- was applied in two incubation experiments in the RM soil and three incubation experiments in the WA soil and the δ^{15}N values of individual AAs and their percentage retained ^{15}N incorporations are shown in Figs. 4.4a, c, 4.5a, b, 4.7a, c, 4.8a and b with the results of linear and non-linear regressions in Tables 4.5, 4.6, 4.7 and 4.8. Differences exist in the patterns of the δ^{15}N values of individual AAs and of their percentage retained ^{15}N incorporations in each incubation experiment, but broadly: (i) there is a lag and, for some AAs, a small dip in δ^{15}N values or percentage ^{15}N incorporations over the first 24 h of the incubation experiment; (ii) the hierarchy in δ^{15}N values at $t = 0$ is maintained throughout, for the most part, but this is not the case for percentage ^{15}N incorporations; and (iii) AA δ^{15}N values do not rise very much over the course of the experiments and this is reflected in low percentage ^{15}N retentions. The general shape of the trends in AA δ^{15}N values and percentage retained ^{15}N incorporations (lags/dips and small rises) resulted in often a poor, very poor or even no fit for the linear and non-linear regressions applied (Tables 4.5, 4.6, 4.7 and 4.8), but in general, considering the data available: (i) zero-order rate constants are low (unsurprising given the initial lags/dips); (ii) first-order rate constants are also low (except where the non-linear regressions fit the data poorly); and (iii) plateau δ^{15}N values/percentages are low (due to the small rises in δ^{15}N values/percentage incorporations), except for the WA $^{15}NO_3^-$ incubation experiment in which ^{15}N incorporation was considerably higher.

The lags/dips in AA δ^{15}N values manifest as flatline/negative percentage ^{15}N incorporations and are most prominent in the RM $^{15}NO_3^-$ incubation experiment (Figs. 4.4a and 4.7a), discernible, but less pronounced in the WA $^{15}NO_3^-$ incubation experiment (Figs. 4.4c and 4.7c), and may be present, but missing due to lower sampling frequency, in the WA $^{15}NO_3^- + 2$ incubation experiment (Figs. 4.5b and 4.8b). An initial lag/dip in AA δ^{15}N values is not observable in the RM $^{15}NO_3^- + 2$ incubation experiment (Fig. 4.5a), but δ^{15}N values and accordingly, percentage ^{15}N incorporations fluctuate close to $t = 0$ values/0% throughout (Figs. 4.5a and 4.8a) indicating that very little ^{15}N incorporation into hydrolysable AAs occurred during this experiment. The lags in $^{15}NO_3^-$ assimilation could result from the uptake and reduction processes (and energy) required to assimilate NO_3^- [44, 69] leading to a slow response.

Depressions in AA δ^{15}N values from $t = 0$ values, however, indicate that substrates with lower N isotopic compositions than previously/at $t = 0$ have been used for AA biosynthesis and, although perhaps surprising, comparable increases in unlabelled inorganic N (which could be responsible for such 'dips' in assimilation of ^{15}N) have been commonly found in many MIT tracer experiments for years [47] and have been ascribed to: the stimulation of net mineralisation by N addition; salt effects; pH changes; and simply the regular operation of MIT, among other things. In this case, since immobilisation into AAs is being directly measured and AA δ^{15}N values do begin to increase after 24 h, a temporary effect of substrate injection, rather than the regular operation of MIT, is inferred and, since similar features are not observed with the addition of the other substrates tested, immediate salt effects, pH changes or a temporary stimulation of net mineralisation as a result of the N addition also seem unlikely.

A proposed possibility is that direct contact with high NO_3^- concentrations at the point of injection caused cell lysis and the release of labile non-proteinaceous substrates of a lower N isotopic composition for AA biosynthesis, but this may not realistically occur within the time-frame of the observed dip (as much of the material released from lysed cells would require mineralisation prior to assimilation into AAs; [115]). Another possibility is that the KNO_3 treatment stimulated the release of clay fixed NH_4^+ (by replacement of NH_4^+ with K^+) and fast assimilation of this apparently ^{15}N-depleted (compared with other N sources for AA biosynthesis, perhaps due to some isotope effect(s) associated with NH_4^+ fixation and subsequent release) NH_4^+ results in the biosynthesis of ^{15}N-depleted AAs [124–126]. Such effects of the added ^{15}N on the soil system would also constitute an 'apparent' ANI or triggering effect as the turnover of stable SOM is not affected (and due to its fast onset and short duration; [114, 115, 123]).

The conservation of the hierarchy of AAs with respect to their various $t = 0$ N isotopic compositions results from the generally low response of AA δ^{15}N values to $^{15}NO_3^-$ addition. This low response is reflected in low percentage ^{15}N incorporations and as a result, small variations in AA concentrations over the course of the incubation experiments have a perceptible effect on the trends in these data. In general, AAs present at higher concentrations (larger moles N) incorporate a larger percentage of retained ^{15}N, but ^{15}N is not distributed absolutely in proportion to moles AA N due to differently responding sub-pools of the AAs and/or the particular biochemical routing of $^{15}NO_3^-$.

The observations from the time course plots are reflected in the regression parameters and, as mentioned above, one result of this is that the selected regressions often did not fit the $^{15}NO_3^-$ experiment data or fitted only poorly. As a result, sufficient regression parameters for a quantitative comparison of the behaviour of different AAs within an incubation experiment are only available for the WA $^{15}NO_3^-$ incubation experiment (Tables 4.6 and 4.8). In this incubation experiment, ^{15}N was initially incorporated/partitioned mostly quickly into Glx and most slowly (not at all) into Hyp (followed by Phe and Tyr), suggesting that Glx is biosynthetically the closest AA to NO_3^-, while Hyp was not initially biosynthesised from the applied ^{15}N. This is consistent with the known biochemistry of NO_3^- assimilation, during which

NO_3^- is reduced to NH_4^+ and then assimilated into Glu, mainly via the GS-GOGAT pathway [44, 52].

The secondary amino structure of Hyp, on the other hand, prevents its participation in transamination, decarboxylation and other reactions involving the coenzyme pyridoxal phosphate, which are common for primary AAs [127, 128]. In fact, Hyp can only be biosynthesised from Pro (mainly via a specialised route that requires peptide-bound Pro residues as substrate) and bacteria are not reliably known to utilise Hyp in proteins or cell structural components, but only in bacterial antibiotics (e.g. etamycin and actinomycin) produced by specialist organisms [127, 128]. Thus, lack of ^{15}N assimilation into Hyp in soil is unsurprising and is supported by the low concentration of Hyp found in both soils (Tables A1.1–A1.11; Fig. 4.3). Further relationships between initial rates and presumed metabolic pathways are illustrated in Fig. 4.15.

The noticeably further biosynthetic proximity of Phe and Tyr compared with other AAs in Fig. 4.15, is likely due to the long (many step) biosynthetic routes to these aromatic AAs—beginning with the condensation of phosphoenolpyruvate (glycolytic intermediate) and erythrose 4-phosphate (pentose phosphate pathway intermediate) and then via the intermediates shikimate, chorismate and prephenate in order to produce the required α-ketoacid skeletons for transamination to Phe and Tyr [129].

Figure 4.15 also depicts the steady state AA percentage retained ^{15}N incorporations or fluxes of ^{15}N (line width) enabling comparison between AAs (flux into Glx is greatest and into Tyr is lowest) and highlighting the relationship between biosynthetic proximity and flux of ^{15}N. There is also a good correlation between initial and overall rates of AA biosynthesis indicating that the AAs with the closest biosynthetic proximity to NO_3^- were also generally biosynthesised most quickly and the flux of ^{15}N into these AA pools was greatest.

Finally, the shading of the AA boxes in Fig. 4.15 represents the steady state proportions of NO_3^--^{15}N. The shading for most of the AAs is quite similar, reflecting a relatively narrow range in $\Delta^{15}N$ values in this incubation experiment (Table 4.6) and the observation that ^{15}N is distributed for the most part in proportion to the moles of N in each AA pool. A greater proportion of the Leu pool than expected (darkest box in Fig. 4.15) has been biosynthesised from the applied $^{15}NO_3^-$ (19.3‰) and Val (6.6‰) contains the lowest steady state proportion of NO_3^--^{15}N. This suggests there may be some preferential routing of the ^{15}N into the Leu pool, and/or this AA has a smaller less active/inactive sub-pool than other AAs and possibly reduced routing of NO_3^--^{15}N into the Val pool or the existence of a (larger) less active sub-pool for Val compared with other AAs.

The main difference between the incubation experiments in which $^{15}NO_3^-$ was applied is the extent to which $^{15}NO_3^-$ was incorporated into the AAs, particularly between the two soils and between the $^{15}NO_3^-$ and $^{15}NO_3^- + \mathcal{C}$ treatments in the WA soil. This difference becomes even more evident when considering the total hydrolysable AA pool as a whole and will therefore be discussed further in Sect. 4.5.5. As already noted, the very low incorporation of $^{15}NO_3^-$ in the two RM incubation

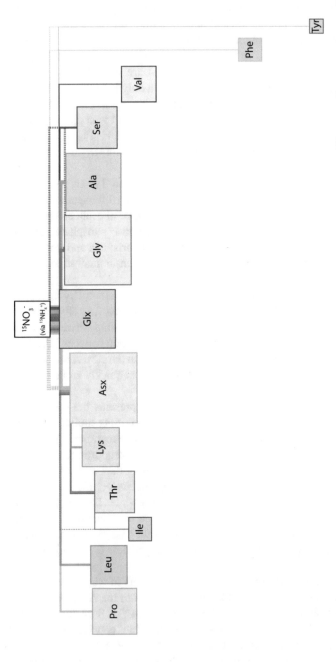

Fig. 4.15 Biosynthetic 'map' of the experimentally determined initial incorporation/partitioning rates and fluxes of ^{15}N from $^{15}NO_3^-$ in the WA $^{15}NO_3^-$ incubation experiment applied to known metabolic pathways [129–131]. Lines illustrate the biochemical connectivity of the substrate and AAs and dotted lines indicate possible, but likely only secondary biosynthetic pathways. Box area is proportional to moles N in that compound/pool; vertical distance from the base of the substrate box is proportional to the reciprocal of the initial rate of incorporation (representative of biosynthetic proximity); line width is proportional to the plateau incorporation of ^{15}N, or the flux of ^{15}N into that AA; and shading is based on plateau AA $\Delta^{15}N$ values translated into percentage opacities, where darker shades represent higher plateau $\Delta^{15}N$ values

experiments and insufficient data points in two of the three WA $^{15}NO_3^-$ experiments leaves only the WA $^{15}NO_3^-$ incubation experiment with enough quantitative regression parameters for a quantitative comparison of the behaviour of different AAs within the incubation experiment and this, by extension, prevents statistical comparison of the behaviour of individual AAs in the different $^{15}NO_3^-$ incubation experiments. Visual similarities in the patterns in AA $\delta^{15}N$ values and percentage retained ^{15}N incorporations are highlighted in the first paragraph of this section and discussed thereafter. In addition to this, there are similarities in the hierarchy of AA percentage retained ^{15}N incorporations between the incubation experiments (e.g. Ala, Glx and Gly commonly have the highest percentage ^{15}N incorporations), but also differences (e.g. 'mid-range' incorporation is found for Pro in all except the RM $^{15}NO_3^- + 2$ incubation experiment in which Pro has the lowest [and a negative] percentage retained ^{15}N incorporation). Since AA percentages are generally close together in each incubation experiment and values result from a combination of AA concentration and only small changes in AA $\delta^{15}N$ values, this is unsurprising and does not particularly provide insights into the different biochemical functioning of the two soils or treatments.

4.5.4.2 Assimilation of ^{15}N-Labelled Ammonium into Hydrolysable Amino Acids

^{15}N-labelled NH_4^+ was applied in one incubation experiment in the RM soil and three incubation experiments in the WA soil. The $\delta^{15}N$ values of individual AAs and their percentage retained ^{15}N incorporations are shown in Figs. 4.4b, d, 4.5c, 4.7b, d and 4.8c with the results of linear and non-linear regressions in Tables 4.5, 4.6, 4.7 and 4.8. There are differences in the patterns of the $\delta^{15}N$ values of individual AAs and of their percentage retained ^{15}N incorporations in each incubation experiment, but generally: (i) the $\delta^{15}N$ values and percentage ^{15}N incorporations of all AAs (except Hyp) rise with little delay from $t = 0$ in a two-phase pattern (fast over the first 2–4 days, then more slowly); (ii) assimilation into Glx occurs more quickly and to a greater extent and Glx $\delta^{15}N$ values and percentage ^{15}N incorporations fall after 2 days while those of other AAs continue to rise; and (iii) AA $\delta^{15}N$ values and percentage ^{15}N incorporations maintain relatively stable, elevated levels during the latter period of the incubation experiments. Accordingly, the linear and non-linear regressions applied generally fit both the AA $\Delta^{15}N$ and percentage retained ^{15}N incorporation data well (Tables 4.5, 4.6, 4.7 and 4.8), with the exception of for Hyp and sometimes for Glx, Gly, Lys and Tyr for the linear regressions and for Glx and Hyp for the non-linear regressions.

The definitive reason for the two-phase incorporation pattern of $^{15}NH_4^+$ into most AAs is not known, but as each time point just provides a snapshot of the dynamic balance between ^{15}N assimilation into and loss from each AA pool, likely results from the combination of the rapid initial assimilation of available and labile $^{15}NH_4^+$ (where assimilation $\gg$ loss) with the slower (temporary) equilibrium turnover of ^{15}N in each AA pool (where assimilation $\approx$ loss). This temporary equilibrium may develop as a result of the declining availability of $^{15}NH_4^+$, or perhaps more likely, given the low plateau percentage retained ^{15}N incorporations found, from the regulation of N uptake/assimilation as limitation by another essential nutrients (e.g. C or P) arises.

Both the clear differences in the behaviour of Glx and Hyp in terms of the patterns in their $\delta^{15}N$ values and percentage retained ^{15}N incorporations compared with other AAs, and the AAs' different rates and extents of ^{15}N incorporation/partitioning can be rationalised based on the known biochemical reactions of microorganisms. Glutamate has the highest initial rate of ^{15}N incorporation/partitioning for both the RM $^{15}NH_4^+$ and WA $^{15}NH_4^+$ incubation experiments (2.61% day^{-1} and 1.99% day^{-1}, respectively) indicating that Glx is 'the AA' closest in biosynthetic proximity to NH_4^+ and indeed, NH_4^+ is assimilated directly into Glu (via either GDH or the GS-GOGAT pathway; Fig. 4.1; [60–63]).

The similarly high overall rate of Glx biosynthesis in both incubation experiments (1.97 day^{-1} and 3.64 day^{-1} in the RM and WA soils, respectively) and the faster rise and earlier, higher peak in $\delta^{15}N$ values and percentage ^{15}N incorporations of Glx compared with other AAs in all three time course $^{15}NH_4^+$ experiments further reflects the importance of Glx/Glu in NH_4^+ assimilation and given the subsequent early fall in Glx $\delta^{15}N$ values and percentage ^{15}N incorporations, as a precursor in other biochemical reactions. Glutamate is an important substrate in the biosynthesis of the AAs Gln, Pro and Arg [129] and may also be converted into other AAs via transamination reactions [75–78, 80]. The continued rise in the $\delta^{15}N$ values and percentage ^{15}N incorporations of the other AAs may result partly from the conversion of Glu-^{15}N into these (^{15}N-)AAs. Glutamate is also the precursor of many other biochemical molecules, which could further/instead account for the fall in ^{15}N-Glu observed, including, for example, δ-aminolevulinate (which forms porphyrins in bacteria), glutathione (a redox buffer in some bacteria) and nucleotides [129].

As a result of the patterns in Glx $\delta^{15}N$ values, the non-linear first-order exponential assimilation equations applied to $\Delta^{15}N$ and percentage retained ^{15}N plots fit poorly. This aside, Glx maintains the highest plateau proportions of ^{15}N and plateau incorporations in both the RM $^{15}NH_4^+$ and WA $^{15}NH_4^+$ incubation experiments (75.1‰ and 2.65%, and 36.3‰ and 1.27% for the two experiments, respectively) indicating that even following initial peak incorporation/partitioning into Glx, the flux of ^{15}N through this AA pool remains important. Using an analogous experimental approach (kinetic flux profiling) on an *Escherichia coli* culture, Yuan et al. [132] similarly found largest fluxes of ^{15}N into Glu and Gln and surmised that Glu N was quickly transferred into other AAs [133]. The small/lack of increase in the N isotopic composition of Hyp indicates that very little/none of the applied $^{15}NH_4^+$ was assimilated into Hyp over the course of the incubation experiment (Figs. 4.4b, d and 4.5c) and this is reflected in almost exclusively 0% incorporation of retained ^{15}N throughout (Figs. 4.7b, d and 4.8c). As discussed in Sect. 4.5.4.1, this is consistent with the known biochemistry of N assimilation and Hyp biosynthesis (or lack thereof).

Further insights into the biochemical processing of $^{15}NH_4^+$ can be gained by applying the experimentally derived initial rates, fluxes and ^{15}N proportions of all AAs in the RM $^{15}NH_4^+$ and WA $^{15}NH_4^+$ incubation experiments (Tables 4.5, 4.6,

4.7 and 4.8) to known metabolic pathways, creating biosynthetic 'maps' (Figs. 4.16 and 4.17). Figures 4.16 and 4.17 reveal that Tyr is biosynthetically considerably further from NH_4^+ than other AAs and, as for $^{15}NO_3^-$, this is likely due to the large number of steps in the biosynthesis of Tyr [129]. It is surprising, however, that the initial rate of Phe biosynthesis from $^{15}NH_4^+$ is so much faster than for Tyr as Phe is biosynthesised by a similarly long route, which diverges from that of Tyr only at prephenate from which the AAs corresponding α-ketoacid C skeletons are formed [129]. Perhaps under the conditions prevailing in the soil, the dehydration and decarboxylation required to form the α-ketoacid C skeleton of Phe (phenylpyruvate) proceeds more easily/quickly than the oxidative decarboxylation of prephenate to *p*-hydroxyphenylpyruvate and/or transamination of phenylpyruvate occurs more easily/quickly.

There is also some similarity in the relative fluxes and in the relationship between biosynthetic proximity and flux (strong positive correlation) in Figs. 4.16 and 4.17— AAs biosynthetically closer to NH_4^+ incorporated larger amounts of ^{15}N from $^{15}NH_4^+$. In addition, in both figures, the Glx, Leu and Ile boxes are noticeably darker and the Gly and Val boxes noticeably lighter than those of other AAs. Thus $^{15}NH_4^+$ seems to be preferentially routed into Glx (unsurprising), Leu and Ile and not into Gly and Val, and/or these AAs have smaller and larger (for each group, respectively) less active/inactive sub-pools compared with other AAs.

Figures 4.16 and 4.17 are generally quite similar, but comparing the RM $^{15}NH_4^+$ and WA $^{15}NH_4^+$ incubation experiments, initial rates of AA biosynthesis were generally faster (except for Tyr), fluxes were greater and the proportions of ^{15}N in AAs were higher in the RM $^{15}NH_4^+$ incubation experiment compared with the WA $^{15}NH_4^+$ incubation experiment. Thus, incorporation of $^{15}NH_4^+$ into AAs seems to have proceeded via the same biochemical pathways in the two soils, but occurred at slightly faster rates and to greater extents in the RM soil.

4.5.4.3 Assimilation of ^{15}N-Labelled Urea into Hydrolysable Amino Acids

^{15}N-labelled urea was applied in one incubation experiment in each soil and the $\delta^{15}N$ values of individual AAs and their percentage retained ^{15}N incorporations are shown in Figs. 4.6 and 4.9. There are differences in the patterns of the $\delta^{15}N$ values of individual AAs and of their percentage retained ^{15}N incorporations in each incubation experiment, but a considerable source of these differences is the low sampling frequency in these experiments and generally: (i) the $\delta^{15}N$ values and percentage ^{15}N incorporations of most AAs (except Hyp) rise with little delay from $t = 0$ and stabilise/fall slightly from $t = 16$ to 32 days; (ii) assimilation into Glx occurs more quickly and to a greater extent than into other AAs and Glx $\delta^{15}N$ values and percentage ^{15}N incorporations fall during the latter part of the incubation experiments.

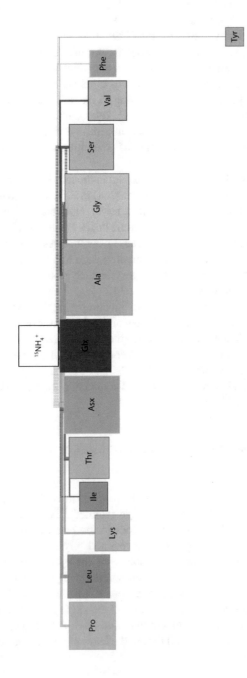

Fig. 4.16 Biosynthetic 'map' of the experimentally determined initial incorporation/partitioning rates and fluxes of ^{15}N from $^{15}NH_4^+$ in the RM $^{15}NH_4^+$ incubation experiment applied to known metabolic pathways [129–131]. Lines illustrate the biochemical connectivity of the substrate and AAs and dotted lines indicate possible, but likely only secondary biosynthetic pathways. Box area is proportional to moles N in that compound/pool; vertical distance from the base of the substrate box is proportional to the reciprocal of the initial rate of incorporation (representative of biosynthetic proximity); line width is proportional to the plateau incorporation of ^{15}N, or the flux of ^{15}N into that AA; and shading is based on plateau AA $\Delta^{15}N$ values translated into percentage opacities, where darker shades represent higher plateau $\Delta^{15}N$ values

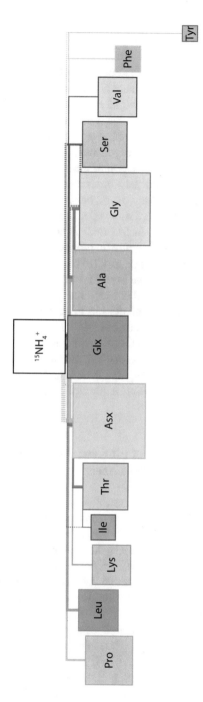

Fig. 4.17 Biosynthetic 'map' of the experimentally determined initial incorporation/partitioning rates and fluxes of ^{15}N from $^{15}NH_4^+$ in the WA $^{15}NH_4^+$ incubation experiment applied to known metabolic pathways [129–131]. Lines illustrate the biochemical connectivity of the substrate and AAs and dotted lines indicate possible, but likely only secondary biosynthetic pathways. Box area is proportional to moles N in that compound/pool; vertical distance from the base of the substrate box is proportional to the reciprocal of the initial rate of incorporation (representative of biosynthetic proximity); line width is proportional to the plateau incorporation of ^{15}N, or the flux of ^{15}N into that AA; and shading is based on plateau AA $\Delta^{15}N$ values translated into percentage opacities, where darker shades represent higher plateau $\Delta^{15}N$ values

Urea-^{15}N was most likely assimilated as ^{15}NH$_4^+$ following hydrolysis [71]. This is supported by similar patterns in the δ^{15}N values and percentage retained ^{15}N incorporations of AAs (particularly Glx) for the ^{15}N-U incubation experiments compared with the ^{15}NH$_4^+$ incubation experiments. The explanations given in Sect. 4.5.4.2 for the trends observed in the ^{15}NH$_4^+$ incubation experiments and the different behaviour of Glx and Hyp thus also apply to the ^{15}N-U incubation experiments. There are however, also differences in the behaviour of individual hydrolysable AAs in the ^{15}N-U compared with the ^{15}NH$_4^+$ incubation experiments (possibly partly exacerbated by the lower sampling frequency of the ^{15}N-U incubation experiments). One noticeable difference is in the RM ^{15}N-U incubation experiment, in which there is a small early peak for Gly, Ile and Val, affecting estimation of the zero-order rates for these AAs. Indeed, due to sampling frequency, the zero-order rate constants of AAs in the ^{15}N-U incubation experiments were calculated over 2 days (compared with 12 h for the ^{15}NH$_4^+$ incubation experiments), which may have resulted in slower estimated zero-order rates (as the window of initial, fastest rates is eclipsed). Considerably slower initial rates (and therefore further biosynthetic proximities) are certainly evident in the RM ^{15}N-U incubation experiment (Fig. 4.18) compared with the ^{15}NH$_4^+$ treatment (Fig. 4.16), but only to a much lesser extent for WA (Figs. 4.17 and 4.19).

In the addition, the overall lower rate of AA biosynthesis from urea-^{15}N in the RM soil compared with the WA soil is observable in Figs. 4.6 and 4.9 and clear in Tables 4.5, 4.6, 4.7 and 4.8. Differences in rates aside, Figs. 4.18 and 4.19 are similar with comparable fluxes (greatest into Glx and smallest in Tyr) and shading patterns (darker for Glx, Leu and Ile and lighter for Val and Gly in both). The range of box shades in Fig. 4.19 is smaller, suggesting that there is less preferential routing of urea-^{15}N and/or a smaller range in the contribution of less active/inactive AA sub-pools in the WA soil.

4.5.4.4   Comparison of the Assimilation of Different ^{15}N-Labelled Substrates into Hydrolysable Amino Acids

Probably the most noticeable difference in the assimilation of the ^{15}N-labelled substrates into hydrolysable AAs is the different rates and extents of ^{15}N incorporation between the substrates—in general ^{15}NH$_4^+$ > ^{15}N-U > ^{15}NO$_3^-$. This is clearly reflected in substrate incorporation into the total hydrolysable AA pool and will be discussed further in Sect. 4.5.5. Consideration of ^{15}N incorporation patterns (and regression parameters) into individual hydrolysable AAs can provide insights into the biochemical pathways of microbial assimilation, particularly in the context of known biochemical reactions (Figs. 4.15, 4.16, 4.17, 4.18 and 4.19). It is not surprising that the three ^{15}N-labelled substrates investigated seem to be processed via similar biochemical pathways, particularly since both ^{15}NO$_3^-$ and ^{15}N-urea are transformed into ^{15}NH$_4^+$ for assimilation into AAs (Fig. 4.20; [44, 71]).

There are however, some differences, including in the relative biochemical proximities of Lys and Thr and Ala and Gly for ^{15}NO$_3^-$ (Fig. 4.15) compared with ^{15}NH$_4^+$ and ^{15}N-urea (Figs. 4.16, 4.17, 4.18 and 4.19). For Lys and Thr this does not affect

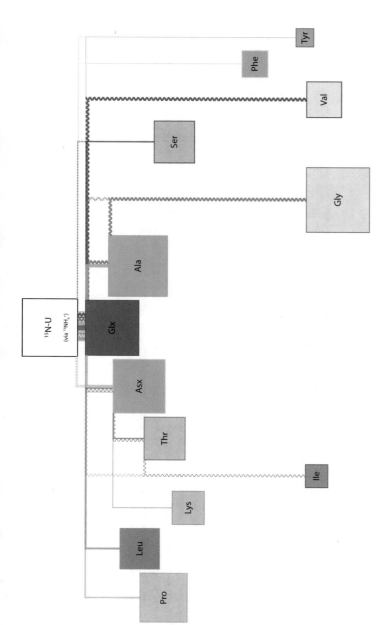

Fig. 4.18 Biosynthetic 'map' of the experimentally determined initial incorporation/partitioning rates and fluxes of ^{15}N from ^{15}N-urea in the RM ^{15}N-U incubation experiment applied to known metabolic pathways [129–131]. Lines illustrate the biochemical connectivity of the substrate and AAs and dotted lines indicate possible, but likely only secondary biosynthetic pathways. Box area is proportional to moles N in that compound/pool; vertical distance from the base of the substrate box is proportional to the reciprocal of the initial rate of incorporation (representative of biosynthetic proximity) - Gly, Ile and Val are shown furthest away, with wavy lines as the zero-order rate constants could not be determined due to poor linear regression fits; line width is proportional to the plateau incorporation of ^{15}N, or the flux of ^{15}N into that AA; and shading is based on plateau AA Δ^{15}N values translated into percentage opacities, where darker shades represent higher plateau Δ^{15}N values

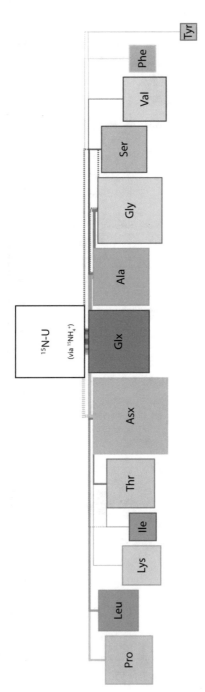

Fig. 4.19 Biosynthetic 'map' of the experimentally determined initial incorporation/partitioning rates and fluxes of ^{15}N from ^{15}N-urea in the WA ^{15}N-U incubation experiment applied to known metabolic pathways [129–131]. Lines illustrate the biochemical connectivity of the substrate and AAs and dotted lines indicate possible, but likely only secondary biosynthetic pathways. Box area is proportional to moles N in that compound/pool; vertical distance from the base of the substrate box is proportional to the reciprocal of the initial rate of incorporation (representative of biosynthetic proximity); line width is proportional to the plateau incorporation of ^{15}N, or the flux of ^{15}N into that AA; and shading is based on plateau AA $\Delta^{15}N$ values translated into percentage opacities, where darker shades represent higher plateau $\Delta^{15}N$ values

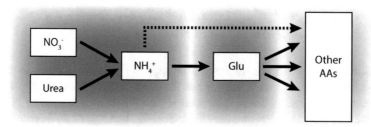

Fig. 4.20 Diagram showing the flow of the N substrates applied in these (left) and related previous ([77]; Glu) incubation experiments during N assimilation into AAs

the presumed routing of ^{15}N, but for Ala and Gly, a closer biochemical proximity for Ala than Gly suggests some ^{15}N-Gly biosynthesis from ^{15}N-Ala (via transamination of glyoxylate; [130] in experiments with [15]NH$_4^+$ and ^{15}N-U, but not [15]NO$_3^-$). Similarly, the large difference in the biosynthetic proximity of Phe and Tyr in the RM [15]NH$_4^+$ incubation experiment (only) is surprising given the similarities in the biosynthetic pathways to these AAs (Sect. 4.5.4.4) and there is also some variation in the biosynthetic proximity of these two AAs relative to those of other AAs across Figs. 4.15, 4.17, 4.18 and 4.19.

Biosynthetic proximities and fluxes are strongly correlated for all incubation experiments presented in Figs. 4.15, 4.17, 4.18 and 4.19 (except RM ^{15}N-U, in which they are weakly correlated) and the relative ordering of fluxes and ^{15}N proportions is similar. In particular, the AA pools containing the highest and lowest proportions of ^{15}N are identical in the [15]NH$_4^+$ and ^{15}N-U incubation experiments—Glx, Leu and Ile with the highest and Gly and Val with the lowest proportions—and also very similar in the WA [15]NO$_3^-$ incubation experiment—Leu is shaded darkest and Gly and Val are lightest. It is not possible to conclusively differentiate whether these differences in the proportions of ^{15}N in different AA pools are due to the preferential biochemical routing of applied ^{15}N-labelled substrates to particular AA pools and/or differences in the contributions of less active/inactive sub-pools for different AAs. Given, however, that N assimilation occurs via Glx, it is not unreasonable to suggest that the high steady state proportions of ^{15}N remaining in Glx in the [15]NH$_4^+$ and ^{15}N-U incubation experiments, (in which fluxes of ^{15}N are appreciable) may be due to preferential routing of ^{15}N through the Glx pool.

The biochemical insights into N assimilation discussed for the incubation experiments in this chapter also compare well (particularly for [15]NH$_4^+$) with published ^{15}N-Glu incubation experiments [77]. Aspartate and Ala, were for example, biosynthetically closest to the applied ^{15}N-Glu substrate, but there was also a larger range in biosynthetic proximities for the ^{15}N-Glu incubation experiments (with Thr furthest away of the AAs shown). In addition, incorporation of Glu-^{15}N into other hydrolysable AAs occurs at higher rates and to greater extents compared with the substrates applied in this chapter. Once again, this is unsurprising since NH$_4^+$ is assimilated into Glu, and Glx comprises the most important source of N in biosynthetic reactions (Fig. 4.20; [134]).

4.5.5   Percentages ^{15}N Incorporated into the Total Hydrolysable Amino Acid Pool

As detailed in Sect. 4.4.5, the percentages of applied ^{15}N incorporated into the total hydrolysable AA pool provide insights into the overall fate of the applied ^{15}N, while the percentages of retained ^{15}N incorporated into the total hydrolysable AA pool reveal the partitioning of ^{15}N into the total hydrolysable AA pool. In this section, the fate and partitioning of ^{15}N applied in different forms to the same soil (Sect. 4.5.5.1) and in the same form to different soils (Sect. 4.5.5.2) are compared and contrasted.

4.5.5.1   Comparison of the Fate and Partitioning of the Different Treatments in Each Soil

For the RM soil, percentage applied and retained ^{15}N incorporations are almost identical and both rates and steady state extents of ^{15}N incorporation/partitioning increase in the order: $^{15}NO_3^- + 2 < ^{15}NO_3^- < ^{15}$N-urea $< ^{15}NH_4^+$ (Fig. 4.10 and Table 4.9). The pattern of percentage applied ^{15}N incorporations is similar for the WA soil (Fig. 4.11a), except that incorporation of $^{15}NO_3^-$ was relatively higher (but with large error bars) compared with the other treatments and differences between the treatments are generally smaller. There are also greater differences between the two percentage ^{15}N incorporation calculations for the WA soil as WA percentage retained ^{15}N incorporations (Fig. 4.11b) are affected by the large range of percentage ^{15}N retentions in the WA incubation experiments (38–115%; Table 4.4; discussed in Sect. 4.5.3). This range is mainly due to more inconsistent treatment applications in the WA soil and the effects of this include changes in the relative order of percentage ^{15}N incorporations from the different treatments and a 'bunching' of all treatments around 6% incorporation of retained ^{15}N.

Considering the general pattern of ^{15}N incorporation/partitioning across the incubation experiments—$^{15}NO_3^- < ^{15}$N-urea $< ^{15}NH_4^+$—the slow uptake (including lags and dips) and low overall enrichment of the total hydrolysable AA pool with $^{15}NO_3^-$ indicates that $^{15}NO_3^-$ was not used extensively as an anabolic source of N. This suggests that an alternative N source was available for this purpose and was consumed in preference—most likely NH_4^+ or perhaps LMW organic N molecules that can be directly assimilated [44]. Ammonium is generally the preferred source of inorganic N for assimilation by soil microorganisms as both NO_3^- uptake and incorporation into cell material (via reduction to NH_4^+) require more energy (and thus C; [44, 48, 59, 67, 68, 135]). Indeed the synthesis of NO_3^- transporter proteins and NO_3^- reductase are thought to be repressed by the presence of NH_4^+ and NO_3^- assimilation in soils has been shown to be strongly and quickly (but not completely) inhibited by even low concentrations of NH_4^+ [44, 59].

Co-application with slurry in the $^{15}NO_3^- + 2$ incubation experiments was carried out in order to investigate $^{15}NO_3^-$ assimilation when supplied with an additional C source, which could be expected to increase $^{15}NO_3^-$ incorporation [68, 88, 135]. The reduced steady state percentage ^{15}N incorporations recorded in these experiments, compared with ^{15}NO$_3^-$ alone, suggest however, that the effect of slurry-NH$_4^+$ in reducing ^{15}NO$_3^-$ assimilation dominated in these experiments [44, 59]. Co-addition with glucose in the WA ^{15}NO$_3^-$ + G incubation experiment was expected to increase ^{15}NO$_3^-$ assimilation compared with ^{15}NO$_3^-$ supplied alone [68, 88, 135], but at 32 days only 4.08% of the applied ^{15}N was present in the total hydrolysable AA pool, compared with a steady state percentage applied ^{15}N incorporation of 6.39% for ^{15}NO$_3^-$ alone (but with large error bars; Fig. 4.11a and Table 4.10). This is partly due to the low mean percentage retention of ^{15}N in the WA ^{15}NO$_3^-$ + G incubation experiment (38%; Table 4.4), which is thought to be due to poor treatment application (Sect. 4.5.3). Indeed, 13.6% of the retained ^{15}N in the WA ^{15}NO$_3^-$ + G incubation experiment is present in the total hydrolysable AA pool at $t = 32$ days compared with a steady state percentage retained ^{15}N incorporation of 6.40% for ^{15}NO$_3^-$ alone (Fig. 4.11b and Table 4.10). Thus, partitioning of NO$_3^-$-^{15}N into the total hydrolysable AA pool was improved when additional C was supplied, as expected [68, 88, 135].

That ^{15}NH$_4^+$ was generally assimilated to the greatest extent, is not surprising as NH$_4^+$ is commonly a preferred N source and can be incorporated directly into Glu and thereby other AAs [44, 48, 59–61, 63, 67, 68]. The steady state percentage applied NH$_4^+$-^{15}N incorporated into the total hydrolysable AA pool in the RM soil is lower (13.4%; Table 4.9), however, than that of Glu-^{15}N in an analogous incubation experiment (22.1%, excluding ^{15}N in the Glx pool; [77]; Fig. 3.4). This might also be expected as Glu is also a preferred microbial N source [44, 48, 134], provides C as well as N nutrition, is a 'biosynthetic step' closer than NH$_4^+$ to most other AAs (Fig. 4.20) and Glu-N is easily transferred to other AAs via transamination reactions [75–78, 80]. The considerably larger incorporation of Glu-^{15}N recorded for the RM soil is nonetheless interesting as AA uptake is usually an active, N regulated process (direct route), which may alternatively proceed via extracellular mineralisation to NH$_4^+$ (MIT route; [42, 44, 45, 136]). The relative contribution of these routes to microbial N nutrition is not well known [44] and while greater (and faster 3.51 day^{-1} compared with 0.73 day^{-1}; Table 4.9) incorporation of Glu-^{15}N compared with NH$_4^+$-^{15}N does not provide unequivocal evidence for a dominance of direct uptake, it certainly suggests that it might be important in the RM soil under the incubation conditions.

Co-application of ^{15}NH$_4^+$ with slurry in the WA ^{15}NH$_4^+$ + S incubation experiment results in higher incorporations of NH$_4^+$-^{15}N compared with ^{15}NH$_4^+$ alone. Dilution of the ^{15}NH$_4^+$ by slurry-NH$_4^+$ (resulting in a lower ^{15}N-labelling of the treatment) was shown to be minimal based on bulk soil δ^{15}N values and mean percentage retentions (Sect. 4.5.3), so incorporation of NH$_4^+$-^{15}N applied with slurry was not expected to be lower than for ^{15}NH$_4^+$ alone due to dilution of the initial treatment. That NH$_4^+$-^{15}N incorporation is higher when applied with slurry, suggests that co-application stimulated immobilisation overall and this dominated any additional mineralisation (supplying additional unlabelled NH$_4^+$) and/or nitrification (consuming NH$_4^+$ and ^{15}NH$_4^+$; [137]). The increased incorporation may be due to the additional C supplied in the slurry [68, 88, 135] or increased microbial activity [108]. Co-addition with glucose in the WA $^{15}NH_4^+ + G$ incubation experiment similarly increased NH_4^+-^{15}N incorporation, but perhaps surprisingly, only to a similar extent as with slurry (higher percentage applied ^{15}N incorporation, but lower percentage retained ^{15}N incorporation for $^{15}NH_4^+ + G$ due to high mean retention of 115%; Fig. 4.11 and Table 4.4), in which there may be expected to be a lower concentration of C as labile/available as the applied glucose.

The intermediate assimilation of urea-^{15}N suggests that urea hydrolysis proceeded more easily than $^{15}NO_3^-$ reduction. This is perhaps unsurprising as urease is ubiquitous in soils and urea hydrolysis can occur extra- or intracellularly [44, 71] providing two routes for uptake. As hydrolysis is required, urea-^{15}N incorporation is slower and lower than NH_4^+-^{15}N incorporation.

Finally, it is worth noting the similar steady state percentage applied and retained ^{15}N incorporations of $^{15}NO_3^-$ and $^{15}NH_4^+$ (*ca.* 6.3%) and generally much more similar incorporations of all treatments in the WA, compared with the RM, soil. As a result, the general patterns of incorporation described above do not always apply as clearly to the WA soil. It is suggested that this may be due to inherent differences between the soils or because the WA soil regularly receives inorganic fertilisers (most recently as sulfur-coated urea and $(NH_4)_2SO_4$) and high soil water NO_3^- concentrations and soil mineral nitrate (SMN) contents have been recorded there in past years, so the soil/soil microbial biomass may be more attuned to inorganic N additions. This will be discussed further in Sect. 4.5.5.2 (below).

4.5.5.2 Comparison of the Fate and Partitioning of Treatments Between the Two Different Soils

Comparing the two soils, assimilation of NO_3^--^{15}N was higher in the WA soil than the RM soil in both the $^{15}NO_3^-$ and $^{15}NO_3^- + G$ incubation experiments (Figs. 4.10 and 4.11 and Tables 4.9 and 4.10). This suggests that there was either a lower availability of NH_4^+ (or other preferred N source) in the WA soil, or as this soil receives regular additions of inorganic N fertiliser, while the RM soil does not, its microbial population is better adapted to NO_3^- anabolism [39, 89]. Ammonium fixation by clay minerals (principally 2:1 phyllosilicates) is well known to influence the availability of NH_4^+ to soil microorganisms and plants—generally, at least temporarily, reducing it, although the magnitude(s) and dynamics of clay fixation (over time, with changes in the soil environment) are complex and remain somewhat unclear [124–126, 138]. The RM and WA soils have relatively similar clay contents (*ca.* 30%), but the calcareous nature and previous potassium (K) fertilisation of the WA soil is likely to have affected NH_4^+ fixation and release in this soil and may have contributed to lower NH_4^+ availabilities during the incubation experiment [124–126].

Previous fertilisation of the WA soil with urea may also explain the faster (initial and overall) assimilation of urea-^{15}N in the WA compared with the RM soil as this could increase the $t = 0$ availability of urease in this soil (compared with cattle depositions alone). That the steady state assimilation of urea-^{15}N, however, and particularly of NH_4^+-^{15}N (a very bioavailable N source) was higher in the RM soil (Figs. 4.10 and 4.11 and Tables 4.9 and 4.10) suggests that this soil clay-fixed less of the applied $^{15}NH_4^+$ and/or has a more active (or larger) soil microbial biomass [32]. Indeed, soil microbial biomass C and/or N or activity (or phospholipid fatty acid content) is commonly positively related to soil % TC, % TOC, % TN and/or SOM contents [22, 139–141] and soil % TOC and % TN contents are higher in the RM soil (6.80% *cf.* 4.17% and 0.63% *cf.* 0.45% for % TOC and % TN contents in the RM *cf.* the WA soil prior to incubation, respectively).

Both observations are also in accordance with those of several studies that have recorded increased SOM contents and microbial biomass (C/N/activity/growth) with manure or organic matter applications compared with inorganic fertilisers [108, 140, 142–145] and others which have found that chronic N deposition or inorganic N fertilisation detrimentally affects the soil microbial community [146–148] as the RM soil has only received cattle slurry for almost 25 years, while the WA soil has received regular additions of $(NH_4)_2SO_4$ or urea since it was converted from spring cropping to a grass ley in 2011. Moreover, the combination of a more active (or larger) soil microbial biomass and greater $t = 0$ soil % TOC and % TN content in the RM soil, as well as a higher total hydrolysable AA N concentration (Tables A1.1–A1.11), which represents a major source of mineralisable N in soil [149], is likely to have resulted in higher NH_4^+ production in the RM soil [22] and may partly explain the observed differences in NO_3^- assimilation between the two soils. In addition, the lower C:N ratio of the RM soil (11.1 *cf.* 16.6 for the WA soil) may have further contributed to higher N mineralisation in this soil [22]. It should be noted however, that repeated long-term slurry applications commonly lead to high soil NO_3^- concentrations and net NO_3^- production rather than NH_4^+ production [108]. Thus, while the different inherent soil types and management histories of the two soils will have affected N assimilation, it is difficult to conclusively determine the underlying mechanisms for the observed differences [108, 150].

4.6 Conclusions

The method presented in Chap. 3 has been used to reveal the biomolecular fate and partitioning of common fertiliser N compounds microbially transformed into hydrolysable soil AAs or the soil protein pool (a relatively labile pool of soil organic N) for the first time. ^{15}N applied in the form of $^{15}NO_3^-$, $^{15}NH_4^+$ and ^{15}N-urea was directly traced into individual hydrolysable AAs using compound-specific ^{15}N-SIP providing insights into the biochemical routing of the applied ^{15}N and enabling quantitative comparisons of the rates and extents of the soil microbial assimilation of different fertiliser N forms applied at environmentally relevant concentrations to two different soils. This quantitative data could also be of use in soil N cycling models.

Important, more specific findings, with reference to the hypotheses presented in Sect. 4.2, include:

(i)   Losses of ^{15}N over the course of most of the incubation experiments were small
      (i.e. mean percentage retentions were generally close to 100%), as expected.
      Poor treatment application rather than ^{15}N loss is the most likely reason for
      low calculated ^{15}N retentions and it is therefore recommended that soil vol-
      umes/packing heights are taken into and account and application volumes no
      greater than 250 μl should be used in future incubation experiments. The trend
      toward increased bulk soil ^{15}N-enrichments over time for all of the ^{15}NO$_3$$^-$
      treatments suggests, however, that some gaseous ^{14}N losses occurred during
      these experiments, most likely via denitrification (and possibly DNRA).

(ii) ^{15}N assimilation patterns reflected known microbial biochemical assimilation
      pathways and all substrates are thought to have been assimilated primarily as
      ^{15}NH$_4$$^+$ into Glu and subsequently converted to other AAs via transamination
      reactions and known metabolic pathways. The lack of ^{15}N incorporation into
      Hyp is consistent with the only specialist biosynthesis and use of this AA in
      soils. Regression fitting and the construction of 'biosynthetic maps' provided
      further insights into the biosynthetic proximities, fluxes and proportional rout-
      ing of ^{15}N into hydrolysable AAs and the relationships between these parame-
      ters. Interestingly the Glx and Leu pools received consistently proportionately
      more ^{15}N than expected, while the Gly and Val pools received less, implicating
      preferential partitioning and/or the existence of differentially active AA sub-
      pools. In general, results compare well with those of Knowles [80] and Knowles
      et al. [77] and indeed, substrate assimilation patterns reflect the sequence of
      transformations through which the substrates are linked (Fig. 4.20).

(iii) Overall, the extents of substrate incorporation into the total hydrolysable AA
      or soil protein pool were ordered as expected and in accordance with N source
      preferences and energy and C assimilation requirements—i.e. ^{15}NO$_3$$^-$ < ^{15}N-
      urea < ^{15}NH$_4$$^+$. The generally (in some cases very) low percentage ^{15}N incor-
      porations of these substrates was somewhat surprising. The faster and greater
      incorporation of ^{15}N-Glu [77] compared with ^{15}NH$_4$$^+$ in the RM soil implies
      that direct uptake, rather than MIT, could be important in the RM soil under
      the incubation conditions.

(iv) Co-additions with slurry reduced ^{15}NO$_3$$^-$ incorporation, presumably due to
      preferential slurry-NH$_4$$^+$ assimilation, but increased ^{15}NH$_4$$^+$ incorporation,
      possibly as a result of the additional C supplied and/or stimulation of the
      microbial biomass. Co-additions with glucose increased ^{15}N assimilation, as
      anticipated.

(v)   Surprisingly a significant difference was found in the AA distributions of the
      two soils. This difference was not reflected in the patterns of ^{15}N assimila-
      tion into individual hydrolysable AAs due, presumably, to the ubiquity of the
      primary metabolic pathways that operate in microorganisms. Differences were
      observed, however, in the rates and extents of assimilation of each ^{15}N-labelled
      substrate between the two soils. This could result from inherent differences
      between the soils (e.g. non-calcareous vs. calcareous, NH$_4$$^+$ fixing capacity
      of soil clays and to some extent, soil N dynamics and microbial biomass

size/activity and community structure) and/or from their different manage-ment histories (e.g. long-term grassland vs. recently converted from arable to grassland, cattle depositions only vs. cattle depositions and synthetic fertiliser) and which will affect soil N dynamics and microbial biomass size/activity and community structure.

These results fit well with the literature, both in terms of the biochemistry of N assimilation (Sect. 4.1.2) and previous studies assessing fertiliser N immobilisation based on bulk measurements (Sect. 4.1.3) and provide vital mechanistic links between theoretical/pure culture derived biochemical expectations and bulk level fertiliser immobilisation studies, bridging these very different scales/levels of detail. Looking beyond the finding that generally the assimilation of $^{15}NO_3^- < {}^{15}N\text{-urea} < {}^{15}NH_4^+$, the insights provided by this work into the effects of various co-additions and differences in N assimilation between different soils highlights the value of the AA ^{15}N-SIP approach applied in improving understanding of microbial N assimilation and soil N cycling at the same time as emphasising the complexity of soil systems and the real need to further investigate feedback relationships to more effectively and sustainably manage these systems. Further work assessing a wider range of amendments and soil types with different management histories would clearly be valuable.

The underlying aim of this work investigating the differential microbial assimi-lation of different N forms is to use the information gained in further understanding soil N cycling in order to develop strategies to reduce NO_3^- leaching. One potential solution is to improve the soil's ability to immobilise and then slowly release applied N, ideally in sync with plant N demand by careful management of the soil sys-tem (with particular reference to the C:N ratio of additions/microbial activity/fixing clays; [8, 51, 125, 151–153]). With this in mind, the ostensible conclusion in this study is to recommend NH_4^+ fertilisation due to the greater microbial incorporation of $^{15}NH_4^+$ into the soil protein pool. In addition, given the extremely low microbial incorporation of the applied $^{15}NO_3^-$, the susceptibility of this mobile ion to leaching is unsurprising.

It is difficult to assess, however, how the findings of this work translate from their small-scale laboratory incubation experiment reality to the real world and such conclusions also do not consider any other effects or feedbacks of such recommen-dations (e.g. 'pollution swapping'; [8]). For example, NO_3^- is commonly the most important N source for plants (for which applied N is actually intended!), due partly to its lack of incorporation by soil microorganisms [24]. Furthermore, the addition of plant roots and their exudates to the system studied could dramatically alter the dynamics of the soil's N cycle, affecting microbial N assimilation and thus giv-ing quite different results from the current work. A repeat, but necessarily slightly larger scale, study including carefully selected plants and further, plant mixtures (as species-specific differences and species interactions could also strongly affect results and conclusions) would therefore be extremely valuable.

References

1. Cameron KC, Di HJ, Moir JL (2013) Nitrogen losses from the soil/plant system: a review. Ann Appl Biol 126:145–173. https://doi.org/10.1111/aab.12014
2. Howden NJK, Burt TP, Worrall F, Mathias SA, Whelan MJ (2013) Farming for water quality: balancing food security and nitrate pollution in UK river basins. Ann Assoc Am Geogr 103:397–407. https://doi.org/10.1080/00045608.2013.754672
3. Sutton MA, Oenema O, Erisman JW, Leip A, van Grinsven H, Winiwarter W (2011) Too much of a good thing. Nature 427:159–161
4. Vitousek PM, Aber JD, Howarth RW, Likens GE, Matson PA, Schindler DW, Schlesinger WH, Tilman DG (1997) Human alteration of the global carbon cycle: sources and consequences. Ecol Appl 7:737–750
5. Smith VH, Tilman GD, Nekola JC (1999) Eutrophication: impacts of excess nutrient inputs on freshwater, marine, and terrestrial ecosystems. Environ Pollut 100:179–196
6. Galloway JN, Aber JD, Erisman JW, Seitzinger SP, Howarth RW, Cowling EB, Cosby BJ (2003) The nitrogen cascade. Bioscience 53:341–356
7. Galloway JN, Dentener FJ, Capone DG, Boyer EW, Howarth RW, Seitzinger SP, Asner GP, Cleveland CC, Green PA, Holland EA, Karl DM, Michaels AF, Porter JH, Townsend AR, Vörösmarty CJ (2004) Nitrogen cycles: past, present, and future. Biogeochemistry 70:153–226
8. Dungait JAJ, Cardenas LM, Blackwell MSA, Wu L, Withers PJA, Chadwick DR, Bol R, Murray PJ, Macdonald AJ, Whitmore AP, Goulding KWT (2012) Advances in the understanding of nutrient dynamics and management in UK agriculture. Sci Total Environ 434:39–50
9. Goulding K, Jarvis S, Whitmore A (2008) Optimizing nutrient management for farm systems. Philos Trans R Soc B 363:667–680. https://doi.org/10.1098/rstb.2007.2177
10. Jenkinson DS (2001) The impact of humans on the nitrogen cycle, with focus on temperate arable agriculture. Plant Soil 228:3–15
11. Bosshard C, Sørensen P, Frossard E, Dubois D, Mäder P, Nanzer S, Oberson A (2009) Nitrogen use efficiency of ^{15}N-labelled sheep manure and mineral fertiliser applied to microplots in long-term organic and conventional cropping systems. Nutr Cycl Agroecosyst 83:271–287. https://doi.org/10.1007/s10705-008-9218-7
12. Hirel B, Le Gouis J, Ney B, Gallais A (2007) The challenge of improving nitrogen use efficiency in crop plants: towards a more central role for genetic variability and quantitative genetics within integrated approaches. J Exp Bot 58:2369–2387. https://doi.org/10.1093/jxb/erm097
13. Spiertz JHJ (2010) Nitrogen, sustainable agriculture and food security: a review. Agron Sustain Dev 30:43–55. https://doi.org/10.1051/agro:2008064
14. Cassman KG, Doberman A, Walters DT (2002) Agroecosystems, nitrogen-use efficiency, and nitrogen management. Ambio 31:132–140
15. Barraclough D, Geens EL, Maggs JM (1984) Fate of fertiliser nitrogen applied to grassland. II. Nitrogen-15 leaching results. J Soil Sci 35:191–199
16. Glendining MJ, Poulton PR, Powlson DS, Jenkinson DS (1997) Fate of ^{15}N-labelled fertilizer applied to spring barley grown on soils of contrasting nutrient status. Plant Soil 195:83–98
17. Hancock JM, McNeill AM, McDonald GK, Holloway RE (2011) Fate of fertiliser N applied to wheat on a coarse textured highly calcareous soil under simulated semi-arid conditions. Plant Soil 348:139–153. https://doi.org/10.1007/s11104-011-0917-5
18. Ladd JN, Amato M (1986) The fate of nitrogen from legume and fertiliser sources in soils successively cropped with wheat under field conditions. Soil Biol Biochem 18:417–425
19. Pilbeam CJ, Hutchison D (1998) Fate of nitrogen applied in different fertilisers to the surface of a calcareous soil in Syria. Nutr Cycl Agroecosyst 52:55–60
20. Sebilo M, Mayer B, Nicolardot B, Pinay G, Mariotti A (2013) Long-term fate of nitrate fertilizer in agricultural soils. Proc Natl Acad Sci USA 110:18185–18189. https://doi.org/10.1073/pnas.1305372110
21. Allison FE (1955) The enigma of soil nitrogen balance sheets. Adv Agron 7:213–250

22. Booth MS, Stark JM, Rastetter E (2005) Controls on nitrogen cycling in terrestrial ecosystems: a synthetic analysis of literature data. Ecol Monogr 75:139–157

23. Monaghan R, Barraclough D (1997) Contributions to N mineralisation from soil macroorganic matter fractions incorporated into two field soils. Soil Biol Biochem 29:1215–1223

24. Schimel JP, Bennett J (2004) Nitrogen mineralisation: challenges of a changing paradigm. Ecology 85:591–602

25. Näsholm T, Kielland K, Ganeteg U (2009) Uptake of organic nitrogen by plants. New Phytol 182:31–48. https://doi.org/10.1111/j.1469-8137.2008.02751.x

26. Neff JC, Chapin FS III, Vitousek PM (2003) Breaks in the cycle: dissolved organic nitrogen in terrestrial ecosystems. Front Ecol Environ 1(4):205–211

27. Chapin FS III, Matson PA, Mooney HA (2002) Principles of terrestrial ecosystem ecology, vol 204. Springer, New York, USA, pp 185–186

28. Macdonald AJ, Poulton PR, Powlson DS, Jenkinson DS (1997) Effects of season, soil type and cropping on recoveries, residues and losses of ^{15}N-labelled fertilizer applied to arable crops in spring. J Agric Sci 129:125–154

29. Murphy DV, Macdonald AJ, Stockdale EA, Goulding KWT, Fortune S, Gaunt JL, Poulton PR, Wakefield JA, Webster CP, Wilmer WS (2000) Soluble organic nitrogen in agricultural soils. Biol Fertil Soils 30:374–387

30. Jones DL, Kielland K (2012) Amino acid, peptide and protein mineralization dynamics in a taiga forest soil. Soil Biol Biochem 55:60–69. https://doi.org/10.2136/sssaj2011.0252

31. Gärdenäs AI, Ågren GI, Bird JA, Clarholm M, Hallin S, Ineson P, Kätterer T, Knicker H, Nilsson SI, Näsholm T, Ogle S, Paustian K, Persson T, Stendahl J (2011) Knowledge gaps in soil carbon and nitrogen interactions—from molecular to global scale. Soil Biol Biochem 43:702–717. https://doi.org/10.1016/j.soilbio.2010.04.006

32. Charteris AF, Knowles TDJ, Michaelides K, Evershed RP (2016) Compound-specific amino acid ^{15}N stable isotope probing of nitrogen assimilation by the soil microbial biomass using gas chromatography/combustion/isotope ratio mass spectrometry. Rapid Commun Mass Spectrom 30:1846–1856. https://doi.org/10.1002/rcm.7612

33. Jones DL, Shannon D, Murphy DV, Farrar J (2004) Role of dissolved organic nitrogen (DON) in soil N cycling in grassland soils. Soil Biol Biochem 36:749–756. https://doi.org/10.1016/j.soilbio.2004.01.003

34. Jackson LE, Schimel JP, Firestone MK (1989) Short-term partitioning of ammonium and nitrate between plants and microbes in an annual grassland. Soil Biol Biochem 21:409–415

35. Rosswall T (1982) Microbiological regulation of the biogeochemical nitrogen cycle. Plant Soil 67:15–34

36. Tahovská K, Kaňa J, Bárta J, Oulehle F, Richter A, Šantrůčková H (2013) Microbial N immobilization is of great importance in acidified mountain spruce forest soils. Soil Biol Biochem 59:58–71

37. Abaas E, Hill PW, Roberts P, Murphy DV, Jones DL (2012) Microbial activity differentially regulates the vertical mobility of nitrogen compounds in soil. Soil Biol Biochem 53:120–123. https://doi.org/10.1016/j.soilbio.2012.05.003

38. Hodge A, Robinson D, Fitter A (2000) Are microorganisms more effective than plants at competing for nitrogen? Trends Plant Sci 5:304–308

39. Inselsbacher E, Hinko-Najera Umana N, Stange FC, Gorfer M, Schüller E, Ripka K, Zechmeister-Boltenstern S, Hood-Novotny R, Strauss J, Wanek W (2010) Short-term competition between crop plants and soil microbes for inorganic N fertilizer. Soil Biol Biochem 42:360–372. https://doi.org/10.1016/j.soilbio.2009.11.019

40. Kaštovská E, Šantrůčková H (2011) Comparison of uptake of different N forms by soil microorganisms and two wet-grassland plants: a pot study. Soil Biol Biochem 43:1285–1291. https://doi.org/10.1016/j.soilbio.2011.02.021

41. Bender SF, van der Heijden MGA (2015) Soil biota enhance agricultural sustainability by improving crop yield, nutrient uptake and reducing nitrogen leaching losses. J Appl Ecol 52(1):228–239. https://doi.org/10.1111/1365-2664.12351

42. Barraclough D (1997) The direct or MIT route for nitrogen immobilisation: a ^{15}N mirror image study with leucine and glycine. Soil Biol Biochem 29:101–108
43. Farrell M, Hill PW, Wanniarachchi SD, Farrar J, Bardgett RD, Jones DL (2011) Rapid peptide metabolism: a major component of soil nitrogen cycling? Global Biogeochem Cycles 25:GB3014. https://doi.org/10.1029/2010gb003999
44. Geisseler D, Horwath WR, Joergensen RG, Ludwig B (2010) Pathways of nitrogen utilization by soil microorganisms—a review. Soil Biol Biochem 42:2058–2067. https://doi.org/10.1016/j.soilbio.2010.08.021
45. Hadas A, Sofer M, Molina JAE, Barak P, Clapp CE (1992) Assimilation of nitrogen by soil microbial population: NH_4^+ versus organic nitrogen. Soil Biol Biochem 24:137–143
46. Hill PW, Farrell M, Jones DL (2012) Bigger may be better in soil N cycling: does rapid acquisition of small L-peptides by soil microbes dominate fluxes of protein-derived N in soil? Soil Biol Biochem 48:106–112. https://doi.org/10.1016/j.soilbio.2012.01.023
47. Jansson SL, Persson J (1982) Mineralisation and immobilisation of soil nitrogen. In: Stevenson FJ (ed) Nitrogen in agricultural soils. American Society of Agronomy, Madison, Wisconsin, USA, pp 229–252
48. Magasanik B (1993) The regulation of nitrogen utilization in enteric bacteria. J Cell Biochem 51:34–40
49. Burger M, Jackson LE (2003) Microbial immobilization of ammonium and nitrate in relation to ammonification and nitrification rates in organic and conventional cropping systems. Soil Biol Biochem 35:29–36
50. Kramer AW, Doane TA, Horwath WR, van Kessel C (2002) Combining fertilizer and organic inputs to synchronize N supply in alternative cropping systems in California. Agr Ecosyst Environ 91:233–243
51. Kramer SB, Reganold JP, Glover JD, Bohannan BJM, Mooney HA (2006) Reduced nitrate leaching and enhanced denitrifier activity and efficiency in organically fertilized soils. Proc Natl Acad Sci USA 103:4522–4527. https://doi.org/10.1073/pnas.0600359103
52. Cabello P, Roldán MD, Moreno-Vivián C (2004) Nitrate reduction and the nitrogen cycle in archaea. Microbiology 150:3527–3546. https://doi.org/10.1099/mic.0.27303-0
53. Moreno-Vivián C, Cabello P, Martínez-Luque M, Blasco R, Castillo F (1999) Prokaryotic nitrate reduction: molecular properties and functional distinction among bacterial nitrate reductases. J Bacteriol 181:6573–6584
54. Moir JWB, Wood NJ (2001) Nitrate and nitrite transport in bacteria. Cell Mol Life Sci 58:215–224
55. González PJ, Correia C, Moura I, Brondino CD, Moura JJG (2006) Bacterial nitrate reductases: molecular and biological aspects of nitrate reduction. J Inorg Biochem 100:1015–1023
56. Murphy MJ, Siegel LM, Tove SR, Kamin H (1974) Siroheme: a new prosthetic group participating in six-electron reduction reactions catalyzed by both sulfite and nitrite reductases. Proc Natl Acad Sci USA 71:612–616
57. Martínez-Espinosa RM, Marhuenda-Egea FC, Bonete MJ (2001) Purification and characterisation of a possible assimilatory nitrite reductase from the halophile archaeon *Haloferax mediterranei*. FEMS Microbiol Lett 196:113–118
58. Martínez-Espinosa RM, Marhuenda-Egea FC, Bonete MJ (2001) Assimilatory nitrate reductase from the haloarchaeon *Haloferax mediterranei*: purification and characterisation. FEMS Microbiol Lett 204:381–385
59. Rice CW, Tiedje JM (1989) Regulation of nitrate assimilation by ammonium in soils and in isolated soil microorganisms. Soil Biol Biochem 21:597–602
60. Meers JL, Tempest DW, Brown CM (1970) 'Glutamine(amide): 2-Oxoglutarate amino transferase oxido-reductase (NADP)', an enzyme involved in the synthesis of glutamate by some bacteria. J Gen Microbiol 64:178–194
61. Miller RE, Stadtman ER (1972) Glutamate synthase from *Escherichia coli*: an iron-sulfide flavoprotein. J Biol Chem 247:7407–7419
62. Santero E, Hervás A, Canosa I, Govantes F (2012) Glutamate dehydrogenases: enzymology, physiological role and biotechnological relevance. In: Canuto RA (ed) Dehydrogenases, pp 289–291. InTech. Published online Nov 2014

63. Tempest DW, Meers JL, Brown CM (1970) Synthesis of glutamate in *Aerobacter aerogenes* by a hitherto unknown route. Biochem J 117:405–407
64. Helling RB (1994) Why does *Escherichia coli* have two primary pathways for synthesis of glutamate? J Bacteriol 176:4664–4668
65. Chávez S, Lucena JM, Reyes JC, Florencio FJ, Candau P (1999) The presence of glutamate dehydrogenase is a selective advantage for the cyanobacterium *Synechocystis* sp. Strain PCC 6803 under nonexponential growth conditions. J Bacteriol 181:808–813
66. Lomnitz A, Calderón J, Hernández G, Mora J (1987) Functional analysis of ammonium assimilation enzymes in *Neurospora crassa*. J Gen Microbiol 133:2333–2340
67. Puri G, Ashman MR (1999) Microbial immobilization of ^{15}N-labelled ammonium and nitrate in a temperate woodland soil. Soil Biol Biochem 31:929–931
68. Recous S, Mary B, Faurie G (1990) Microbial immobilisation of ammonium and nitrate in cultivated soils. Soil Biol Biochem 22:913–922
69. Bengtson P, Bengtsson G (2005) Bacterial immobilization and remineralization of N at different growth rates and N concentrations. FEMS Microbiol Ecol 54:13–19. https://doi.org/10.1016/j.femsec.2005.02.006
70. Glibert PM, Harrison J, Heil C, Seitzinger S (2006) Escalating worldwide use of urea: a global change contributing to coastal eutrophication. Biogeochemistry 77:441–463
71. Mobley HLT, Island MD, Hausinger RP (1995) Molecular biology of microbial ureases. Microbiol Rev 59:451–480
72. Sumner JB, Hand DB, Holloway RG (1931) Studies of the intermediate products formed during hydrolysis of urea by urease. J Biol Chem 91:333–341
73. Arkoun M, Sarda X, Jannin L, Laîne P, Etienne P, Garcia-Mina J, Yvin J, Ourry A (2012) Hydroponics versus field lysimeter studies of urea, ammonium, and nitrate uptake by oilseed rape (*Brassica napus* L.). J Exp Bot 63:5245–5258
74. Salazar F, Martínez-Lagos J, Alfaro M, Misselbrook T (2012) Ammonia emissions from urea application to permanent pasture on a volcanic soil. Atmos Environ 61:395–399
75. Braunstein AE, Kritzmann MG (1937) Formation and breakdown of amino-acids by intermolecular transfer of the amino group. Nature 140:503–504
76. Feldman LI, Gunsalus IC (1950) The occurrence of a wide variety of transaminases in bacteria. J Biol Chem 187:821–830
77. Knowles TDJ, Chadwick DR, Bol R, Evershed RP (2010) Tracing the rate and extent of N and C flow from ^{13}C,[15]N-glycine and glutamate into individual de novo synthesised soil amino acids. Org Geochem 41:1259–1268. https://doi.org/10.1016/j.orggeochem.2010.09.003
78. Rudman D, Meister A (1953) Transamination in Escherichia coli. J Biol Chem 200:591–604
79. Jones DL, Hodge A (1999) Biodegradation kinetics and sorption reactions of three differently charged amino acids in soil and their effects on plant organic nitrogen availability. Soil Biol Biochem 31:1331–1342
80. Knowles TDJ (2009) Following the fate of proteinaceous material in soil using a compound-specific [13]C- and ^{15}N-labelled tracer approach, Unpublished Ph.D. thesis, University of Bristol, Bristol, UK
81. Vieublé Gonod L, Jones DL, Chenu C (2006) Sorption regulates the fate of the amino acids lysine and leucine in soil aggregates. Eur J Soil Sci 57:320–329. https://doi.org/10.1111/j.1365-2389.2005.00744.x
82. Farrell M, Hill PW, Farrar J, DeLuca TH, Roberts P, Kielland K, Dahlgren R, Murphy DV, Hobbs PJ, Bardgett RD, Jones DL (2013) Oligopeptides represent a preferred source of organic N uptake: a global phenomenon? Ecosystems 16:133–145. https://doi.org/10.1007/s10021-012-9601-8
83. Hill PW, Farrell M, Roberts P, Farrar J, Grant H, Newsham KK, Hopkins DW, Bardgett RD, Jones DL (2011) Soil- and enantiomer-specific metabolism of amino acids and their peptides by Antarctic soil microorganisms. Soil Biol Biochem 43:2410–2416. https://doi.org/10.1016/j.soilbio.2011.08.006
84. Pilbeam CJ, McNeill AM, Harris HC, Swift RS (1997) Effect of fertilizer rate and form on the recovery of ^{15}N-labelled fertilizer applied to wheat in Syria. J Agric Sci 128:415–424

85. Powlson DS, Hart PBS, Poulton PR, Johnston AE, Jenkinson DS (1992) Influence of soil type, crop management and weather on the recovery of ^{15}N-labelled fertilizer applied to winter wheat in spring. J Agric Sci 118:83–100
86. Recous S, Machet JM, Mary B (1992) The partitioning of fertilizer-N between soil and crop: comparison of ammonium and nitrate applications. Plant Soil 144:101–111
87. Wickramasinghe KN, Rodgers GA, Jenkinson DS (1985) Transformations of nitrogen fertilisers in soil. Soil Biol Biochem 17:625–630
88. Christie P, Wasson EA (2001) Short-term immobilization of ammonium and nitrate added to a grassland soil. Soil Biol Biochem 33:1277–1278
89. Bunch ND, Bernot MJ (2012) Nitrate and ammonium uptake by natural stream sediment microbial communities in response to nutrient enrichment. Res Microbiol 163:137–141
90. Ladd JN, Parsons JW, Amato M (1977) Studies of nitrogen immobilisation and mineralisation in calcareous soils—I. Distribution of immobilised nitrogen amongst soil fractions of different particle size and density. Soil Biol Biochem 9:309–318
91. Morvan T, Leterme P, Arsene GG, Mary B (1997) Nitrogen transformations after the spreading of pig slurry on bare soil and ryegrass using ^{15}N-labelled ammonium. Eur J Agron 7:181–188
92. Liang B, Yang XY, Murphy DV, He XH, Zhou JB (2013) Fate of ^{15}N-labeled fertilizer in soils under dryland agriculture after 19 years of different fertilizations. Biol Fertil Soils 49:977–986. https://doi.org/10.1007/s00374-013-0789-3
93. Nannipieri P, Ciardi C, Palazzi T, Badalucco L (1990) Short-term nitrogen reactions following the addition of urea to a grass-legume association. Soil Biol Biochem 22:533–549
94. Nannipieri P, Falchini L, Landi L, Benedetti A, Canali S, Tittarelli F, Ferri D, Convertini G, Badalucco L, Grego S, Vittori-Antisari L, Raglione M, Barraclough D (1999) Nitrogen uptake by crops, soil distribution and recovery of urea-N in a sorghum-wheat rotation in different soils under Mediterranean conditions. Plant Soil 208:43–56
95. Jensen LS, Pedersen IS, Hansen TB, Nielsen NE (2000) Turnover and fate of ^{15}N-labelled cattle slurry ammonium-N applied in the autumn to winter wheat. Eur J Agron 12(1):23–35
96. Sulçe S, Palma-Lopez D, Jacquin F, Vong PC, Guiraud G (1996) Study of immobilization and remobilization of nitrogen fertilizer in cultivated soils by hydrolytic fractionation. Eur J Soil Sci 47:249–255
97. He HB, Li XB, Zhang W, Zhang XD (2011) Differentiating the dynamics of native and newly immobilized amino sugars in soil frequently amended with inorganic nitrogen and glucose. Eur J Soil Sci 62:144–151. https://doi.org/10.1111/j.1365-2389.2010.01324.x
98. Geisseler D, Horwath WR (2014) Investigating amino acid utilisation by soil microorganisms using compound specific stable isotope analysis. Soil Biol Biochem 74:100–105. https://doi.org/10.1016/j.soilbio.2014.02.024
99. Dörr N, Kaiser K, Sauheitl L, Lamersdorf N, Stange CF, Guggenberger G (2012) Fate of ammonium ^{15}N in a Norway spruce forest under long-term reduction in atmospheric N deposition. Biogeochemistry 107:409–422. https://doi.org/10.1007/s10533-010-9561-z
100. Redmile-Gordon MA, Evershed RP, Hirsch PR, White RP, Goulding KWT (2015) Soil organic matter and the extracellular microbial matrix show contrasting responses to C and N availability. Soil Biol Biochem 88:257–267. https://doi.org/10.1016/j.soilbio.2015.05.025
101. Low AJ, Armitage ER (1970) The composition of the leachate through cropped and uncropped soils in lysimeters compared with that of the rain. Plant Soil 33:393–411
102. Defra (2015) The British survey of fertiliser practice 2014: fertiliser use on farm crops for crop year 2014. Report available from https://www.gov.uk/government/collections/fertiliser-usage. Accessed 15/05/2016
103. Defra (2016) The British survey of fertiliser practice 2015: statistical notice. Report available from https://www.gov.uk/government/collections/fertiliser-usage. Accessed 27/01/2017
104. Derrien D, Marol C, Balesdent J (2004) The dynamics of neutral sugars in the rhizosphere of wheat: an approach by ^{13}C pulse-labelling and GC/C/IRMS. Plant Soil 267:243–253
105. Kuzyakov Y (2010) Priming effects: interactions between living and dead organic matter. Soil Biol Biochem 42:1363–1371

106. Defra (2010) Fertiliser Manual (RB209), p 60, 65. Report available from http://www.ahdb. org.uk/projects/CropNutrition.aspx. Accessed 26/03/2014
107. Heaton THE (1986) Isotopic studies of nitrogen pollution in the hydrosphere and atmosphere: a review. Chem Geol (Isotope Geoscience Section) 59:87–102
108. Müller C, Laughlin RJ, Christie P, Watson CJ (2011) Effects of repeated fertilizer and cattle slurry applications over 38 years on N dynamics in a temperate grassland soil. Soil Biol Biochem 43:1362–1371
109. Ge T, Yuan H, Roberts P, Jones DL, Qin H, Tong C, Huang D (2012) Amino acid and peptide dynamics in horticultural soils under conventional and organic management strategies. J Soils Sediments 12:323–333
110. Friedel JK, Scheller E (2002) Composition of hydrolysable amino acids in soil organic matter and soil microbial biomass. Soil Biol Biochem 34:315–325
111. Jones DL, Kermit SJ, Wright D, Cuttle SP, Bol R, Edwards AC (2005) Rapid intrinsic rates of amino acid biodegradation in soils are unaffected by agricultural management strategy. Soil Biol Biochem 37:1267–1275
112. Schulten H-R, Schnitzer M (1998) The chemistry of soil organic nitrogen: a review. Biol Fertil Soils 26:1–15
113. Harrod TR, Hogan DV (2008) The soils of North Wyke and Rowden. Revised edition of original report by Harrod TR (1981) Soil survey of England and Wales (now the National Soil Resources Institute, Cranfield University, UK). Report available from Rothamstead Research, http://www.rothamsted.ac.uk/northwyke. Accessed 05/06/2015
114. Jenkinson DS, Fox RH, Rayner JH (1985) Interactions between fertiliser nitrogen and soil nitrogen—the so-called 'priming' effect. J Soil Sci 36:425–444
115. Kuzyakov Y, Friedel JK, Stahr K (2000) Review of mechanisms and quantification of priming effects. Soil Biol Biochem 32:1485–1498
116. Giles M, Morley N, Baggs EM, Daniell TJ (2012) Soil nitrate reducing processes—drivers, mechanisms for spatial variation, and significance for nitrous oxide production. Front Microbiol 3. https://doi.org/10.3389/fmicb.2012.00407
117. Knowles R (1982) Denitrification. Microbiol Rev 46:43–70
118. Bouwman AF, Boumans JM, Batjes NH (2002) Estimation of global NH_3 volatilization loss from synthetic fertilizers and animal manure applied to arable lands and grasslands. Global Biogeochem Cycles 16:1024. https://doi.org/10.1029/2000GB001389
119. Carozzi M, Ferrara RM, Rana G, Acutis M (2013) Evaluation of mitigation strategies to reduce ammonia losses from slurry fertilisation. Sci Total Environ 449:126–133. https://doi. org/10.1016/j.scitotenv.2012.12.082
120. Thompson RB, Ryden JC, Lockyer DR (1987) Fate of nitrogen in cattle slurry following surface application or injection to grassland. J Soil Sci 38:689–700
121. Sexstone AJ, Parkin TB, Tiedje JM (1985) Temporal response of soil denitrification rates to rainfall and irrigation. Soil Sci Soc Am J 49:99–103
122. Tiedje JM, Sexstone AJ, Parkin TB, Revsbech NP (1984) Anaerobic processes in soil. Plant Soil 76:197–212
123. Blagodatskaya E, Kuzyakov Y (2008) Mechanisms of real and apparent priming effects and their dependence on soil microbial biomass and community structure: critical review. Biol Fertil Soils 45:115–131
124. Juang TC, Wang MK, Chen HJ, Tan CC (2011) Ammonium fixation by surface soils and clays. Soil Sci 166:345–352
125. Nieder R, Benbi DK, Scherer HW (2011) Fixation and defixation of ammonium in soils: a review. Biol Fertil Soils 47:1–14. https://doi.org/10.1007/s00374-010-0506-4
126. Scherer HW, Feils E, Beuters P (2014) Ammonium fixation and release by clay minerals as influenced by potassium. Plant, Soil and Environment 60:325–331
127. Adams E (1970) Metabolism of proline and hydroxyproline. In: Hall DA, Jackson DS (eds) International review of connective tissue research, vol 5. Academic Press, New York, USA, pp 2–82

128. Adams E, Frank L (1980) Metabolism of proline and the hydroxyprolines. Annu Rev Biochem 49:1005–1061
129. Nelson DL, Cox MM (2013) Biosynthesis of amino acids, nucleotides, and related molecules. Lehninger principles of biochemistry, 6th edn. Macmillan Higher Education, Basingstoke, UK, pp 881–928
130. Caspi R, Foerster H, Fulcher CA, Kaipa P, Krummenacker M, Latendresse M, Paley S, Rhee SY, Shearer AG, Tissier C, Walk TC, Zhang P, Karp PD (2007) The MetaCyc Database of metabolic pathways and enzymes and the BioCyc collection of Pathway/Genome Databases. Nucleic Acids Res 36:D623–D631. https://doi.org/10.1093/nar/gkm900
131. Berg JM, Tymoczko JL, Gatto GJ Jr, Stryer L (2015) The biosynthesis of amino acids. Biochemistry, 8th edn. W. H. Freeman and Company, New York, USA, pp 713–742
132. Yuan J, Fowler WU, Kimball E, Lu W, Rabinowitz JD (2006) Kinetic flux profiling of nitrogen assimilation in *Escherichia coli*. Nat Chem Biol 2:529–530. https://doi.org/10.1038/nchembio816
133. Reitzer L (2003) Nitrogen assimilation and global regulation in *Escherichia coli*. Annu Rev Microbiol 57:155–176. https://doi.org/10.1146/annurev.micro.57.030502.090820
134. Merrick MJ, Edwards RA (1995) Nitrogen control in bacteria. Microbiol Rev 59:604–622
135. Ahmad Z, Kai H, Harada T (1972) Effect of nitrogenous forms on immobilization and release of nitrogen in soil. J Fac Agric Kyushu Univ 17:49–65
136. Geisseler D, Horwath WR, Doane TA (2009) Significance of organic nitrogen uptake from plant residues by soil microorganisms as affected by carbon and nitrogen availability. Soil Biol Biochem 41:1281–1288
137. Müller C, Stevens RJ, Laughlin RJ, Ottow JCG, Jäger H-J (2003) Ammonium immobilisation during chloroform fumigation. Soil Biol Biochem 35:651–665. https://doi.org/10.1016/S0038-0717(03)00014-2
138. Kowalenko CG, Cameron DR (1976) Nitrogen transformations in an incubated soil as affected by combinations of moisture content and temperature and absorption-fixation of ammonium. Can J Soil Sci 56:63–70
139. Börjesson G, Menichetti L, Kirchmann H, Kätterer T (2012) Soil microbial community structure affected by 53 years of nitrogen fertilisation and different organic amendments. Biol Fertil Soils 48:245–257
140. Černý J, Balík J, Pavlíková D, Zitková M, Sýkora K (2003) The influence of organic and mineral nitrogen fertilisers on microbial biomass nitrogen and extractable organic nitrogen in long-term experiments with maize. Plant Soil Environ 49:560–564
141. Chakraborty A, Chakrabarti K, Chakraborty A, Ghosh S (2011) Effect of long term fertilisers and manure application on microbial biomass and microbial activity of a tropical agricultural soil. Biol Fertil Soils 47:227–233
142. Edmeades DC (2003) The long-term effects of manures and fertilisers on soil productivity and quality: a review. Nutr Cycl Agroecosyst 66:165–180
143. Ge T, Nie S, Huang DF, Xiao H, Jones DL, Iwasaki K (2010) Assessing soluble organic nitrogen pools in horticultural soils: a case study in the suburbs of Shanghai (China). Acta Agric Scand Sect B Soil Plant Sci 60:529–538. https://doi.org/10.1080/09064710903246427
144. Kaur K, Kapoor KK, Gupta AP (2005) Impact of organic manures with and without mineral fertilisers on soil chemical and biological properties under tropical conditions. J Plant Nutr Soil Sci 168:122–177
145. Kirchmann H, Haberhauer G, Kandeler E, Sessitisch A, Gerzabek MH (2004) Effects of level and quality of organic matter input on carbon storage and biological activity in soils: synthesis of a long-term experiment. Global Biogeochem Cycles 18:GB4011. https://doi.org/10.1029/2003gb002204
146. Compton JE, Watrud LS, Porteous LA, DeGrood S (2004) Response of soil microbial biomass and community composition to chronic nitrogen additions at Harvard forest. For Ecol Manage 196:143–158
147. Kamble PN, Rousk J, Frey SD, Bååth E (2013) Bacterial growth and growth-limiting nutrients following chronic nitrogen additions to a hardwood forest soil. Soil Biol Biochem 59:32–37

148. Söderström B, Bååth E, Lundgren B (1983) Decrease in soil microbial activity and biomasses owing to nitrogen amendments. Can J Microbiol 29:1500–1506
149. Mengel K (1996) Turnover of organic nitrogen in soils and its availability to crops. Plant Soil 181:83–93
150. Mallory EB, Griffin TS (2007) Impacts of soil amendment history on nitrogen availability from manure and fertiliser. Soil Sci Soc Am J 71:964–973
151. Drinkwater LE, Wagoner P, Sarrantonio M (1998) Legume-based cropping systems have reduced carton and nitrogen losses. Nature 396:262–265
152. Kirchmann H, Johnston AEJ, Bergström LF (2002) Possibilities for reducing nitrate leaching from agricultural land. Ambio 31:404–408
153. Vinten AJA, Whitmore AP, Bloem J, Howard R, Wright F (2002) Factors affecting N immobilisation/mineralisation kinetics for cellulose-, glucose- and straw-amended sandy soils. Biol Fertil Soils 36:190–199

Chapter 5
Microbial Fertiliser Nitrogen Assimilation in the Field as Compared with the Laboratory Incubation Experiments

5.1 Introduction

The issue of scale is well-recognised in the determination and modelling of NO_3^- leaching and indeed, in many other areas of research as almost all observations are strongly affected by the scale(s) at which they are made [1, 2]. Processes described and validated at one scale often do not translate to other scales and/or become relatively much more/less important [1, 3, 4]. Major causes of scale-issues with respect to NO_3^- uptake/leaching are the changes in the heterogeneity of soil/subsoil/the system and in the relative importance of NO_3^- leaching/cycling processes depending on the scale considered (e.g. plot/field/catchment).

This chapter presents the results of a ^{15}N-SIP incubation experiment with $^{15}NO_3^-$ conducted under field conditions at the WA site. This work extends that of Chap. 4 by repeating the WA $^{15}NO_3^-$ laboratory incubation experiment on a larger scale and in a field setting, such that the influences of 'normal' processes (plant uptake, leaching, volatile losses etc.) and conditions (soil structure, weather) are included. Although the laboratory incubation experiment (Chap. 4) and the field incubation experiment described in this chapter are probably both best considered as 'point scale' experiments [3, 5], the additional heterogeneity and processes operating in the field experiment are expected to affect NO_3^- assimilation measurements analogously to a larger scale change. A short review of microbial N uptake and assimilation was included in Chap. 4 (Sect. 4.1.2). While more ^{15}N losses (via various processes, e.g. plant N uptake, gaseous N losses and leaching) are likely to occur in this experiment, they are only considered in the respect that they reduce the amount of ^{15}N available in the soil for microbial incorporation. The physical partitioning and transport of ^{15}N away from the ^{15}N-treated area are discussed in Chap. 6.

© Springer Nature Switzerland AG 2019
A. F. Charteris, *^{15}N Tracing of Microbial Assimilation, Partitioning and Transport of Fertilisers in Grassland Soils*, Springer Theses,
https://doi.org/10.1007/978-3-030-31057-8_5

5.2 Objectives

The aim of this work was to investigate the microbial assimilation of $^{15}NO_3^-$ into hydrolysable AAs under field conditions and to compare this with the findings of the laboratory incubation experiments (Chap. 4). Nitrate was selected as it is a commonly applied fertiliser compound and the N form which is often most available to plants [6–9] and particularly susceptible to leaching, resulting in groundwater NO_3^- pollution [10]. The experiment was carried out for 64 days, beginning in October, with the date of $^{15}NO_3^-$ application coinciding with the last fertilisation of the rest of the field before winter. This period was selected for study as the weather usually remains sufficiently warm during October to maintain soil microbial activity and NO_3^- leaching over late-autumn, winter and early spring is generally greatest [10, 11].

The specific objectives of this work were to:

(i)   Monitor bulk soil $\delta^{15}N$ values in the top 15 cm of soil following ^{15}N application of $^{15}NO_3^-$ as an indication of the ^{15}N retained in this portion of soil for microbial and plant uptake.

(ii) Determine and compare the patterns, rates and fluxes of the conversion of $^{15}NO_3^-$ into individual soil AAs via their $\delta^{15}N$ values and percentage ^{15}N incorporations as an indication of the primary biochemical pathways operating in the soil microbial assimilation of $^{15}NO_3^-$.

(iii) Determine and compare the patterns, rates and fluxes of the conversion of $^{15}NO_3^-$ into the total hydrolysable AA pool as a whole.

(iv) Compare the results of the experiment under field versus laboratory conditions.

Potential findings include:

(i)   Insights into the biochemical pathways, rates and extents of microbial assimilation of $^{15}NO_3^-$ under field conditions (compared with the laboratory).

(ii) A greater understanding of the biomolecular fate of fertiliser $^{15}NO_3^-$ under field conditions (compared with the laboratory).

(iii) Measures of newly synthesised soil protein from the applied $^{15}NO_3^-$ under field conditions (compared with the laboratory).

The hypotheses tested in this work are:

(i)   Losses of ^{15}N from the ^{15}N-treated area over the course of the experiment will be considerable.

(ii) There will be higher variability between samples taken at the same time points due to the heterogeneity of the field environment.

(iii) The patterns in $^{15}NO_3^-$ incorporation into hydrolysable AAs will be very similar to those observed in the WA $^{15}NO_3^-$ laboratory incubation experiment.

(iv) As a result of greater ^{15}N losses from the treated area during the field incubation experiment, the percentage of the *applied* NO_3^--^{15}N incorporated into the total hydrolysable AA pool in the field experiment will be lower than that recorded in the WA $^{15}NO_3^-$ laboratory incubation experiment. The percentage of *retained*

$NO_3^- $-^{15}N (^{15}N remaining in the treated area soil, based on bulk soil δ^{15}N values and soil % TN contents) incorporated into the total hydrolysable AA pool in the field experiment will be higher than that recorded in the WA $^{15}NO_3^-$ laboratory incubation experiment as ^{15}N not held in soil AAs will be more susceptible to loss.

5.3 Site, Experimental Design, Sampling and Analytical Methods

5.3.1 Site

The field experiment was carried out in the 'Little Broadheath' field of Longlands Dairy Farm, WA (Fig. 5.1), described further in Sect. 2.2.2. The site is a Wessex Water catchment management area (Sect. 1.5) due to its close proximity to the WA PWS source and high groundwater NO_3^- concentrations (Sect. 2.2.2). As part of Wessex Water's established monitoring activities, a set of ten porous ceramic cups transects the field (at 90 cm depth; Fig. 5.1).

The valley in which the WA PWS source resides is relatively steep, with land elevation ranging from 110 to 190 m above sea level. The dry valley to the north-west of the catchment is characterised by fine, silty over clay and fine, loamy over

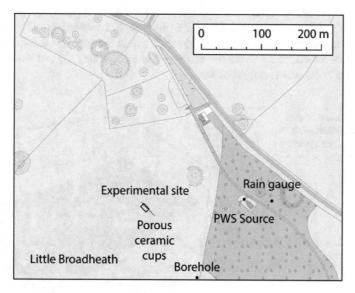

Fig. 5.1 Annotated map of the field experiment site. © Crown Copyright and Database Right [01/02/2016]. Ordnance Survey (Digimap Licence)

clay soils, whilst shallow, well drained, calcareous silty soils predominate elsewhere. Relatively high mean annual rainfall totals of approximately 900 mm are typical and localised flooding is not uncommon along the valley base in winter. Rainfall rates are generally low, however, and despite relatively steep slopes in the catchment, overland flow is not usually observed.

5.3.2 Experimental Design

The field experiment was designed with the dual purpose of: (i) providing a direct comparison with the laboratory incubation experiments (Chap. 4); and (ii) investigating NO_3^- partitioning and transport (Chap. 6). The experimental design is described briefly here and more detail is only included where relevant to the field incubation experiment. Little Broadheath was surveyed and shown to exhibit a gently sloping hollow in the area selected for the experiment (Fig. 5.2). An area of approximately 50 m^2 around the hollow was fenced off from the rest of the field and the site was prepared for the field experiment (including porous ceramic cup/suction cup installation, see Sect. 6.3.2) in July 2014.

The site was allowed to settle and maintained (by grass cutting and partial removal of the cuttings) until the start of the experiment in October 2014. At 10:00 on 17th October 2014 5.5 g ^{15}N-labelled N as K^{15}NO$_3$, (46 atom % in 500 ml DDW) was spray-applied to three adjoining 1 × 0.33 m areas overlying three porous ceramic cups/suction cups, which were located at depths of 0.8 m (sampling area A), 0.2 m (sampling area B) and 0.5 m (sampling area C; Fig. 5.3). This is equivalent to 55 kg N ha^{-1}, which, based on five to six fertiliser applications between February and October is approximately 300 kg N ha^{-1} year^{-1}, within the range (140–340 kg N ha^{-1} year^{-1}) recommended by Defra for dairy-grazed grasslands [12] and similar to the 40 kg N ha^{-1} the site usually receives.

Fig. 5.2 Surface map of Little Broadheath, with the experimental area marked with a rectangle

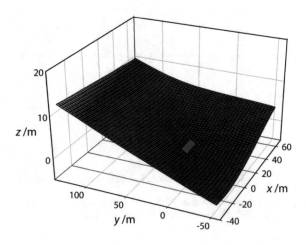

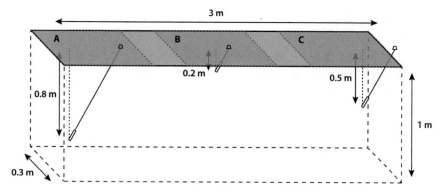

Fig. 5.3 Scale diagram of the treated and sampling areas (A, B and C) for the field incubation experiment including the three porous ceramic cups/suction cups located beneath the treated area. Soil was randomly sampled to a depth of 10 cm within each of the areas A, B and C at time points matching those of the laboratory incubation experiments (Sect. 5.3.3). (Note that the buried cup of the porous ceramic cups is not drawn to scale)

The following factors were considered in the experimental design:

(i)   The application date was selected to coincide with the last fertilisation of the rest of the field before winter and based on weather forecasts of some rain (not excessive or heavy) only later in the day to help the applied $^{15}NO_3^-$ penetrate and disperse into the soil more fully. Rainfall was automatically telemetered by a rain gauge located approximately 200 m from the experimental site and this data is presented in Sect. 5.4.1.

(ii) A relatively high substrate ^{15}N-enrichment (compared with the laboratory incubation experiments) was selected in order to ensure that the ^{15}N label would not be diluted in the much larger and more dynamic field system (the potential effects this could have on biological ^{15}N discrimination and isotopic fractionation are acknowledged; [13, 14]).

(iii) A spray, rather than injection (as in the laboratory experiments), application method was selected to better represent usual farming practices (although advantageous in many respects, injection applications are less convenient and more expensive and therefore less commonly used; [15]). It could be argued that this adds a difference between the laboratory and field incubation experiments, but part of the purpose of carrying out both experiments is to compare the laboratory results with those obtained in a field setting and the use of spray application is arguably part of this.

5.3.3 Sampling

In order to generate samples that were as comparable as possible with those of the laboratory incubation experiments, soil samples were taken at the same time points (1.5, 3, 6 and 12 h and 1, 2, 4, 8, 16, 32 and 64 days) and immersed in liquid N_2 prior to temporary -20 °C storage and freeze-drying. Soil was randomly sampled to a depth of 10 cm from the areas highlighted in Fig. 5.3. The resulting void was refilled with soil collected from elsewhere in the field, close to the experimental area and the sampled point was clearly marked.

5.3.4 Analytical Methods

Any foliage, roots or stones were removed from freeze-dried samples and soils ground. Hydrolysable AAs were extracted, isolated and derivatised as described in Sect. 2.5 and soils analysed (bulk % TC and % TN contents and N isotopic compositions and AA concentrations and N isotopic compositions) as described in Sect. 2.6. Calculations were carried out as described in Sect. 2.7. For the purpose of bulk soil percentage ^{15}N retentions and percentage applied ^{15}N incorporation calculations, the mass of soil to which the $^{15}NO_3^-$ was applied was calculated based on a 3×0.3 m treated area, 0.1 m sampling depth and a soil density of 1.1 g cm^{-3}.

5.4 Results

5.4.1 Prevailing Weather Conditions

UK weather in October 2014 was described as unsettled and particularly mild due to the predominance of (often strong) southerly winds [16, 17]. Overall, both rainfall and mean temperature in south-west England and Wales were above average (118% and 2.0 °C, respectively, compared with regional monthly means 1910–2013; [17]), with October 31st being widely reported as the warmest Halloween on record [18]. Sunshine hours in south-west England and Wales were 85% of average (compared with regional monthly means 1910–2013; [17]).

The mild and unsettled theme continued into November 2014 and rainfall was 119% of average, mean temperature 1.5 °C above average and sunshine hours 106% of average in south-west England and Wales (compared with regional monthly means 1910–2013; [17]). The UK in December 2014 was dominated by westerlies and alternated between wet and dry and bright days and mild and cold temperatures [17]. Compared with monthly means in south-west England and Wales 1910–2013, rainfall was 75% of average, mean temperature 0.5 °C above average and sunshine hours 139% of average [17].

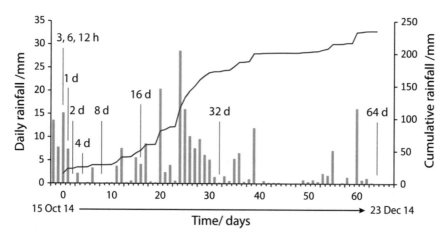

Fig. 5.4 Automatically telemetered daily rainfall totals (measured from 09:00 on each day until 09:00 the next day) and cumulative total rainfall over the experimental period. Sampling dates are highlighted

Daily rainfall totals and cumulative rainfall over the experimental period are displayed in Fig. 5.4. The day of treatment application was dry until 17:50, after which 5.4 mm rain fell between then and 21:10. The remaining 9.8 mm rain recorded on 17th October 2014 fell between 02:00 and 09:00 on 18th October. The highest daily rainfall total is 28.6 mm on 10th November and the highest hourly rainfall rate of 6 mm h^{-1} was also recorded on this date, between 19:00 and 20:00 (24 days after the start of the experiment). The largest rainfall volume between sampling points is 125.6 mm from $t = 16$–32 days and the cumulative rainfall total following treatment application until the last sample was taken on 20th December is 234.8 mm.

Mean weekly air temperatures and weekly sunshine hours for the area are shown in Fig. 5.5 (data obtained from Wessex Water based on Met Office Rainfall and Evaporation Calculation System [MORECS] data). Mean weekly temperatures are highest at the start of the experimental period (14.1 °C) in October and lowest (5 °C) at the beginning of December. Total weekly sunshine hours vary from 3.9 to 27.4 h.

Soil water potentials were not measured directly over the course of the field incubation experiment, but estimates of daily soil moisture deficits (SMDs) were generated by Wessex Water's Catchment Science Team using IRRIGUIDE, a field-scale water balance model developed by ADAS ([19–21]; Fig. 5.6). Modelled SMDs fall below 0 mm on 9th November 2014 (23 days after the start of the experiment), indicating that the soil has reached field capacity, and remain at or below 0 mm throughout the field incubation experiment.

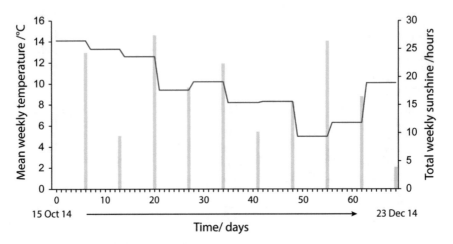

Fig. 5.5 Mean weekly air temperatures and total weekly sunshine hours for the MORECS square in which the experimental site resided

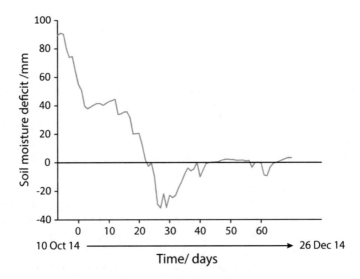

Fig. 5.6 IRRIGUIDE modelled daily SMDs during the field incubation experiment

5.4.2 *Bulk Soil Percentage Total Carbon and Percentage Total Nitrogen Contents*

Bulk soil % TC and % TN contents vary over the course of the experiment (Table 5.1). There is no observable trend in the bulk soil % TC content with time, but there is a slight rise in the bulk soil % TN content (linear regression gives $m = 0.0021$, $R^2 = 0.56$).

Table 5.1 Bulk soil % TC and % TN contents for every time point sampled (n = 3 for each) alongside mean values (for all samples at every time point, including $t = 0$ pre-application values) and their associated SEs

Time/days												Mean	SE
	0	0.125	0.25	0.5	1	2	4	8	16	32	64		
% TC	5.58	6.56	6.08	6.62	6.05	7.48	6.92	5.45	5.96	6.39	7.87	6.45	0.181
% TN	0.53	0.60	0.55	0.62	0.64	0.67	0.63	0.60	0.63	0.66	0.73	0.62	0.013

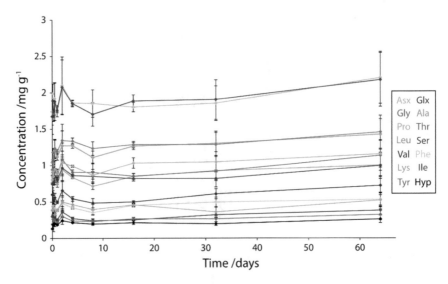

Fig. 5.7 Concentrations of individual hydrolysable soil AAs with time over the course of the field experiment. Error bars are ± SE (n = 3)

5.4.3 Hydrolysable Soil Amino Acid Concentrations

The concentrations of individual hydrolysable soil AAs range from 0.13 mg g^{-1} soil (Hyp at $t = 0$ days) to 2.20 mg g^{-1} soil (Asx at $t = 64$ days) during the experiment (Fig. 5.7). The concentration of each AA fluctuates over the course of the experiment, but changes little overall, with the exception of possibly a slight rise toward the end of the experiment (Fig. 5.7). Accordingly, there is little change in the hierarchy of AA concentration and the AAs that are present in highest concentrations at $t = 0$ (e.g. Glx and Asx) remain so throughout, as did those at intermediate (e.g. Pro, Thr, Leu and Ser) and low concentrations (e.g. Ile, Tyr and Hyp).

The total concentration of hydrolysable soil AAs ranges from 10.6 mg g^{-1} soil at $t = 0$ to 14.3 mg g^{-1} soil at $t = 64$ days (Fig. 5.8). Between these time points, total hydrolysable AA concentrations fluctuate, but overall rise slightly over time (linear regression gives $m = 0.039$, $R^2 = 0.49$). The contribution of total hydrolysable AA N to TN, however, does not change much overall and fluctuates around a mean of $23.7 \pm 0.46\%$ (n = 33; Fig. 5.8).

5.4.4 Bulk Soil Nitrogen Isotopic Composition

Prior to treatment application, the bulk soil N isotopic composition at the site was determined as $6.4 \pm 0.2\permil$. The mean N isotopic composition of the soil at each time point following treatment application is presented in Fig. 5.9 using

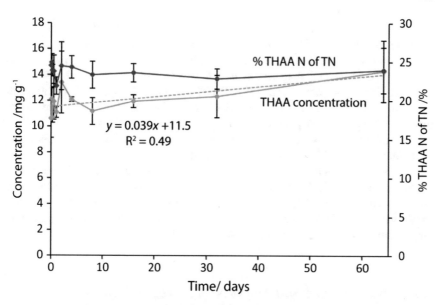

Fig. 5.8 Total hydrolysable AA (THAA) concentrations and the percentage contribution of total hydrolysable AA N to TN with time over the course of the field experiment. Error bars are ± SE (n = 3)

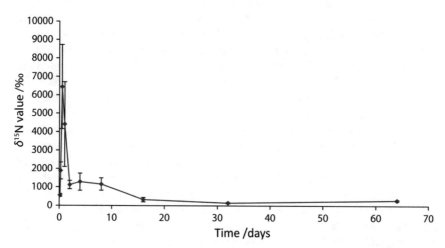

Fig. 5.9 Bulk soil N isotopic compositions in per mil over the course of the field experiment. Error bars are ± SE (n = 3)

$\delta^{15}N$ values (‰). Mean bulk soil $\delta^{15}N$ values rise between the first post-treatment sampling point at $t = 3$ (559‰) and 12 h (6440‰) and then fall sharply until $t = 2$ days (1120‰). Values are subsequently fairly stable until $t = 8$ days and then fall until $t = 16$ days, restabilising at around 250‰ over the rest of the experiment. Mean bulk soil $\delta^{15}N$ values range from 145 ± 32.0‰ at $t = 32$ days to 6440 ± 2300‰ at $t = 12$ h. This is equivalent to 0.420 ± 0.387 and 2.66 ± 1.12 atom %, respectively. Error bars, especially at peak bulk soil $\delta^{15}N$ values, are relatively large and the lowest and highest individual $\delta^{15}N$ values are 81.6 and 10,700‰ (0.396 and 4.13 atom %) at $t = 32$ days and 12 h, respectively.

The mean percentage retentions to which these $\delta^{15}N$ values correspond (based on the application of 5.5 g N at 46 atom % to a $3 \times 0.33 \times 0.1$ m treated volume with a soil density of 1.1 g cm^{-3}, which equates to 1.76×10^{-6} mol ^{15}N g^{-1} soil) are shown in Table 5.2. Percentage ^{15}N retentions reflect bulk soil $\delta^{15}N$ values and range from 14% at $t = 32$ days to 571% at $t = 12$ h.

5.4.5 Hydrolysable Soil Amino Acid Nitrogen Isotopic Compositions and Percentage ^{15}N Incorporations

The N isotopic compositions of individual hydrolysable soil AAs over the course of the experiment are shown in Fig. 5.10. In general, AA $\delta^{15}N$ values co-vary (rising and falling together) over the course of the experiment. Not all AAs, however, follow this pattern throughout and particularly Glx, Asx and Hyp and initially, Pro, Gly, Ser and Val exhibit slightly different trends. At $t = 3$ h, Glx $\delta^{15}N$ values are highest (25.7 ± 3.1‰) and those of most other AAs have risen to some extent, with only Pro showing lower $\delta^{15}N$ values than at $t = 0$ (Pro $\delta^{15}N$ values only begin to rise above those at $t = 0$ in samples from $t = 12$ h and 1 day). By $t = 6$ h, Asx $\delta^{15}N$ values are highest (66.9 ± 3.1‰) and an early peak in Asx (to 233 ± 76.5‰; as well as Ser, Glx, Gly and Hyp to a lesser extent) occurs at $t = 12$ h. At the end of this peak at $t = 2$ days there is a dip in the $\delta^{15}N$ values of almost all AAs. After this, Glx $\delta^{15}N$ values are consistently highest, followed by Asx and the $\delta^{15}N$ values of these AAs are considerably higher than those of other AAs until $t = 16$ days for Asx and $t = 32$ days for Glx. Hydroxyproline $\delta^{15}N$ values, on the other hand, are consistently lowest from $t = 2$ days (after the initial peak at $t = 12$ h) and rise gently rather than tracking the pattern of the other AAs.

The percentages of applied and retained ^{15}N incorporated into individual hydrolysable soil AAs are shown in Fig. 5.11. The graph of the percentage of applied ^{15}N incorporated into individual hydrolysable AAs (Fig. 5.11a) is very similar in general shape to that of AA $\delta^{15}N$ values (Fig. 5.10). The wider spread of the data and differences in the hierarchy of AAs in Fig. 5.11a compared with Fig. 5.10 result from the effects of AA concentration (AAs present at higher concentrations need to incorporate a larger amount of ^{15}N to raise the overall $\delta^{15}N$ value of the AA by a

Table 5.2 Bulk soil percentage ^{15}N retentions and their associated SEs (n = 3 for individual time points) over the course of the experiment

Time/days	0.125	0.25	0.5	1	2	4	8	16	32	64	Mean
% Retention ^{15}N	48	153	571	427	112	119	100	30	14	30	160
SE	5.3	38.6	205	216	26.9	43.9	26.7	12	3.2	5.0	41.7

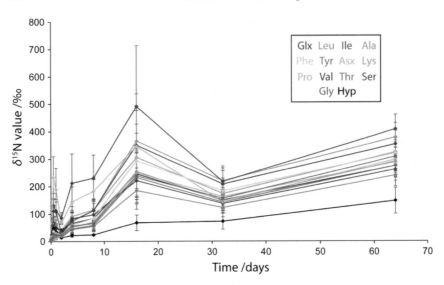

Fig. 5.10 $\delta^{15}N$ values of individual hydrolysable soil AAs over the course of the field experiment. Error bars are $\pm$ SE ($n = 3$)

certain amount) and those AAs present in higher concentrations have larger percentage ^{15}N incorporations. The highest percentage applied ^{15}N incorporated occurs into Glx at $t = 16$ days (1.26%), closely followed by this AA at $t = 64$ days (1.24%). The percentage applied ^{15}N incorporations of most other AAs are also highest at $t = 64$ days (ranging from Hyp at 0.07% to Ala at 1.05%).

Percentage retained ^{15}N incorporations are similarly affected by AA concentration, but also take into account the moles ^{15}N available in the soil for assimilation into hydrolysable AAs (based on bulk soil $\delta^{15}N$ values; Fig. 5.9), which changes the shape of the graph considerably. With the exception of a small initial peak for Glx, Asx and Ala at $t = 1$ day (0.70%, 0.26% and 0.20%, respectively), AA percentage retained ^{15}N incorporations rise throughout the experiment, most sharply from $t = 8$–16 days and only very gently from $t = 16$–64 days. Highest percentage retained ^{15}N incorporations occur at $t = 64$ days for most AAs (except Ala, Leu, Pro, Phe and Ile which peak at $t = 32$ days) and range from 0.23% for Hyp to 4.22% for Glx.

Linear and non-linear regressions of the form described in Sect. 2.7.6 were fitted to time course plots of AA $\Delta^{15}N$ values (Fig. 5.10) and to the data presented in Fig. 5.11 in order to obtain quantitative information about: the rates at which AA N isotopic compositions change; the overall change in AA N isotopic compositions at steady state; the rates of applied and retained ^{15}N incorporation/partitioning into hydrolysable AAs; and the steady state percentages of ^{15}N incorporation/partitioning into hydrolysable AAs (Table 5.3). Linear regressions were applied to data from both the first 12 h and the first 4 days of the experiment because data for the first 12 h display an approximately linear trend for time course plots of AA $\Delta^{15}N$ values and percentage applied ^{15}N incorporations, while linear regressions over 4 days could

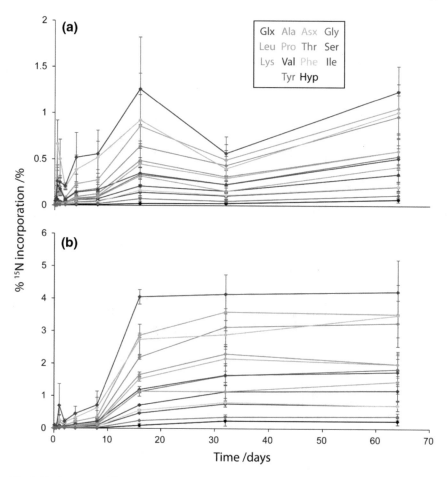

Fig. 5.11 Percentage ^{15}N incorporations into individual hydrolysable soil AAs over the course of the field experiment: **a** applied and, **b** retained. Error bars are $\pm$ SE (n = 3)

be fitted to a larger proportion of AA percentage retained ^{15}N incorporation data and also facilitate comparison with the regression parameters generated for the WA ^{15}NO$_3^-$ laboratory incubation experiment (Chap. 4). As might be expected, 12-h zero-order rate constants are generally higher than those calculated over 4 days.

Amino acid Δ^{15}N values initially increased fastest for Asx (390‰ day^{-1}) and decreased from $t = 0$ for Pro (with a very poor fit) over the first 12 h. Pro also has the lowest (measurable) initial rate over the first 4 days of the experiment (9.4‰ day^{-1}), while over this period, the linear regression does not fit the time course Δ^{15}N data for Asx and Glx has the highest zero-order rate (52.9‰ day^{-1}). Amino acid Δ^{15}N values increase most quickly for Asx (poor fit) and Glx overall (2.77 and 0.21 day^{-1}, respectively) and most slowly for Hyp (0.02 day^{-1}). Glutamate has the highest plateau

Table 5.3 Results of linear (zero-order rate constants) and non-linear single first-order exponential assimilation (first-order rate constants and plateau levels) regressions (alongside measures of the equation fits, R^2 and S, which is the standard error of the regression) applied to time course plots of AA $\Delta^{15}N$ values, percentage applied ^{15}N incorporations and percentage retained ^{15}N incorporations for the field incubation experiment

	WA $^{15}NO_3^-$ $\Delta^{15}N$ values							WA $^{15}NO_3^-$: % applied ^{15}N incorporated							WA $^{15}NO_3^-$: % retained ^{15}N incorporated						
	0–0.5 days		0–4 days		0–64 days			0–0.5 days		0–4 days		0–64 days			0–0.5 days		0–4 days		0–64 days		
	Zero-order[a]	R^2	Zero-order[b]	R^2	First-order	Plateau	S	Zero-order[a]	R^2	Zero-order[b]	R^2	First-order	Plateau	S	Zero-order[a]	R^2	Zero-order[b]	R^2	First-order	Plateau	S
Ala	74.2	0.98	20.9	0.58	0.10	267	48.5	0.20	0.95	0.06	0.66	0.06	0.93	0.15	–	–	0.06	0.41	0.05	4.03	0.47
Asx	390	0.88	–	–	2.77	193	84.7	1.14	0.87	–	–	2.53	0.58	0.26	0.27	0.73	0.09	0.45	0.05	3.72	0.33
Glx	182	0.91	52.9	0.68	0.21	350	78.0	0.46	0.91	0.13	0.71	0.17	1.00	0.22	0.11	0.71	0.14	0.18	0.06	4.62	0.60
Gly	149	0.88	–	–	0.06	216	43.4	0.50	0.86	–	–	0.04	0.99	0.16	0.01	0.89	0.03	0.59	0.04	3.82	0.38
Hyp	79.7	0.84	–	–	0.02	214	19.9	0.02	0.85	–	–	–	–	–	0.01	0.02	0.00	0.61	0.03	0.31	0.03
Ile	70.2	0.91	17.6	0.62	0.08	314	53.7	0.03	0.87	0.01	0.70	0.04	0.21	0.02	0.01	0.61	0.01	0.94	0.04	0.85	0.09
Leu	33.8	0.87	16.1	0.92	0.08	341	56.3	0.05	0.87	0.02	0.92	0.05	0.57	0.08	–	–	0.02	0.95	0.05	2.37	0.32
Lys	32.0	0.90	11.7	0.85	0.06	262	42.9	0.03	0.94	0.01	0.91	0.03	0.43	0.06	–	–	0.01	0.88	0.04	1.61	0.18
Phe	54.3	0.96	13.5	0.63	0.07	288	46.1	0.03	0.97	0.01	0.68	0.04	0.21	0.03	–	–	0.01	0.78	0.05	0.84	0.11
Pro	–5.41	0.07	9.35	0.87	0.05	266	39.2	0.01	0.09	0.02	0.87	0.04	0.61	0.08	–	–	0.01	0.93	0.04	2.35	0.29
Ser	244	0.94	–	–	0.15	203	60.8	0.39	0.94	–	–	0.09	0.40	0.11	0.10	0.24	0.03	0.45	0.04	2.00	0.17
Thr	23.9	0.90	9.77	0.87	0.05	256	38.8	0.04	0.91	0.02	0.89	0.02	0.65	0.06	0.01	0.02	0.01	0.88	0.03	2.19	0.19
Tyr	38.0	0.94	12.5	0.83	0.05	293	39.2	0.01	0.94	0.00	0.86	0.02	0.14	0.01	–	–	0.00	0.92	0.03	0.46	0.04
Val	136	0.79	–	–	0.08	230	44.8	0.13	0.84	–	–	0.03	0.36	0.04	–	–	0.02	0.31	0.04	1.40	0.12

Units of the zero-order rate constant are $‰\ day^{-1}$ and $\%\ day^{-1}$ and units of the first-order rate constant are day^{-1}. Poor and very poor fits are highlighted in orange and red, respectively and a dash (–) indicates either that a horizontal line would have produced a better fit than the linear regression or the non-linear regression selected did not fit the data

[a] zero-order rate constant over 0.5 days

[b] zero-order rate constant over 4 days

AA Δ^{15}N value (350‰) while Asx (poor fit) and Ser have the lowest (193 and 203‰, respectively).

^{15}N was initially incorporated into Asx and Glx fastest and into Pro and Tyr most slowly when considered as a percentage of the applied ^{15}N. As a percentage of retained ^{15}N, it was incorporated into Glx most rapidly, but the linear regressions fit poorly for many AAs. Overall, incorporation of applied ^{15}N was fastest for Asx (poor fit; 2.53 day^{-1}) and Glx (0.17 day^{-1}) and slowest for Thr and Tyr (0.02 day^{-1}) and of retained ^{15}N was fastest for Glx (0.06 day^{-1}) and slowest for Hyp, Thr and Tyr (0.03 day^{-1}). Glutamate maintains the highest percentages of ^{15}N (1.00 and 4.62% of applied and retained, respectively) while Tyr and Hyp have the lowest percentages of applied and retained ^{15}N, respectively (0.14 and 0.31%, respectively).

5.4.6 Percentage ^{15}N Incorporations into the Soil Total Hydrolysable Amino Acid Pool

The percentages of applied and retained ^{15}N incorporated into the soil hydrolysable AA pool as a whole are shown in Fig. 5.12. Until $t = 4$ days and while percentage ^{15}N incorporations are low (up to 1.85%), there is little difference between the results of applied and retained calculations. Following this, percentage retained ^{15}N incorporations are consistently higher than those of applied ^{15}N with the largest increase in retained ^{15}N incorporated occurring from $t = 8$ to 16 days (2.86–20.6%). For both calculations, percentage ^{15}N incorporations are highest at $t = 64$ days (7.86 and 26.7% for applied and retained ^{15}N, respectively).

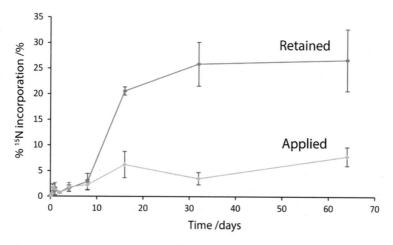

Fig. 5.12 Percentage applied and retained ^{15}N incorporated into the total hydrolysable AA pool over the course of the field experiment. Error bars are $\pm$ SE (n = 3)

Table 5.4 Results of linear (zero-order rate constants) and non-linear single first-order exponential assimilation (first-order rate constants and plateau levels) regressions (and measures of the equation fits, R^2 and S, which is the standard error of the regression) applied to the percentage applied and retained ^{15}N incorporation data shown in Fig. 5.12

	Zero-order[a]	R^2	Zero-order[b]	R^2	First-order	Plateau	S
Applied ^{15}N	3.01	0.91	0.53	0.08	0.07	6.62	1.25
Retained ^{15}N	0.78	0.07	0.45	0.49	0.05	30.3	3.19

Units of the zero-order rate constant are % day^{-1} and units of the first-order rate constant are day^{-1}.
Very poor fits are highlighted in red
[a]zero-order rate constant over 0.5 days
[b]zero-order rate constant over 4 days

Table 5.4 shows the results of linear and non-linear regressions (also of the form described in Sect. 2.7.6) fitted to the data presented in Fig. 5.12 in order to obtain quantitative information about the rates and steady state extents of applied and retained ^{15}N incorporation/partitioning into the total hydrolysable AA pool. The zero order rate constants calculated over the first 12 h of the experiment reveal a strong linear increase in the percentage of applied ^{15}N incorporated into the total hydrolysable AA pool as the amount of ^{15}N reaching the soil also increases, while the rate constant for retained ^{15}N shows the opposite effect of the sharp peak in bulk soil $\delta^{15}N$ values at $t = 12$ h (Fig. 5.9). Over 4 days, this peak has subsided and the fit of the linear regression for the percentage of retained ^{15}N incorporated improves, while the variability in the availability of ^{15}N in the defined soil volume affects the calculated rate at which applied ^{15}N is incorporated. Both initially and overall, the rate of ^{15}N incorporation is higher when calculated based on the moles ^{15}N applied, rather than the moles present, probably due to the $t = 0$–12 h increase in bulk soil $\delta^{15}N$ values (Fig. 5.9) outweighing any increases in ^{15}N incorporation into hydrolysable AAs. As ^{15}N is lost from the system over the course of the experiment, the steady state percentage retained ^{15}N incorporated into hydrolysable AAs is understandably much higher (30.3%) than the applied (6.62%).

5.5   Discussion

The site was likely slightly warmer and wetter than average over the experimental period based on historical Met Office data for south-west England and Wales, but no exceptional weather events occurred during the field experiment to strongly affect the results. Field experiment results are discussed below in the context of the weather and compared with the laboratory incubation results from Chap. 4 in each section.

5.5.1 Bulk Soil Carbon and Nitrogen Contents and Amino Acid Concentrations

The insensitivity of bulk soil % TC and % TN contents and AA concentrations to the addition of a small amount of fertiliser N are highlighted in Chaps. 3 and 4. Thus, rather than acting as measurements to assess responses to, or processing of, the applied N, these results provide background information about the system. In the laboratory incubation experiments where losses were minimised and conditions kept constant, no changes over time were expected and indeed, no notable changes over time were observed (Sect. 4.5.2). In the field experiment, however, the experimental system was not isolated, C and N could be lost or gained and many processes (e.g. leaching or plant uptake/excretion) were available to affect changes in the soil. In addition, the system was exposed to precipitation and temperature variations, which in turn influence the rates and magnitudes of these soil processes.

No overall changes were observed in the % TC content of the soil over the course of the field experiment, but % TN contents rose slightly with time (Table 5.1), as did the concentrations of many hydrolysable soil AAs (and accordingly the total hydrolysable AA concentration; Figs. 5.7 and 5.8). That the percentage contribution of total hydrolysable AA N to TN remained relatively constant throughout, at 23.7%, suggests that the later samples contain more N, but the balance of the system has not been altered. The reason for the rise over the course of the experiment is not known and, as it is small, could be an artefact of sample variability.

5.5.2 Bulk Soil $\delta^{15}N$ Values and Corresponding Percentage ^{15}N Retentions

Following treatment application, mean bulk soil $\delta^{15}N$ values and percentage ^{15}N retentions did not peak immediately (to theoretical values of 1090‰ and 100%, respectively) and then fall as a result of ^{15}N losses from the system over the course of the experiment, as might theoretically be expected. This is partly due to the spray application of $^{15}NO_3^-$ in 500 ml DDW, which resulted in the coating of grass swards, clover leaves and moss patches, with only a portion of the $^{15}NO_3^-$ being applied directly to the soil. Although much of the N applied to the plant cover could eventually reach the soil, some direct foliar absorption of $^{15}NO_3^-$ by perennial ryegrass is likely to have occurred [22], reducing the ^{15}N available for the soil. In the context of this experiment and these calculations, a true measured 100% ^{15}N retention in the soil could only result from the perfect distribution of all the applied ^{15}N within the soil volume (i.e. with no losses from the soil due to, for example, plant uptake or leaching). This is clearly unattainable and in reality, sampled percentage ^{15}N retentions reflect the distribution of the ^{15}N within a heterogeneous soil system as well as any losses. Absolutely separating the influence of these factors based on the data available is, however, impossible and the following assumptions are considered reasonable:

(i) ^{15}N will be irreversibly 'lost' from the defined soil volume from the moment of application, as not all of the applied ^{15}N will ever even reach the soil; (ii) further losses of ^{15}N from the defined soil volume will occur continuously over the course of the experiment; (iii) the defined soil volume may initially appear to gain ^{15}N as a result of the application method, due to ^{15}N originally applied to foliage reaching the soil; (iv) uneven distribution of ^{15}N in the soil volume will result in high variability between individual samples at particular sampling points, as has been recorded in other field studies (e.g. [23]).

Following ^{15}NO$_3$$^-$ application, a portion of the applied ^{15}N would have begun to absorb into the soil and probably a little more ^{15}N would have been added by leaf-drip from foliage over time. A mean percentage ^{15}N retention at $t = 3$ h of 48 $\pm$ 5.3% indicates that roughly half of the applied ^{15}N was found in the defined soil volume at this time (although some influence of an uneven ^{15}N distribution on these values is likely). By $t = 6$ h this had risen to 153 $\pm$ 38.6% suggesting that more of the applied ^{15}N had found its way to the defined soil volume and, while this is possible due to the delivery of ^{15}N from foliage to the soil, a percentage ^{15}N retention of over 100% indicates that the ^{15}N is not distributed evenly in the defined soil volume. The further large rise in mean bulk soil δ^{15}N values and percentage ^{15}N retentions between $t = 6$–12 h (Fig. 5.9 and Table 5.2) coincides with the first rainfall event (5.4 mm) after treatment application and while this is likely to have washed more of the applied ^{15}NO$_3$$^-$ into the soil, the very high mean percentage ^{15}N retention of 571% at $t = 12$ h indicates that the rise is also likely due to an uneven distribution of the ^{15}N within the defined soil volume. The plant cover and soil characteristics (texture, structure and moisture content) as well as the intensity and duration of rainfall events following application will have affected the percolation of rainwater, and hence the transport of dissolved ^{15}NO$_3$$^-$, into and through the soil [10, 24–27]. The predominance of low intensity rainfall over the course of the field experiment is likely to have favoured slow percolation of soil water (and hence the applied ^{15}N) into and through the soil rather than preferential flow (e.g. macropore flow) and hence a dominance of vertical over lateral transport [24, 26]; Chap. 6).

The fall in the mean bulk soil δ^{15}N value and percentage ^{15}N retention from $t = 12$ h to 1 day may be due to further rainfall during the night on 17th October (9.8 mm, or 10 mm prior to the $t = 1$ day sampling), which, as well as potentially leaching ^{15}N out of the defined soil volume, could have also aided distribution of the applied ^{15}NO$_3$$^-$ within the defined soil volume, although the percentage ^{15}N retention still remains well above 100%. This pattern is continued with the precipitation of 7.2 mm rainfall from $t = 1$ to 2 days and a further decrease in the mean bulk soil δ^{15}N value and percentage ^{15}N retention (which reaches a value close to 100%, indicating that the applied ^{15}N is still poorly distributed in the defined soil volume). It is worth noting also that the SEs of the mean bulk soil δ^{15}N values and percentage ^{15}N retentions at shorter sampling periods are large, supporting the suggestion that the ^{15}N is less evenly dispersed in the defined soil volume at earlier sampling points. Treatment application method and weather effects aside, the distribution of ^{15}N is also likely to be further complicated, compared with the laboratory experiments, by the heterogeneity of the soil system.

The influence of rainfall on mean bulk soil $\delta^{15}N$ values and percentage ^{15}N retentions is further confirmed by the relative stability of these parameters from $t = 2$–8 days, during which there was low rainfall (Figs. 5.4 and 5.9). Mean bulk soil $\delta^{15}N$ values fall from $t = 8$–16 days and the percentage ^{15}N retention drops well below 100% ($30 \pm 12\%$). During this time, 18.2 mm rainfall was recorded and the low percentage ^{15}N retention indicates considerable losses of ^{15}N from the defined soil volume (note, however, that it does not indicate that the applied ^{15}N is now well distributed). Large rainfall volumes (Fig. 5.4) and the fall of SMDs below 0 mm (Fig. 5.6) between $t = 16$ and 32 days appear to have further reduced the mean bulk soil $\delta^{15}N$ value and the percentage ^{15}N retention and it is highly likely that ^{15}N was leached from the defined soil volume during this time. Little rainfall occurs throughout the remainder of the experiment and the apparent recovery of the mean bulk soil $\delta^{15}N$ value and the percentage ^{15}N retention at $t = 64$ days is more likely to result from differences in ^{15}N distribution than from any gain of ^{15}N to the system. Mean percentage ^{15}N retentions of 14–30% in the top 10 cm of soil between $t = 16$ and 64 days correspond well with percentage ^{15}N recoveries recorded in $^{15}NO_3^-$ field experiments in the literature—e.g. [28] recovered 25% NO_3^--^{15}N in the top 23 cm of soil under a grass-clover ley 48 days after application, while [23] recovered 17% NO_3^--^{15}N in the top 10 cm of soil under a mustard seed crop 42 days after application.

As well as affecting the rate of leaching, rainfall also has the potential to increase denitrification losses of the applied ^{15}N via the increased development of anaerobic microsites at higher soil MCs [28–31]. Denitrification losses are also controlled by soil temperature, however, and denitrification rates are reduced at lower soil temperatures [10, 32]. De Klein and van Logtestijn [33], for example, recorded tenfold lower denitrification rates at 10 than 20 °C in a dry grassland soil in the Netherlands. In a study of the denitrification losses from a UK grassland soil, Ryden [29] found that denitrification was only appreciable when NO_3^- concentrations were $>5\ \mu g\ N\ g^{-1}$ soil, the soil water content was >20% (by weight) and soil temperature was >5–8 °C. Thus, although both leaching and denitrification losses may increase with higher soil MCs as a result of rainfall, the contribution of denitrification to ^{15}N losses is likely to have deteriorated over the course of the experiment as the availability of the applied ^{15}N declined and soil temperatures fell[1].

Declining soil temperatures from October to December could also have affected plant N uptake, which, as the 'system' considered in this experiment is a defined

---

[1]Although only mean weekly air temperatures for the region are available, soil temperature is strongly related to air temperature; generally, fluctuations are dampened and responses at 10 cm depth lag 4–6 h behind air temperatures [34]. Thus, as averaging nullifies the greater amplitude of fluctuations in air temperature and the weekly timescale is large in comparison to the response lag, mean weekly soil temperatures are likely to be approximately similar to mean weekly air temperatures and have been considered so in this analysis. Furthermore, it should be noted that local variability may well have more of an impact on the 'true' soil temperatures at the site than any differences between air and soil temperatures on a regional scale. As a result, the data presented can only be used to provide some indication of the trends in air/soil temperatures at the site and the actual values recorded may not be representative.

volume of soil, represents a (potentially considerable) 'loss' pathway for the applied [15]N. The relationship between plant N uptake, growth and temperature is complicated, but in general, plant N uptake and growth are interdependent and both are thought to decrease as temperatures decline in late autumn/winter [10, 35, 36]. More specifically, soil temperatures above 5 and 8 °C are commonly considered necessary for grass and clover growth, respectively [34, 37]. As mean weekly air temperatures for the region were only below 8 °C for 2 weeks from 3rd to 16th December 2014 (47–60 days after the start of the experiment) and were never below 5 °C, it is possible, on this basis, that some grass and clover growth (and concomitant N uptake) was maintained during the experiment. It has also been suggested, however, that the suspension in plant growth and N uptake usually observed during late autumn/winter occurs in response to fixed seasonal cues such as day length, rather than temperature, so that plants are not 'caught out' by sudden frosts, for example [38]. Thus, it is likely that the contribution of plant N uptake to ^{15}N losses declined over the course of the experiment, although the mechanism for this is uncertain.

Finally, mean bulk soil δ^{15}N values and percentage ^{15}N retentions could also be reduced over the course of the experiment by inputs of N at natural isotopic abundance. The mechanisms for such inputs include: lateral transport of N (fertiliser or cow-derived) from elsewhere in the field; N_2-fixation by the clover (and possibly, but much less probably, by free living N_2-fixing microorganisms, which are not usually considered important in agricultural soils; [39]); mixing by soil macrofauna; and wet and dry N deposition. As the predominance of low intensity rainfall is likely to have favoured vertical leaching rather than horizontal transport [24, 26], N input via the first of these mechanisms was most probably low. The second, clover N_2-fixation, is affected by temperature (fixation declines with temperature) and strongly related to clover growth patterns and thus, the temperature/seasonal growth effects discussed above, which favour low growth are likely to have also promoted low N_2-fixation [40–44]. Dilution by inputs of clover-fixed N therefore probably played less of a role over time in the experiment and, given that: the clover content of the sward was quite low in the experimental area; clover N_2-fixation rates have been observed to decline from flowering in summer onwards in some species [42, 43]; and fixation can be negatively impacted by N applications (particularly of NO_3^-, although the reported application rates at which inhibition occurs vary widely, positive effects have also been observed and timing may be more important; [42, 45, 46]), it is possible that N inputs via fixation were not extensive throughout the experiment.

Macrofaunal activities in the soil are likely to have both transported applied ^{15}N out of the defined volume and natural abundance N into it [47], but the extent to which this occurred and affected mean bulk soil δ^{15}N values and percentage ^{15}N retentions is difficult to assess. Soil mixing by macrofauna is also likely to have contributed to the redistribution of the applied ^{15}N within the defined soil volume [47] and this activity will also have affected mean bulk soil δ^{15}N values and percentage ^{15}N retentions. As the last, but probably the most important input of natural abundance N mentioned above, wet and dry N deposition in the area is estimated to have been *ca.* 1 g N m^{-2} year^{-1} (based on 2012 grid-average estimates for multiple land classes from the Concentration Based Estimated Deposition [CBED] model; [48]). Based on

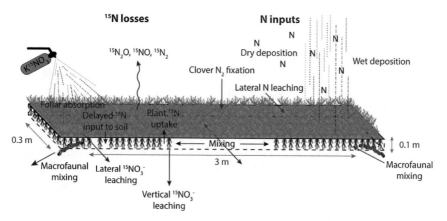

Fig. 5.13 Main applied ^{15}N loss and natural abundance N dilution processes presumably responsible for the trend in bulk soil δ^{15}N results over the course of the field experiment

the concentration of N measured in rainfall, however, *ca.* 320 mg N m^{-2} is estimated to have been deposited in rainfall collected at the field experiment over the course of the experiment (Sect. 6.4.1.1). This would be equivalent to *ca.* 1.9 g N m^{-2} year^{-1} of wet deposition, considerably higher than the wet *and* dry estimate of 1 g N m^{-2} year^{-1} and this is thought to be due to the regular close proximity of large numbers of cows to the sampling area.

Figure 5.13 summarises the main applied ^{15}N loss and natural abundance N dilution processes discussed in this section. Unsurprisingly, in contrast to the controlled laboratory incubation experiments discussed in Chap. 4, the less direct method of application and open and more heterogeneous system in which the field experiment was carried out resulted in large changes and high variabilities (SEs range from 32.0–2300‰) in bulk soil δ^{15}N values over the course of the incubation experiment. As bulk soil δ^{15}N values are indicative of the amount of ^{15}N available in the defined soil volume for incorporation into hydrolysable AAs, this is likely to affect AA δ^{15}N values and the percentages of applied and retained ^{15}N incorporated into the total hydrolysable AA pool.

5.5.3 Nitrogen Isotopic Compositions of Individual Amino Acids

The δ^{15}N values of individual hydrolysable AAs rise, with some fluctuations over the course of the field experiment (Fig. 5.10) and these patterns are reflected in AA percentage applied ^{15}N incorporations, (Fig. 5.11a) although differences in AA concentration alter the hierarchy of AAs and give a wider spread of results. Amino acid δ^{15}N values rise much more than in the WA ^{15}NO$_3^-$ laboratory incubation experiment (Fig. 4.4c) due to the much higher ^{15}N-labelling of the NO$_3^-$ applied in the field experiment (46 atom % *cf.* 10 atom %) and there are also differences in the hierarchy of AAs between the two experiments. The percentages of applied ^{15}N incorporated into individual hydrolysable AAs, however, are fairly comparable, ranging from almost nothing (for Hyp) to around 1% (for Glx) and the ordering of enrichment in AAs is similar (Fig. 5.14).

The rise in AA δ^{15}N values in Fig. 5.10 is less consistent than in Fig. 4.4c, but in contrast to the WA ^{15}NO$_3^-$ laboratory incubation experiment, the δ^{15}N values and percentage ^{15}N incorporations of all AAs except Pro rise within 3 h from $t = 0$. The absence of an 'initial dip' for most AAs in this experiment could be due to one or several factors: (i) the different method of treatment application (spray vs. injection) perhaps reducing the direct contact of soil microorganisms with high concentrations of NO$_3^-$ (note however that the concentration of NO$_3^-$ in the field experiment was higher than that applied in the laboratory incubation experiment; 0.011 g N ml^{-1} *cf.* 0.000275 g N ml^{-1}, respectively); (ii) the undisturbed soil structure in the field experiment may have been more effective at preventing direct contact of microorganisms with high concentrations of NO$_3^-$, or perhaps at maintaining any fixed NH$_4^+$ and/or restricting access of microorganisms to any NH$_4^+$ released; (iii) any initial dip may be obscured by the much higher ^{15}N-labelling of the NO$_3^-$ applied in the field experiment (46 atom % *cf.* 10 atom %). The reason for the contrary behaviour of Pro compared with all other AAs in this experiment is unknown.

The uneven rise in AA δ^{15}N values and percentage applied ^{15}N incorporations in the field experiment is partly due to variability in bulk soil δ^{15}N values and therefore in the amount of ^{15}N present in the soil for incorporation into hydrolysable AAs. The early peak in AA δ^{15}N values and percentage applied ^{15}N incorporations, particularly for Asx, at $t = 12$ h coincides with the peak in bulk soil δ^{15}N values, while the dip in values at $t = 32$ days coincides with the lowest bulk soil δ^{15}N value recorded over the course of the experiment (145‰; Figs. 5.9, 5.10 and 5.11a). Since percentage retained ^{15}N incorporations are calculated based on bulk soil δ^{15}N values, the influence of variations in the amount of ^{15}N present in the soil for incorporation into hydrolysable AAs is removed and Fig. 5.11b shows a somewhat different pattern. The early peak, particularly for Glx, at $t = 1$ day likely reflects the additional N remaining in this AA pool following greater assimilation during peak ^{15}N availability at $t = 12$ h (which is masked at $t = 12$ h in Fig. 5.11b by exceptionally high bulk soil δ^{15}N values).

As the influence of both differences in substrate ^{15}N-labelling and variations in bulk soil δ^{15}N values are removed, percentage retained ^{15}N incorporations represent a good choice for comparing ^{15}N incorporation into individual AAs between the laboratory and the field. Examining Figs. 4.7c and 5.11b, there is a noticeable difference in the time taken for percentage retained ^{15}N incorporations to reach their maximum plateau levels—4 days in the laboratory incubation experiment compared with 16 days in the field. Accordingly, 4-day zero-order rate constants are 1.4–6.6 times higher (on average 3.4 ± 0.4 times) and overall first order rate constants are 1.7–11.7 times higher (on average 5.7 ± 0.7 times) for individual AAs in the laboratory incubation experiment compared with the field experiment (Tables 4.19 and 5.3). The hierarchy/ordering of percentage retained ^{15}N incorporation into AAs is very similar between the two experiments, but the levels of ^{15}N incorporation are

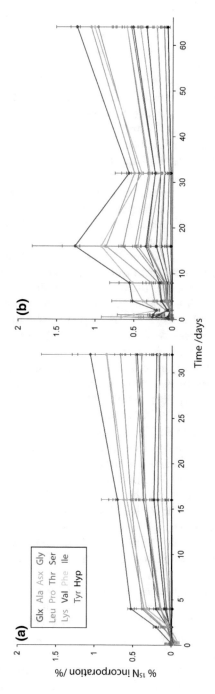

Fig. 5.14 Percentage applied ^{15}N incorporation graphs for: **a** the WA $^{15}NO_3^-$ laboratory incubation experiment and, **b** the field incubation experiment, shown alongside one another to highlight similarities in the hierarchy and percentages of applied ^{15}N incorporated into individual hydrolysable AAs. Error bars are $\pm$ SE ($n = 3$)

3.1–5.8 times higher (on average 4.7 ± 0.7 times) in the field experiment compared with the WA $^{15}NO_3^-$ laboratory incubation experiment (Tables 4.19 and 5.3). Possible reasons for differences in the rates and extents of ^{15}N incorporation between the laboratory and the field are discussed in more detail in Sect. 5.5.4 using results for the total hydrolysable AA pool.

The important observation for this section is that AAs behave similarly in both experiments. Thus, the slightly different patterns in the $\delta^{15}N$ values of Glx, Asx and Hyp compared with other AAs in Fig. 5.10 and their relative percentage ^{15}N incorporations (highest and lowest for Glx and Hyp, respectively) are also observed in the laboratory incubation experiments, although to a much lesser extent, presumably due to the lower ^{15}N-labelling of the NO_3^- applied in the field experiment leading to less clear changes in AA $\delta^{15}N$ values. As already discussed in detail in Sect. 4.5.4, these differences in AA behaviour reflect common biochemical reactions shared by almost all microorganisms. Briefly, microorganisms mainly assimilate N directly into Glu, primarily via the GS-GOGAT pathway [49, 50], N can be directly assimilated into Asp via Asp ammonia-lyase [51] and Knowles et al. [52] found that Glu-^{15}N was incorporated into Asx most quickly and to the greatest extent, whereas Hyp can only be formed via a specialised route and is not known in microbial proteins or cell structural components [53, 54]. In order to gain further insights and make further comparisons with the WA $^{15}NO_3^-$ laboratory incubation experiment, the experimentally derived rate and steady state ^{15}N proportion and flux parameters for the field experiment (Table 5.3) were applied to known metabolic pathways to create a biosynthetic 'map' of ^{15}N routing (Fig. 5.15).

Comparing Figs. 4.15 and 5.15, there is a much greater spread in biosynthetic proximity for the field experiment and also some differences in the relative biosynthetic proximities of AAs to NO_3^- between the laboratory and field experiments. Lysine is biosynthetically further from NO_3^- than Leu and Thr and Gly is biosynthetically further from NO_3^- than Ala in the field experiment compared with the WA $^{15}NO_3^-$ laboratory incubation experiment. These differences may result from the initial dip in the laboratory incubation experiment versus the lack of it in the field experiment. Four days is a relatively long time (in N assimilation/biochemical reaction terms) and is not ideal to investigate biosynthetic proximity, but unfortunately is the time-frame necessary to avoid the initial dip in the WA $^{15}NO_3^-$ incubation experiment and is therefore also used for the field experiment to enable comparison. It is reassuring however, that broadly, the relative ordering of biosynthetic proximities between the laboratory and field experiment is similar, with Glx biosynthetically closest and Phe and Tyr furthest.

Relative steady state percentage retained ^{15}N incorporations or fluxes are also very similar, but of quite different magnitudes between the experiments, (as already noted when comparing Figs. 4.7c and 5.11b) and the relationship/positive correlation between biosynthetic proximity and ^{15}N flux is clear in both Figs. 4.15 and 5.15. Finally, although a much wider range in plateau $\Delta^{15}N$ values was recorded in the field experiment, AA boxes in each experiment are similar relative shades reflecting ^{15}N distribution approximately in proportion to the moles of N in each AA pool. In addition, the same AAs have the darkest shading in both

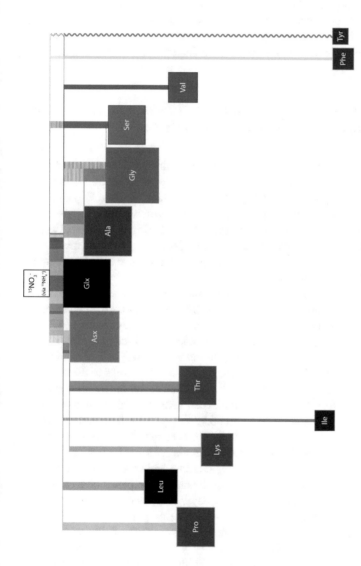

Fig. 5.15 Biosynthetic 'map' of the experimentally determined initial incorporation/partitioning rates and fluxes of $NO_3^- - {}^{15}N$ in the field experiment applied to known metabolic pathways [51, 55, 56]. Lines illustrate the biochemical connectivity of the substrate and AAs and dotted lines indicate possible, but likely secondary biosynthetic pathways. Box area is proportional to moles N in that compound/pool; vertical distance from the base of the substrate box is proportional to the reciprocal of the (4-day) initial rate of retained ${}^{15}N$ incorporation (representative of biosynthetic proximity), Tyr has been represented using a wavy line as the vertical distance for this AA should be twice that of Phe and did not fit on the diagram; line width is proportional to the plateau percentage retained ${}^{15}N$ incorporation, or the flux of ${}^{15}N$ into that AA; and shading is based on plateau AA $\Delta{}^{15}N$ values translated into percentage opacities (349.7‰ for Glx = 100%), where darker shades represent larger plateau $\Delta{}^{15}N$ values

Figs. 4.15 and 5.15 (Glx, Leu, Ile, Tyr and Phe) indicating that these AA pools contain a greater proportion of the applied ^{15}N than expected based on pool size. This supports suggestions in Sect. 4.5.4 of either some preferential ^{15}N routing into these AA pools, and/or smaller less active/inactive sub-pools for these AAs compared with the other AAs. Similarities between the laboratory and field experiments in which AA pools contain the smallest proportions of the applied ^{15}N are less clear. It is possible that the longer sampling period of the field experiment (64 cf. 32 days) affected the fit of the non-linear regression and resultant plateau Δ^{15}N values.

5.5.4   Percentage Applied and Retained ^{15}N Incorporations into the Total Hydrolysable Amino Acid Pool

There is little difference in the percentage applied and retained ^{15}N incorporated into the total hydrolysable AA pool until $t = 4$–8 days, despite wide variations in bulk soil δ^{15}N values (including the peak at $t = 12$ h) during this time (Fig. 5.9) and the differences in the fit of the 12-h and 4-day linear regressions this causes. Variations in bulk soil δ^{15}N values are also thought to be responsible for the higher calculated rates of applied ^{15}N incorporation compared with retained ^{15}N incorporation as increases in bulk soil δ^{15}N values from $t = 0$–12 h outweigh increases in ^{15}N incorporation into hydrolysable AAs during this time.

As expected, as ^{15}N is lost from the system the percentage retained ^{15}N incorporated into the total hydrolysable AA pool becomes much higher than the percentage of applied ^{15}N incorporated. By steady state it is almost five times greater (30.3 cf. 6.62%). The main divergence between percentage applied and retained ^{15}N incorporations occurs between $t = 8$ and 16 days, during which time 18.2 mm rainfall was recorded and the bulk soil mean percentage ^{15}N retention drops to $30 \pm 12\%$ creating a large difference between the moles ^{15}N retained versus the moles applied.

Compared with the WA ^{15}NO$_3{}^-$ incubation experiment, applied ^{15}N 4-day zero-order rate constants are very similar (0.55 and 0.53% day^{-1} for the laboratory and field [poor fit] experiments, respectively), whilst the retained ^{15}N 4-day zero-order rate constant for the laboratory incubation experiment is 2.7 times higher than in the field (1.20 cf. 0.45% day^{-1}). This difference is due to a higher retained ^{15}N 4-day zero-order rate constant for the laboratory experiment (as the percentage retained ^{15}N incorporated increases to over 5% over 4 days, compared with to 2.5% for applied ^{15}N—mainly the result of inconsistent treatment applications) combined with a lower retained ^{15}N 4-day zero-order rate constant for the field experiment (due to variations/the $t = 12$ h peak in bulk soil δ^{15}N values). It is interesting that applied ^{15}N 4-day zero-order rate constants are so similar between the laboratory and the field. As already discussed in Sect. 5.5.3, however, the lack of initially negative ^{15}N assimilation in the field experiment and the relatively long period of 4 days for an initial rate constant somewhat complicates and reduces the reliability/usefulness of comparisons and interpretations.

First-order rate constants for the incorporation of applied and retained ^{15}N into the total hydrolysable AA pool are higher for the laboratory incubation experiment than the field. As noted in Sect. 5.5.3, there is a substantial difference in the time taken for percentage ^{15}N incorporations to reach their maximum plateau levels—4 days in the laboratory incubation experiment compared with 16 days in the field. Possible reasons for this difference in overall incorporation rate are: (i) the different application method used in the field compared with the laboratory (spray vs. injection) leading to delays in ^{15}N reaching the soil (and in ^{15}N concentrations peaking in the soil) and thereby affecting the overall rate of ^{15}N assimilation; (ii) greater spatial and temporal variations in ^{15}N availability in the field experiment due to treatment application method, greater soil heterogeneity and ^{15}N losses, which were strongly affected by rainfall; (iii) temperature—the laboratory incubation experiments were carried out at a constant temperature of 20 °C, whereas the field experiment was exposed to diurnal and seasonal (declining) temperature variations; (iv) soil moisture—constantly 50% WHC in the laboratory incubation experiment, but much more variable in the field; and (v) differences in soil structure (maintained vs. disturbed) between the field and laboratory experiments affecting microbial access to the applied $^{15}NO_3^-$ and activity.

As bulk soil $\delta^{15}N$ values maximise relatively early (at $t = 12$ h) and assimilation into hydrolysable AAs constitutes only a minor fate of $^{15}NO_3^-$, lower/delayed substrate availability is unlikely to have been rate limiting. The more uneven and less direct spray application method did, however, contribute to greater variability in all field experiment results and differences in the timing of substrate access between different areas of the defined soil volume could have affected experimentally derived regression parameters (via additional variability in the time course of ^{15}N incorporation). The same is true for any other factors contributing to greater spatial and temporal variations in ^{15}N availability.

Diurnal variations and much lower (and declining) soil temperatures are likely to have led to lower microbial growth rates and activities in the field compared with the laboratory incubation experiment [57, 58]. In general, the square root of bacterial growth rate constants increase linearly with temperature, while soil mineralisation rates increase by 1.5–2.5 times for every 10 °C rise in temperature between 0 and 30 °C [57, 58]. The mean of the mean weekly air temperatures shown in Fig. 5.5 is 9.8 °C, approximately half the incubation temperature used in the laboratory, but first order incorporation rate constants are 0.14 and 5.4 times higher for applied and retained ^{15}N, respectively—much higher and lower, respectively than the literature multiples of 1.5–2.5 times. This is likely partly because the overall average air temperature of 9.8 °C does not represent the more intricate effects of fluctuating soil temperatures well, but also because such literature multiples are usually generated with all other factors, such as soil moisture potential, being equal. As soil moisture potential and variations in soil moisture potential can also affect microbial growth and activity, it is necessary to also consider the effects of soil moisture potential, although in reality, it is the compound effect (and variations thereof) of varying temperatures and soil moisture potentials over the course of the field experiment that is important [57, 59–61].

Typically, a log-linear relationship exists between microbial activity and soil water potential from -0.01 (wet, but not saturated) to -8.5 MPa (dry), but large, transitory fluxes of CO_2 (released through microbial respiration and thus related to microbial activity) are observed upon rewetting of dry soils [57, 61]. Soil water potentials were not measured directly over the course of the field experiment, but estimates of daily SMDs are shown in Fig. 5.6. Modelled SMDs fall below 0 mm on 9th November 2014 (23 days after the start of the experiment), indicating that the soil is likely to have reached field capacity around this time. Modelled SMDs then fall to a minimum of -31.8 mm on 13th November, rise and remain close to 0 mm for the remainder of the field incubation experiment. Microbial N assimilation rates in the field may therefore be expected to have increased, based on falling SMDs, over the first 16 days of the experiment (potentially outweighing any declines in rates due to decreasing temperatures). It is also possible that drier soil conditions at the start of the experiment may be partly responsible for the lower first-order rate constants and the greater lags to plateau in the field compared with the laboratory.

It is worth noting that it is the soil moisture potential, rather than soil moisture content that is important in affecting microbial activity as soil moisture potentials provide an indication of the energy required to draw water from the soil [61]. There is generally a log-linear relationship between soil water potential and gravimetric soil moisture content, but the quantitative relationship between these parameters varies between different soils [61]. Indeed, soil structural disturbances, such as sieving can change this relationship and the water potential of the sieved and homogenised laboratory soil at 50% WHC is likely to be different from that of the field soil at 50% WHC [61].

Soil disturbance has also been propounded to lead to reduced NO_3^- assimilation rates as a result of the destruction of NH_4^+-poor microsites in which more mobile NO_3^- is the only source of N available to microorganisms [62], but the opposite has been observed between the laboratory and field experiments in this work. This may be due to the dominance of other effects (e.g. substrate application or temperature), or it is possible that disturbances in soil structure led to artificially enhanced NO_3^- incorporation rates in the laboratory incubation experiments via NH_4^+ depletion by increased immobilisation and/or increases in microbial biomass [63] during the 4-day pre-incubation period. Thus, the experimentally-determined first order rate constants for the incorporation of NO_3^--^{15}N into the total hydrolysable AA pool result from the combination of independently, temporally and spatially fluctuating and interacting factors, the relative importance of which are likely to vary over the course of the experiment. It is therefore impossible to precisely (and quantitatively) determine the cause(s) of the difference in first order rate constants beyond the general discussion of contributing factors already given.

That the steady state percentage of applied ^{15}N incorporated into the total hydrolysable AA pool in the field experiment was slightly higher than that recorded in the laboratory (6.62 cf. 6.39%, respectively; Tables 4.21 and 5.4), despite slower incorporation rates and greater ^{15}N losses, implies that field $^{15}NO_3^-$ incorporation was not limited by the additional competing (loss) processes operating under field conditions (e.g. plant uptake or leaching). Moreover, NO_3^- assimilation constituted a quantitatively slightly more important fate for the applied ^{15}N. As with first order assimilation rates discussed above, however, this could be an artefact of temporal and spatial variability of ^{15}N or it could be due to greater NO_3^- assimilation in the field experiment, perhaps due to soil structure and the diffusion of mobile $^{15}NO_3^-$ ions to NH_4^+-poor microsites [62], facilitated by increasing soil moisture potentials over the course of the field experiment.

The steady state percentage retained ^{15}N incorporated into the total hydrolysable AA pool is almost five times higher in the field experiment compared with the laboratory (30.3 *cf.* 6.40%, respectively; Tables 4.21 and 5.4). This indicates that a much greater proportion of ^{15}N present in the soil has been partitioned into the total hydrolysable AA pool in the field compared with the laboratory. Large ^{15}N losses (especially via leaching) during the field experiment greatly reduced the moles of ^{15}N present (based on bulk soil δ^{15}N values) and as ^{15}N incorporated into the soil protein pool is less susceptible to leaching, total hydrolysable AA-^{15}N constitutes an increasing proportion of retained ^{15}N. Finally, as noted in Sect. 5.5.3, it is possible that the longer sampling period of the field experiment (64 *cf.* 32 days) affected the fit of the non-linear regression and thus regression-derived first order rate constants and plateau percentage incorporations.

5.6 Conclusions

The WA $^{15}NO_3^-$ laboratory incubation experiment presented in Chap. 4 has been repeated in a field experiment in order to assess how representative the results of a mesocosm laboratory incubation experiment are of those obtained in the field and the impact of field heterogeneity and uncontrolled environmental conditions. In general, bulk soil results compare well, but reflect much greater ^{15}N losses, as would be expected. It is encouraging that the operation of known biochemical pathways is reflected in the incorporation of ^{15}N into individual hydrolysable AAs, as in the laboratory, despite the influence of field conditions (e.g. diurnal and seasonal temperature variation) and additional processes (e.g. plant uptake and leaching). While overall, the rates of ^{15}N incorporation into the soil protein pool were slower and steady state percentage applied and retained ^{15}N incorporations were greater in the field, differences between the laboratory and field results are generally within ranges that can be accounted for by discrepancies in experimental conditions (although elucidating specific control mechanisms is not straightforward). Indeed, that steady state percentage applied ^{15}N incorporations were quantitatively very similar between the laboratory and the field suggests that the extent of $^{15}NO_3^-$ assimilation was not affected by the different conditions and additional variability and processes, but that a relatively constant proportion of applied $^{15}NO_3^-$ is incorporated into the total hydrolysable AA pool.

Important, more specific findings, with reference to the hypotheses presented in Sect. 5.2, include:

(i)   Losses of ^{15}N over the course of the experiment were considerable and loss pathways are summarised in Fig. 5.13. While good synchrony between rainfall and patterns in bulk soil δ^{15}N values was observed, it is difficult to conclusively demonstrate that ^{15}NO$_3^-$ leaching was the main ^{15}N loss process based on the data presented in this chapter and the physical partitioning of ^{15}N applied to the treated area will be discussed further in Chap. 6.

(ii)   High variability was observed between samples taken at the same time points, as expected, but this was only partly due to the heterogeneity of the field environment and the uneven application and subsequent distribution (and thereby processing) of ^{15}N contributed considerably. This caused somewhat unforeseen difficulty in separating the influence of ^{15}N distribution from ^{15}N loss.

(iii) Patterns in ^{15}NO$_3^-$ incorporation into hydrolysable AAs reflected known biochemical pathways as in the laboratory, but were affected by variations in bulk soil δ^{15}N values. In addition, the lag/initial dip in AA δ^{15}N values observed in the WA ^{15}NO$_3^-$ incubation experiment was absent for all AAs except Pro and complicated comparison of, particularly initial, ^{15}N incorporation rates. Three possible reasons for this are discussed in Sect. 5.5.3, but further work is required to establish a definitive answer. The presence of similar discrepancies in AA steady state bulk soil Δ^{15}N values (with the Glx, Leu, Ile, Tyr and Phe pools containing a greater proportion of the applied ^{15}N than expected based on pool size) in the field and laboratory experiments is particularly interesting. Further work (perhaps employing a greater range of ^{15}N substrates) is required to establish whether this might be due to preferential ^{15}N routing or the existence of smaller less active/inactive sub-pools for these AAs.

(iv)   The percentage of applied NO$_3^-$-^{15}N incorporated into the total hydrolysable AA pool was slightly higher than that recorded in the WA ^{15}NO$_3^-$ laboratory incubation experiment (6.62 cf. 6.39%, respectively; Tables 4.21 and 5.4). This suggests that the extent of ^{15}NO$_3^-$ assimilation in the field experiment was not affected or limited by the different conditions and additional variability and processes in the field, but that a relatively constant proportion of applied ^{15}NO$_3^-$ is incorporated into the total hydrolysable AA pool. The percentage of retained NO$_3^-$-^{15}N incorporated into the total hydrolysable AA pool was greater because as ^{15}N was lost from the soil system, that incorporated, and thereby less susceptible to loss, constituted a greater proportion of the retained ^{15}N.

The work presented in this chapter suggests that the laboratory incubation experiments presented in Chap. 4 can provide a reasonable indication of ^{15}N incorporation into hydrolysable AAs, at least for ^{15}NO$_3^-$. Other ^{15}N-labelled substrates are expected to behave comparably, but it could be worthwhile to conduct analogous comparative field experiments with other substrates in order to confirm this—any differences, for example, could provide extremely interesting insights into the relative importance of different processes for different substrates at a different scale/under field conditions. There is a need also for some intermediate-scale/level of control experiments in order to elucidate the controlling factors behind the differences between the laboratory and field experiments (e.g. the reasons for the absence of a lag/initial dip in AA $\delta^{15}N$ values in the field experiment and for differences in overall ^{15}N incorporation rates, particularly the delay in the field experiment to reach steady state).

In addition, since NO_3^- incorporation into hydrolysable AAs is not an important fate, it is clearly worth investigating the (relative) importance of other fates. As has already been noted, the susceptibility of NO_3^- to leaching is well known and this is expected to represent an important fate for field-applied $^{15}NO_3^-$. While many studies have been carried out to assess NO_3^- leaching, uncertainty remains as to the origin of diffuse NO_3^- pollution in groundwater and the timing and mechanisms of fertiliser NO_3^- transport to groundwater. Finally, the desired fate of fertiliser NO_3^- is plant/crop uptake, so it is necessary to consider the results of this work with this in mind—as highlighted in the conclusions of Chap. 4, the low microbial incorporation of NO_3^- likely contributes to its importance as a plant N source [9].

References

1. Dungait JAJ, Cardenas LM, Blackwell MSA, Wu L, Withers PJA, Chadwick DR, Bol R, Murray PJ, Macdonald AJ, Whitmore AP, Goulding KWT (2012) Advances in the understanding of nutrient dynamics and management in UK agriculture. Sci Total Environ 434:39–50
2. Levin SA (1992) The problem of pattern and scale in ecology. Ecology 73:1943–1967
3. Addiscott TM (1996) Measuring and modelling nitrogen leaching: parallel problems. Plant Soil 181:1–6
4. Beven K (1989) Changing ideas in hydrology—the case of physically-based models. J Hydrol 105:157–172
5. Refsgaard JC, Thorsen M, Jensen JB, Kleeschulte S, Hansen S (1999) Large scale modelling of groundwater contamination from nitrate leaching. J Hydrol 221:117–140
6. Hodge A, Robinson D, Fitter A (2000) Are microorganisms more effective than plants at competing for nitrogen? Trends Plant Sci 5:304–308
7. Jackson LE, Schimel JP, Firestone MK (1989) Short-term partitioning of ammonium and nitrate between plants and microbes in an annual grassland. Soil Biol Biochem 21:409–415
8. Kaštovská E, Šantrůčková H (2011) Comparison of uptake of different N forms by soil microorganisms and two wet-grassland plants: a pot study. Soil Biol Biochem 43:1285–1291. https://doi.org/10.1016/j.soilbio.2011.02.021
9. Schimel JP, Bennett J (2004) Nitrogen mineralisation: challenges of a changing paradigm. Ecology 85:591–602
10. Cameron KC, Di HJ, Moir JL (2013) Nitrogen losses from the soil/plant system: a review. Ann Appl Biol 126:145–173. https://doi.org/10.1111/aab.12014
11. Redmile-Gordon MA, Armenise E, Hirsch PR, Brookes PC (2014) Biodiesel co-product (BCP) decreases soil nitrogen (N) losses to groundwater. Water, Air & Soil Pollution 225. https://doi.org/10.1007/s11270-013-1831-7
12. Defra (2010) Fertiliser Manual (RB209), p 60, 65. Report available from http://www.ahdb.org.uk/projects/CropNutrition.aspx. Accessed 26 Mar 2014
13. Mathieu O, Lévêque J, Hénault C, Ambus P, Milloux M, Andreux F (2007) Influence of ^{15}N enrichment on the net isotopic fractionation factor during the reduction of nitrate to nitrous oxide in soil. Rapid Commun Mass Spectrom 21:1447–1451. https://doi.org/10.1002/rcm.2979

14. Tang FHM, Maggi F (2012) The effect of ^{15}N to ^{14}N ratio on nitrification, denitrification and dissimilatory nitrate reduction. Rapid Commun Mass Spectrom 26:430–442. https://doi.org/10.1002/rcm.6119

15. Newell Price JP, Harris D, Taylor M, Williams JR, Anthony SG, Duethmann D, Gooday RD, Lord EI, Chambers BJ, Chadwick DR, Misselbrook TH (2011) An inventory of mitigation methods and guide to their effects on diffuse water pollution, greenhouse gas emissions and ammonia emissions from agriculture: user guide. Prepared as part of Defra Project WQ0106

16. CEH (2014) Hydrological Summary for the United Kingdom, October 2014. A full catalogue of past Hydrological Summaries can be accessed and downloaded at: http://www.ceh.ac.uk/data/nrfa/nhmp/nhmp.html. Accessed 31 Mar 2016

17. Met Office (2014) Climate summaries. http://www.metoffice.gov.uk/climate/uk/summaries. Accessed 31 Mar 2016

18. BBC (2014) Warmest UK Halloween on record. http://www.bbc.co.uk/news/uk-29851285. Accessed 06 Jan 2016

19. Bailey RJ, Spackman E (1996) A model for estimating soil moisture changes as an aid to irrigation scheduling and crop water-use studies: I. Operation details and description. Soil Use Manage 12:122–128

20. Bailey RJ, Groves SJ, Spackman E (1996) A model for estimating soil moisture changes as an aid to irrigation scheduling and crop water-use studies: II. Field test of the model. Soil Use Manage 12:129–133

21. Silgram M, Hatley D, Gooday R (2007) IRRIGUIDE: a decision support tool for drainage estimation and irrigation scheduling. ADAS UK Ltd

22. Bowman CB, Paul JI (1992) Foliar absorption of urea, ammonium, and nitrate by perennial ryegrass turf. J Am Soc Hortic Sci 117:75–79

23. Holbeck B, Amelung W, Wolf A, Südekum K-H, Schloter M, Welp G (2013) Recoveries of ^{15}N-labelled fertilisers (chicken manure, mushroom compost and potassium nitrate) in arable topsoil after autumn application to winter cover crops. Soil Tillage Res 130:120–127

24. Armstrong AC, Leeds-Harrison PB, Harris GL, Catt JA (1999) Measurement of solute fluxes in microporous soils: techniques, problems and precision. Soil Use Manag 15:240–246

25. Bachmair S, Weiler M, Nützmann G (2009) Controls of land use and soil structure on water movement: lessons for pollutant transfer through the unsaturated zone. J Hydrol 369:241–252

26. Jarvis NJ (2007) A review of non-equilibrium water flow and solute transport in soil macropores: principles, controlling factors and consequences for water quality. Eur J Soil Sci 58:523–546

27. Gazis C, Feng X (2004) A stable isotope study of soil water: evidence for mixing and preferential flow paths. Geoderma 119(1–2):97–111

28. Goulding KWT, Webster CP, Powlson DS, Poulton PR (1993) Denitrification losses of nitrogen fertiliser applied to winter wheat following ley and arable rotations as estimated by acetylene inhibition and ^{15}N balance. J Soil Sci 44:63–72

29. Ryder JC (1983) Denitrification loss from a grassland soil in the field receiving different rates of nitrogen as ammonium nitrate. J Soil Sci 34:355–365

30. Sexstone AJ, Parkin TB, Tiedje JM (1985) Temporal response of soil denitrification rates to rainfall and irrigation. Soil Sci Soc Am J 49:99–103

31. Tiedje JM, Sexstone AJ, Parkin TB, Revsbech NP (1984) Anaerobic processes in soil. Plant Soil 76:197–212

32. Luo J, Tillman RW, Ball PR (1999) Factors regulating denitrification in a soil under pasture. Soil Biol Biochem 31:913–927

33. de Klein CAM, van Logtestijn RSP (1996) Denitrification in grassland soils in the Netherlands in relation to irrigation, N-application rate, soil water content and soil temperature. Soil Biol Biochem 28:231–237

34. Zheng D, Hunt ER Jr, Running SW (1993) A daily soil temperature model based on air temperature and precipitation for continental applications. Climate Res 2:183–191

35. Di HJ, Cameron KC (2002) Nitrate leaching in temperate agroecosystems: sources, factors and mitigating strategies. Nutr Cycl Agroecosyst 46:237–256

36. Jarvis SC, Macduff JM (1989) Nitrate nutrition of grasses from steady-state supplies in flowing solution culture following nitrate deprivation and/or defoliation. J Exp Bot 40:965–975
37. EBLEX (2013) Improving pastures for better returns. Agriculture and Horticulture Development Board (AHDB)
38. Gutschick VP (1981) Evolved strategies in nitrogen acquisition by plants. Am Nat 118:607–637
39. Olivares J, Bedmar EJ, Sanjuán J (2013) Biological nitrogen fixation in the context of global change. Mol Plant Microbe Interact 26:486–494
40. Bolger TP, Pate JS, Unkovich MJ, Turner NC (1995) Estimates of seasonal nitrogen fixation of annual subterranean clover-based pastures using the ^{15}N natural abundance technique. Plant Soil 175:57–66
41. Grant DA, Lambert MG (1979) Nitrogen fixation in pasture V.: unimproved North Island hill country, "Ballantrae". New Zealand J Exp Agric 7:19–22
42. Liu Y, Wu L, Baddeley JA, Watson CA (2011) Models of biological nitrogen fixation of legumes: a review. Agron Sustain Develope 31:155–172. https://doi.org/10.1051/agro/2010008
43. Rice WA (1980) Seasonal patterns of nitrogen fixation and dry matter production by clovers grown in the Peace River region. Can J Plant Sci 60:847–858
44. Unkovich M (2012) Nitrogen fixation in Australian dairy systems: review and prospect. Crop Pasture Sci 63:787–804
45. Elgersma A, Schlepers H, Nassiri M (2000) Interactions between perennial ryegrass (*Lolium perenne* L.) and white clover (*Trifolium repens* L.) under contrasting nitrogen availability: productivity, seasonal patterns of species composition, N_2 fixation, N transfer and N recovery. Plant Soil 221:281–299
46. Høgh-Jensen H, Schjoerring JK (1997) Interactions between white clover and ryegrass under contrasting nitrogen availability: N_2 fixation, N fertiliser, N transfer and water use efficiency. Plant Soil 197:187–199
47. Coleman D, Wall D (2007) Fauna: The engine for microbial activity and transport. In: Paul E (ed) Soil microbiology, ecology, and biochemistry. Academic Press, New York, pp 163–191
48. CEH (2016) Nitrogen atmospheric Concentration Based Estimated Deposition (CBED) data for the UK 2012. Accessed 18 Apr 2016
49. Cabello P, Roldán MD, Moreno-Vivián C (2004) Nitrate reduction and the nitrogen cycle in archaea. Microbiology 150:3527–3546. https://doi.org/10.1099/mic.0.27303-0
50. Geisseler D, Horwath WR, Joergensen RG, Ludwig B (2010) Pathways of nitrogen utilization by soil microorganisms—a review. Soil Biol Biochem 42:2058–2067. https://doi.org/10.1016/j.soilbio.2010.08.021
51. Caspi R, Foerster H, Fulcher CA, Kaipa P, Krummenacker M, Latendresse M, Paley S, Rhee SY, Shearer AG, Tissier C, Walk TC, Zhang P, Karp PD (2007) The MetaCyc database of metabolic pathways and enzymes and the BioCyc collection of Pathway/Genome Databases. Nucleic Acids Res 36:D623–D631. https://doi.org/10.1093/nar/gkm900
52. Knowles TDJ, Chadwick DR, Bol R, Evershed RP (2010) Tracing the rate and extent of N and C flow from ^{13}C,^{15}N-glycine and glutamate into individual de novo synthesised soil amino acids. Org Geochem 41:1259–1268. https://doi.org/10.1016/j.orggeochem.2010.09.003
53. Adams E (1970) Metabolism of proline and hydroxyproline. In: Hall DA, Jackson DS (eds) International review of connective tissue research, vol 5. Academic Press, New York, pp 2–82
54. Adams E, Frank L (1980) Metabolism of proline and the hydroxyprolines. Annu Rev Biochem 49:1005–1061
55. Berg JM, Tymoczko JL, Gatto GJ Jr, Stryer L (2015) The biosynthesis of amino acids. Biochemistry, 8th edn. W. H. Freeman and Company, New York, pp 713–742
56. Nelson DL, Cox MM (2013) Biosynthesis of amino acids, nucleotides, and related molecules. Lehninger principles of biochemistry, 6th edn. Macmillan Higher Education, Basingstoke, pp 881–928
57. Jamieson N, Barraclough D, Unkovich M, Monaghan R (1998) Soil N dynamics in a natural calcareous grassland under a changing climate. Biol Fertil Soils 27:267–273
58. Ratkowsky DA, Olley J, McMeekin TA, Ball A (1982) Relationship between temperature and growth rate of bacterial cultures. J Bacteriol 149:1–5

59. Howard DM, Howard PJA (1993) Relationship between CO_2 evolution, moisture content and temperature for a range of soil types. Soil Biol Biochem 25:1537–1546
60. Steinweg JM, Dukes JS, Wallenstein MD (2012) Modelling the effects of temperature and moisture on soil enzyme activity: linking laboratory assays to continuous field data. Soil Biol Biochem 55:85–92
61. Orchard VA, Cook FJ (1983) Relationship between soil respiration and soil moisture. Soil Biol Biochem 15:447–455
62. Booth MS, Stark JM, Rastetter E (2005) Controls on nitrogen cycling in terrestrial ecosystems: a synthetic analysis of literature data. Ecol Monogr 75:139–157
63. Černohlávková J, Jarkovský J, Nešporová M, Hofman J (2009) Variability of soil microbial properties: effects of sampling, handling and storage. Ecotoxicol Environ Saf 72:2102–2108. https://doi.org/10.1016/j.ecoenv.2009.04.023

Chapter 6
^{15}N Tracing of the Partitioning and Transport of Nitrate Under Field Conditions

6.1 Introduction

6.1.1 Context

The results of Chaps. 4 and 5 suggest that processes other than microbial incorporation are more important for NO_3^- cycling in the soil system. Indeed, the propensity of NO_3^- to leaching is well known (Sect. 1.2) and NO_3^- is an important plant N source (Sects. 6.1.2 and 6.1.3). Furthermore, clear indications of other processes were recorded in the field incubation experiment (Chap. 5) and there were even some signs of gaseous N production processes in the laboratory incubation experiments in which plant uptake and leaching were excluded and gaseous N losses limited as far as possible (Chap. 4). It is therefore important to view microbial NO_3^- incorporation in the context of its wider partitioning in the plant-soil system, particularly since the desired fate of fertiliser N is plant/crop uptake. Clearly, this partitioning (and the intended crop uptake) will be affected by the transport of NO_3^- away from the target/application area.

Although there has already been a considerable amount of research into the partitioning of NO_3^-, particularly in the context of NUE and N balance studies, a complete N balance remains to be achieved [1]. In addition, there are many challenges to accurately recording all the fates of ^{15}N under field conditions, limiting progress in understanding. Uncertainty also remains in the mechanisms and rates of NO_3^- leaching, particularly in undisturbed, heterogeneous soil systems (Sect. 1.2). Accordingly, the work described in this chapter further investigates the partitioning and transport of $^{15}NO_3^-$ in a unique plot-scale field experiment and in the context of its microbial incorporation (Chap. 5). Whilst several aspects of this work are reflected in previous studies, the experiment presented in this chapter is believed to be unique in its design and combined assessment of the partitioning and (vertical and lateral) transport of NO_3^- in an undisturbed soil under field conditions. An early goal when planning the experiment was to attempt a full N balance of the system under field conditions, but

© Springer Nature Switzerland AG 2019
A. F. Charteris, ^{15}N Tracing of Microbial Assimilation, Partitioning
and Transport of Fertilisers in Grassland Soils, Springer Theses,
https://doi.org/10.1007/978-3-030-31057-8_6

due to time, money, sampling, instrumentation and analysis constraints this simply was not possible. As a result, while the experiment moves towards an N balance of an unconfined and relatively undisturbed system under field conditions, the balance is incomplete as gaseous N losses were not measured.

NO_3^- leaching is reviewed in Sect. 1.2 and the importance of alternative fates for NO_3^- in controlling/reducing NO_3^- leaching are briefly described in Sect. 1.2.2. Plant uptake is the desired fate of applied NO_3^- and plant NO_3^- uptake and assimilation are reviewed in the following sections. Finally, denitrification is likely to represent the other major potential fate of field applied fertiliser NO_3^- and N losses via this process may even be greater than those via leaching [2]. The process is introduced in Sect. 1.2.2 and discussed briefly in Sects. 4.5.3 and 5.5.2, but is not reviewed in further detail here as denitrification (or indeed any gaseous N losses) was not measured in any of the experiments described in this thesis and detailed consideration of the process beyond that already presented was deemed to be of no advantage.

6.1.2   Plant Nitrogen Uptake and Assimilation

Despite pre-1900 reports and much debate during the latter part of the 20th century for the utilisation of organic N by plants, the traditional focus has remained, until perhaps recently, on inorganic N species as the only (and now the major) source of plant N nutrition [3–5]. While it is now accepted that plants (both with and without mycorrhiza and in a range of ecosystems) can directly take up LMW organic N compounds (including urea, AAs and small peptides) as well as NO_3^- and NH_4^+ [3, 6–12], the technical difficulty of unequivocally demonstrating and quantifying intact uptake of organic N [4, 13–16], the effects of interactions between plants and microorganisms (with varying N concentrations) and species, as well as ecosystem differences complicate assessments of the relative importance of different N forms in plant nutrition [3, 17, 18]. Thus, the ecological significance of organic N in plant nutrition remains a matter of debate [3, 13, 16], but it is likely to be most important for mycorrhizal plants and in N-limited ecosystems (e.g. arctic tundra and boreal and taiga forests; [9, 12, 17–19]).

In the absence of any competition effects, most plants preferentially take up NH_4^+ (followed by AAs) when all forms of N are equally available [3, 19]. Inter-species differences have been recorded, however (e.g. [20]), and for some species, N preferences may have some temperature dependence, with lower temperatures increasing the importance of organic N compared with NH_4^+ and of NH_4^+ compared with NO_3^- [21, 22]. In addition, while dependent mainly on the abundance and operation of transporter systems, uptake preferences may also be related to the energy/C cost of incorporating the N form (as it is for soil microorganisms)—AAs may be utilised directly or transaminated, NH_4^+ requires a C skeleton, while NO_3^- must first be reduced to NH_4^+ [3, 19, 23]. High NH_4^+ concentrations, however, are toxic to plants (sensitivity between species varies) and can inhibit growth [3, 24–26]. These effects can be alleviated by pH buffering or NO_3^- supply [24–26] and indeed, the benefits of supplying mixed N sources have been proclaimed in several studies [24, 26]. This is purportedly as a result of the complementary effects of the different sources: NO_3^- encourages organ elongation, whilst NH_4^+ can increase yields [24, 27]. The presence of NH_4^+ has also been shown to increase NO_3^- uptake in several species (although the reverse effect has been recorded for others; [21, 26]).

In an agricultural context, NO_3^- is usually considered the main source of plant N (e.g. [28]), probably due to the effect of competition from microorganisms in controlling the supply of N to plants, which commonly leads to highest NO_3^- availabilities and consequently, plant dependence on NO_3^- (e.g. [3, 5, 29]; Sect. 6.1.3). The higher mobility of NO_3^- (compared with NH_4^+) is also valuable for plant N uptake as NO_3^- can diffuse directly to roots, replenishing often N-depleted rhizosphere zones [30].

A wide variety of transporters for the uptake of the various N forms have been identified in plants and are commonly regulated by N availability, although expression can also be affected by plant growth stage [3, 8, 31, 32]. Following uptake, NH_4^+ and AAs are more likely to temporarily remain in roots, while a proportion of the absorbed NO_3^- is translocated directly to shoots, where it may be reduced [3, 19]. Ammonium (either taken up or derived from the reduction/hydrolysis of NO_3^-/urea) is assimilated into Glu and Gln via GDH or the GS-GOGAT pathway [8, 33]. Amino acids may be incorporated directly into cell matter, transaminated, or probably less commonly, de-aminated or decarboxylated [3].

6.1.3 Competition Between Plants and Soil Microorganisms for Nitrogen

As the most commonly limiting nutrient in terrestrial ecosystems, competition for N between plants and microorganisms is likely to be fierce, particularly in rhizosphere zones where N may be depleted by plant uptake and microbial activity stimulated by higher C availability [30, 34]. Conventionally, it has been understood that microorganisms outcompete plants for all forms of N in the short term and that plants must satisfy their N demand with what remains [5, 34, 35]. Thus, preferred microbial N sources, such as NH_4^+ and organic N, are expected to be less available to plants and, in the presence of other N sources, competition for NO_3^- will be lower [34–36]. In this way, competition, as well as N availability, can play an important role in determining the apparent plant preferences for N substrates [5, 34].

The superiority of microorganisms has been attributed to their larger surface area to volume ratios, wider spatial distributions, greater substrate affinities and higher growth rates [5, 34, 35]. These advantages, however, cannot always be of assistance and where N (in any form) is accessible by plant roots, it may be taken up, reducing the N available for microbial consumption [5]. In addition, the mycorrhizal infections of some plants complicate assessments of plant-microbe competition and can play a role in improving plant competitiveness and N uptake [3, 5, 30, 34].

Moreover, while microbial N uptake is initially extremely rapid, it plateaus relatively quickly (possibly due to C-limitation) and the fast turnover of the microbial biomass re-releases N into the soil for plant acquisition [34, 35, 37]. Since plant biomass turnover is much slower, plants can gradually accumulate N and effectively outcompete microorganisms in the longer term [30, 34, 35, 37]. Thus, while microbes might win the battle for N, plants win the war. This separation in the timescales of microbial and plant N utilisation and turnover is important for retaining biologically available N within the soil and reducing any volatilisation and leaching losses [30].

6.2 Objectives

This chapter focuses on the physical partitioning and transport of ^{15}N over the course of a 32-week plot-scale field experiment. Nitrate is a commonly applied N fertiliser compound, and the N form which is often most available to plants ([5, 29, 34, 35]; Sects. 6.1.2 and 6.1.3) and particularly susceptible to leaching ([38]; Sect. 1.2). The experiment started in October 2014 and continued until June 2015 (32 weeks) in order to capture the late-autumn, winter and early spring periods during which NO_3^- leaching is generally greatest [38, 39], as well as spring and early summer plant growth, during which uptake of remaining $^{15}NO_3^-$ may be expected to increase.

The specific objectives of this work were to:

(i)   Conduct a ^{15}N partitioning and transport tracer study in undisturbed and unisolated soil under field conditions.
(ii) Explore the patterns in soil, root, foliage and soil water δ^{15}N values over the course of the 32-week ^{15}N tracer field experiment.
(iii) Investigate the partitioning of ^{15}N into different components of the plant-soil system (soil, roots, foliage and soil water) in the treated area over the course of 32 weeks (and in the context of the microbial NO_3^- assimilation considered in Chap. 5).
(iv) Obtain estimates of the relative proportions and velocities of vertical and lateral NO_3^- leaching at the site.
(v)   Investigate soil water flow at the site and gain a greater understanding of soil water NO_3^- transport in relation to soil water flow paths and residence times.

Potential findings include:

(i)   Developments in experimental design recommendations for conducting ^{15}N tracer studies under field conditions.
(ii) Improved understanding of $^{15}NO_3^-$ distribution, proportional representation of particular N pools and transport under field conditions immediately following fertiliser NO_3^- application and over the course of the following 32 weeks.
(iii) Insights into the isotopic mass-balanced/pool-size-based partitioning of $^{15}NO_3^-$ amongst different components of the plant-soil system (soil, roots, foliage and soil water) in the treated area over the course of 32 weeks (and in the context of the microbial NO_3^- assimilation considered in Chap. 5).

(iv) Estimates of the relative proportions and velocities of vertical and lateral NO_3^- leaching at the site.

(v)   A greater understanding of NO_3^- leaching mechanisms and transport pathways at the site.

The hypotheses tested in this work are:

(i)   Following application, $^{15}NO_3^-$ will not be (or ever become) evenly distributed in the plant-soil system due to the difficulty of spray applying it completely evenly and the heterogeneity of the plot.

(ii) Due to losses, $NO_3^- \text{-}^{15}N$ will represent a declining proportion of the N in each component of the treated area plant-soil system over the course of the 32-week experiment.

(iii) Due to the spray application of $^{15}NO_3^-$ to the foliage and soil in the treated area, direct foliar ^{15}N uptake will occur.

(iv) Due to the only very gently sloping hollow in which the experiment is sited, vertical $^{15}NO_3^-$ leaching will occur faster and to a greater extent than lateral and due to stones and soil structure, some preferential flow is likely to occur at the site.

6.3 Site, Experimental Design, Sampling and Analytical Methods

6.3.1 Site

The field experiment was carried out in the 'Little Broadheath' field of Longlands Dairy Farm, near Winterbourne Abbas and this site is described further in Sects. 2.2.2 and 5.3.

6.3.2 Experimental Design

The location of the field experiment was selected as described in Sect. 5.3 (Figs. 5.1 and 5.2). The three adjoining 1×0.33 m treated areas sited above three porous ceramic cups (at depths of 0.8, 0.2 and 0.5 m; Fig. 5.3) is also described in Sect. 5.3 and actually forms a small part of the larger field experimental design (Fig. 6.1). In order to trace the vertical and lateral transport of applied $NO_3^- \text{-}^{15}N$ in soil water, an array of further porous ceramic cups (than described in Sect. 5.3) were installed at various depths and three downslope distances (rows 1 m apart) from the treated area in July 2014. The sampling array (Fig. 6.1) was designed to include:

(i)   Porous ceramic cups at three different depths below the treated area in order to maximise the chances of detecting vertical transport of ^{15}N and to facilitate

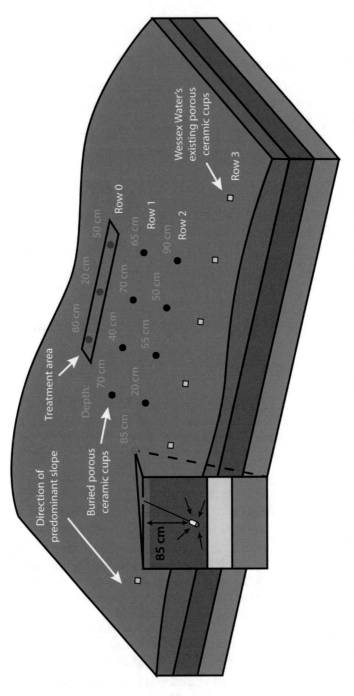

Fig. 6.1 Diagram of the plot-scale field experiment showing the treated area and the array and depths of buried porous ceramic cups

assessment of the contribution of vertical transport to NO_3^- leaching, vertical transport processes (uniform or preferential flow) and calculation of vertical leaching rates.

(ii) Porous ceramic cups at three downslope distances from the treated area (minimum distance of 1 m determined by recommendations for minimum spacing of porous ceramic cups) and at a range of depths in order to maximise the chances of detecting lateral transport of ^{15}N and to facilitate assessment of the contribution of lateral transport to NO_3^- leaching, lateral transport processes (uniform or preferential flow), the depths at which lateral transport occurs and calculation of lateral transport rates.

(iii) Increasing array width with distance from the treated area in order to maximise chances of detecting any plumes in the transport of ^{15}N.

The original intention was to install porous ceramic cups at depths of 0.3, 0.6 and 0.9 m, but difficulties during installation due to the stoniness of the site rendered this impossible. The range in the distribution of depths eventually achieved was deemed to satisfactorily meet the requirements of the original design.

Porous ceramic cup installation was carried out according to manufacturer's instructions and Wessex Water's Standard Operating Procedure for the Installation of Porous Pots (Fig. 6.2). Briefly, for each porous ceramic cup installation, Wessex Water's Envirocorer Spiker 1000 coring rig was used to auger a channel at a 30° angle and of an appropriate length to give the desired vertical depth below

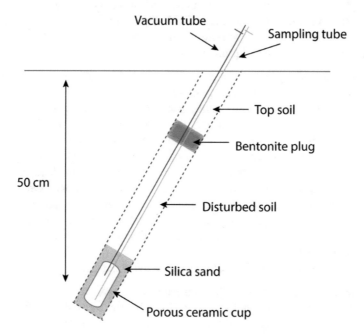

Fig. 6.2 Porous ceramic cup installation

the surface (using trigonometry). Installation at an angle avoids disturbance of the soil directly above the porous ceramic cup [40]. The porous ceramic cup was soaked in water for 10 min prior to insertion into the channel and a silica sand and water mixture poured around the porous ceramic cup once in place using a plastic pipe and funnel. Soaking rinses away any contaminants left by the production process and preconditions the ceramic to reduce any sorption of the soil solution [41].

The channel was then backfilled with a little subsoil and a vacuum applied in the porous ceramic cup to ensure good contact between the cup and the silica. More soil was then gently compacted (using a wooden pole) into the channel to a level of 20 cm below the surface at which point a 5 cm thick plug of bentonite pellets was added to prevent surface infiltrating water from running directly down the tubes leading to the porous ceramic cup. The remainder of the channel was then (over-) filled with topsoil (to allow for settling). The porous ceramic cup tubes that remained sticking out of the soil were passed through a hole in the base of a small screw-topped, plastic pot and clips added to enable them to be sealed. At this point a syringe was used to vacuum extract and discard any water from the newly installed porous ceramic cup (infiltrating from the disturbed soil and silica sand mixture). Finally, the tubes were clipped shut, coiled in the plastic pot and the lid screwed on. As well as protecting the sampling tubes from faunal interference, the sealed pot produces a slight greenhouse effect, helping to prevent them from freezing over winter.

Soil solution sampling using porous ceramic cups has been developed and very widely used in research, industry and agriculture since its inception by Briggs and McCall [42]. The technique has the advantages of simple and inexpensive installation and maintenance, low soil disturbance, flexibility in spatial and temporal sampling patterns and negligible NO_3^- sorption in the ceramic cups, but is also associated with several well-documented problems [40, 41, 43–45], which relate mainly to:

(i)   Point ceramic cup samples' representativeness of the soil solution at field-scale, particularly in more heterogeneous systems.

(ii) The use of porous ceramic cup sampling in well-structured soils in which macropore or fissure flow is likely to bypass cups, or alternatively, but less likely, to 'over-contribute' to a particular porous ceramic cup.

(iii) The effect of the vacuum-sampling of water on the normal infiltration of soil water (and NO_3^-) through the soil system—the potential field of a porous suction cup may extend over 1 m and while some authors have raised this concern (e.g. [44]), others have suggested that it is unlikely to present a problem with the suction pressures and sample volumes typically employed (usually $\leq$80 kPa and 20 ml; e.g. [40]), although ultimately this is likely to depend on edaphic conditions.

(iv) The indeterminate origin and/or nature of the water sampled:

      – firstly, the variable and unknown radius of the soil volume from which the water is extracted (usually between 0.1 and 0.5 m, depending on: the diameter of the ceramic cup; its installation depth; the vacuum pressure in the cup; the pore size distribution of the soil; the capillary pressure in the soil; and the depth of the water table; [41])

- secondly, whether/to what extent soil water is preferentially extracted from particular soil pore size ranges, which may have differing NO_3^- concentrations (Grossmann and Udluft [41] argued that extraction of all pore sizes should occur in perfectly homogenous soils, but that heterogeneity is likely to favour the sampling of water seeping through larger pores, as does greater suction pressures) and to what extent the NO_3^- concentrations of this extracted water are representative of soil water NO_3^- concentrations generally
- and thirdly, and somewhat contentiously, whether porous ceramic cups extract 'mobile' water (generally considered to be that seeping through larger pores), which provides a good indication of the NO_3^- concentrations of percolating water or perhaps by ill-timed sampling, 'immobile' water [46] or due to the well-structured nature of the soil, less mobile water, which does not represent drainage NO_3^- concentrations [43, 47].

(v) The 'snapshot' nature of discontinuous sampling, which may affect quantification of NO_3^- leaching [40, 41, 43, 44, 46, 47].

Accordingly, the use of porous ceramic cups to estimate NO_3^- leaching losses is recommended mainly in homogenous, unstructured sandy soils in which many of the issues noted above are minimised. The soil in Little Broadheath has a reasonable clay content (*ca.* 30%) and is quite stony, so is unlikely to conform entirely to the homogenous ideal required for the accurate use of porous ceramic cups. Elements of the experiment objectives, however, (e.g. the aim to trace the vertical *and* lateral transport of $^{15}NO_3^-$ at different depths through undisturbed soil upslope of a PWS point) and the constraints of conducting a study in a small (*ca.* 50 m^2) section of a field on a working dairy farm, made porous ceramic cups the best (or the only practicable) method of sampling in situ soil water. In addition, it was hoped that the combined measurement of soil water $\delta^{15}N$, δ^2H and $\delta^{18}O$ values would enable assessment of the history of the water in which sampled $^{15}NO_3^-$ was present (e.g. [48, 49]). It is acknowledged that the use of porous ceramic cup samplers may affect accurate quantification of $^{15}NO_3^-$ leaching, as will the use of the IRRIGUIDE model [50–52], rather than direct measurements, to estimate drainage volumes, but this is only one aspect of the experiment and estimates obtained in this way will be presented as such. Consistent and apparently satisfactory results have previously been obtained in medium soils overlying chalk, so it is possible that these estimates will be more reliable than the worst-case scenario presented above suggests [40].

Following porous ceramic cup installation, the site was allowed to settle and maintained (by grass cutting and partial removal of the cuttings) until the start of the experiment in October 2014. At 10:00 on 17th October 2014 5.5 g ^{15}N-labelled N as $K^{15}NO_3$, (46 atom % in 500 ml DDW) was spray-applied to the three 1 × 0.33 m treatment areas (Figs. 5.3 and 6.1). This is equivalent to 55 kg N ha^{-1}, which, based on five to six fertiliser applications between February and October is *ca.* 300 kg N ha^{-1} year^{-1}, within the range (140–340 kg N ha^{-1} year^{-1}) recommended by Defra for dairy-grazed grasslands [53]. About an hour later (to avoid contamination with any wind-borne $^{15}NO_3^-$ spray), three rainfall collectors of the designs depicted in Fig. 6.3

were installed in the experimental area and weighed down with rocks. Collectors were designed to prevent evaporative fractionation of rainwater based on United States Geological Survey (USGS) recommendations [54]. An annotated photograph of the final field experimental set-up is shown in Fig. 6.4.

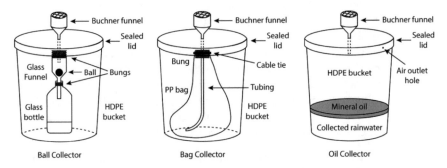

Fig. 6.3 Designs of the three rainfall collectors installed at the site to collect rainwater for N concentration and water isotopic analysis. HDPE; high-density polyethylene; PP; polypropylene

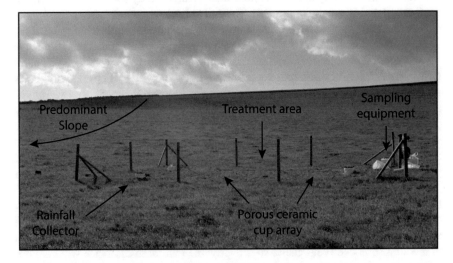

Fig. 6.4 Annotated photograph of the plot-scale field experiment. The experimental area was fenced off from the wider field in order to prevent cows from accessing the area and adding any considerable amounts of excreta N and further complexity

6.3.3 Sampling

6.3.3.1 Soil

Soil was sampled from the treated areas to a depth of 10 cm as described in Chap. 5 for the 64-day laboratory-field comparison experiment and some of the data generated from these samples will be briefly re-presented in this chapter. In addition, soil was sampled to a depth of 20–30 cm using a 2 cm diameter hand auger from the area around each porous ceramic cup (including in the treated area) at 4-weekly intervals throughout the 32-week field experiment. Sampling points were marked and holes refilled with soil collected from nearby in the field.

6.3.3.2 Foliage

Foliage (of grass-clover ley, predominantly grass) was sampled by hand from an area around each porous ceramic cup at 4-weekly intervals throughout the 32-week field experiment. At each sampling date the full length of above ground plant matter was sampled.

6.3.3.3 Rainfall

Rainfall was automatically telemetered by a rain gauge located approximately 200 m from the experimental site. Rainwater was collected at the experimental site in the three different vessels (Sect. 6.3.2; Fig. 6.3) and sampled every 2 weeks for the first 24 weeks of the experiment and then every 4 weeks until the end of the 32-week experiment (2nd June 2015). At each sampling, the entire contents of each collector was poured into separate acid washed and ultrapure water (resistivity of 18.2 MΩ cm) rinsed high-density polyethylene (HDPE) bottles. Samplers were cleaned and mineral oil was replaced at each sampling. The bottle and bag in the respective collectors were replaced every 4 weeks.

6.3.3.4 Soil Water

Soil waters were sampled from buried porous ceramic cups from the treated areas at weekly intervals for the first 4 weeks and thereafter, and from elsewhere on the site, at two weekly intervals for the first 24 weeks and then every 4 weeks until the end of the experiment (32 weeks). The sampling tube was clipped shut and a suction of *ca.* 60–70 kPa was applied to the vacuum tube using a clean, empty syringe. The vacuum tube was then clipped shut and the porous ceramic cup left under vacuum for 2 h. Soil waters were sampled by unclipping both tubes and withdrawing water from

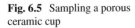
Fig. 6.5 Sampling a porous
ceramic cup

the sampling tube using a clean, empty syringe (Fig. 6.5). Sample volumes were
recorded and samples stored in individual, 30 ml leak-proof polystyrene containers.

6.3.3.5   Groundwater

Groundwater was sampled every 8 weeks from a borehole located *ca.* 100 m away
from the experimental area, to the south-east (Fig. 5.1). At each sampling the full
volume of the borehole was pumped out and allowed to refresh before a sample was
taken (Fig. 6.6). Groundwater samples were collected and stored in acid washed and
ultrapure water rinsed HDPE bottles.

Fig. 6.6 Groundwater
sampling

6.3.4 Analytical Methods

6.3.4.1 Sample Preparation and Summary of Analyses Conducted

Figure 6.7 summarises the processing and analyses carried out by sample type. In some cases, however, only a subset of the samples underwent each/all analyses. In particular, because porous ceramic cups yield only small volumes of sample (22 ml at most in this experiment), it was not possible to conduct all of the desired analyses on every sample. Porous ceramic cup samples from within the treated area and downslope locations most likely to receive some of the applied $^{15}NO_3^-$ were prioritised for $\delta^{15}N$ analysis in order to trace applied ^{15}N transport rates. In addition, based on ceramic cup $\delta^{15}N$ results, preliminary and spot-check analyses, and due to time constraints, not all of the soil and foliage samples collected were analysed, instead potentially interesting transects (in time and space) were chosen. Finally, although included in Fig. 6.7, no further soil samples (to those already discussed in Chap. 5) and only a very small selection of foliage samples were analysed for AA concentrations and $\delta^{15}N$ values. This is partly because the procedure is very time-consuming, but also because the data discussed in Chap. 5 is already available for comparison and ^{15}N incorporation into AAs is not the focus of the work presented in this chapter.

Soil and foliage samples were frozen within 3 h of collection and stored frozen until freeze-drying. Freeze-dried soil samples were gently ground using a pestle and mortar and roots separated out and stones removed. Roots were cleaned of soil and as grinding (even with the assistance of liquid N_2) proved ineffective, roots were very finely chopped with washed and solvent-rinsed scissors. Freeze-dried foliage

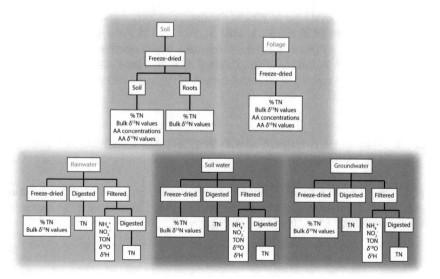

Fig. 6.7 Schematic summary of sample processing and analyses

was similarly difficult to grind and foliage samples were also very finely chopped to generate representative bulk samples.

Immediately after sampling aliquots (2 ml) of rainwater from the bottle and bag collectors, selected soil water samples and groundwater samples were syringe-filtered through Whatman^{TM} PURADISC^{TM} 25 PP 0.45 μm polypropylene disposable filters (previously tested to be isotopic fractionation-free) and sealed in air-less 2 ml glass sample vials for hydrogen (H) and oxygen (O) isotopic analysis [55]. Rainwater from the oil collector was separated from mineral oil using an acid and ultrapure water-rinsed glass separating funnel within 4 h of collection and a 2 ml sample immediately syringe filtered and sealed as above for H and O isotopic analysis. Within 4 h of collection portions of each rainwater sample, selected soil water samples and groundwater samples were hand filtered through Whatman^{TM} 0.45 μm cellulose nitrate membrane filters. Although Whatman^{TM} 0.45 μm cellulose nitrate membrane filters are advertised as having 'very low extractable levels' and are commonly used in such applications, clear NO_3^- contamination was observed in some samples (apparently at random) following filtration and as a result, filtered aqueous sample NO_3^- and TN concentrations are treated with some caution. All aqueous samples were refrigerated at 4 °C and NH_4^+, NO_2^-, total oxidised nitrogen (TON) and TN concentrations determined for filtered (and unfiltered for TN) samples within 48 h (or occasionally 72 h) of sample collection (Sect. 6.3.4.4).

In order to determine the δ^{15}N values of constituents dissolved in soil water samples, which could only be collected in small volumes (porous ceramic cups yielded 22 ml at most in this experiment), and therefore relatively small amounts of N, recorded volumes of selected samples were freeze-dried and the % TN content and δ^{15}N value of the residue produced determined by EA-IRMS (Sects. 2.5.2 and 6.3.4.2). Although available in greater volumes, the low N concentration of rainwater samples resulted in similar sample limitations and these samples, as well as groundwater samples (for consistency) were also freeze-dried to assess suspended and dissolved N δ^{15}N values. Where determined for the same sample, aqueous TN concentrations and residue % TN contents are of the same order of magnitude and generally agree well.

6.3.4.2 Percentage Nitrogen Contents and Bulk Nitrogen Isotopic Compositions

The % TN contents and bulk δ^{15}N values of soil, roots, foliage and freeze-dried rainwater, selected soil water and groundwater samples were determined at the Lancaster Node of the NERC LSMSF, as described in Sect. 2.5.2.

6.3.4.3 Amino Acid Concentrations and Nitrogen Isotopic Compositions

Foliage samples (15 mg) were lipid extracted three times by sonication (20 min) with 5 ml chloroform and MeOH (2:1 v/v) and allowed to dry overnight prior to hydrolysable AA extraction. Hydrolysable AAs were extracted from soil and foliage as described in Sect. 2.4, except that 50 µl of Nle (800 µg ml^{-1} in 0.1 M HCl) was added to the lipid extracted foliage samples. AA concentrations were determined by GC (Sect. 2.5.3) and AA δ^{15}N values by GC-C-IRMS (Sect. 2.5.5).

6.3.4.4 Aqueous Sample Nitrogen Concentrations

The dissolved NH_4^+, NO_2^- and TON concentrations of filtered water samples were determined by air-segmented continuous flow colourimetry using a Bran+Luebbe Autoanalyser 3 controlled by Bran+Luebbe Autoanalyser Control and Evaluation Software (AACE) for Windows.

Ammonium concentrations were determined by reaction to an indophenol via a modified version of the Berthelot reaction [56, 57]. Three reagent mixtures were employed for the NH_4^+ channel: (i) sodium hydroxide (3.5 g) and sodium dichloroisocyanurate (SDIC) dihydrate (0.2 g) dissolved in 100 ml ultrapure water; (ii) sodium salicylate (75 g) dissolved in ultrapure water (250 ml); and (iii) a complexing and catalytic reagent mixture containing disodium ethylenediaminetetraacetate (EDTA; 7.5 g), trisodium citrate dihydrate (60 g) and sodium nitroprusside (0.25 g) dissolved in 500 ml ultrapure water and with 1.5 ml Brij-35 subsequently added. Hydrolysis of SDIC produces hypochlorous acid, which reacts with NH_3 to produce monochloramine (NH_2Cl; [56]). Monochloramine and salicylate react to produce 5-aminosalicylate and oxidation (possibly by a pentacyanoferrate (III) complex formed from sodium nitroprusside due to the presence of chlorine) and oxidative coupling of this compound to another salicylate molecule produces indosalicylate, a blue indophenol compound [56, 57]. The final solution, however, is a blue-green colour due to excess, yellow sodium nitroprusside and maximum absorption occurs at a wavelength, λ, of $ca.$ 660 nm [56]. Sodium salicylate was used as the phenolic compound for the reaction because it is much less toxic than phenol itself, sodium nitroprusside acts as a catalyst (via a range of complex cyanides), while disodium EDTA and trisodium citrate are complexing agents which prevent precipitation of calcium and magnesium hydroxides [56, 57].

Nitrite concentrations were determined via a modified version of the Griess reaction [58]. Samples were combined in the autoanalyser with a reagent mixture containing HCl (75 ml), sulfanilamide (5 g) and N-1-napthylethylenediame dihydrochloride (NED; 0.25 g) in 500 ml ultrapure water. Nitrite reacts with the primary amine group of sulfanilamide under the acidic reaction conditions to form a diazonium cation [59, 60]. The diazonium cation couples to NED in the $para$ position forming a pink azo dye with a strong absorbance at 545 nm [59, 60].

Total oxidised nitrogen ($NO_2^- + NO_3^-$) concentrations were determined as NO_2^- as above following reduction of NO_3^- to NO_2^- using a copperised cadmium column [61–63]. As well as the acidic sulfanilamide and NED solution described above, the TON channel also employs a pH 8.5 ammonium chloride solution—5 g NH_4Cl and ca. 0.5 ml 25% NH_3 for pH adjustment in 500 ml ultrapure water with 0.25 ml Brij-35 and 0.5 ml 1.25 mM copper (II) sulfate solution added. The NH_4Cl solution maintains pH conditions such that NO_3^- is not reduced further than to NO_2^- and prevents the precipitation of cadmium (II) hydroxides in the column, which would deactivate it [61, 63]. Nitrate concentrations were calculated by subtracting NO_2^- concentrations from TON concentrations.

The TN concentrations of filtered and unfiltered samples were also determined by air-segmented continuous flow colourimetry, as TON following microwave-assisted persulfate digestion [64]. The digestion reagent consisted of 2.375 g NaOH and 11.25 g potassium persulfate ($K_2S_2O_8$) dissolved in 250 ml ultrapure water (by stirring at 40 °C). It was stored at 4 °C and refreshed every 2 days. Digestion reagent (5 ml) was added to each sample (10 ml) in acid washed and ultrapure water rinsed MARSXpress^{TM} digestion vessels (CEM) and the tubes were inserted (symmetrically) in the microwave turntable. The MARS 6 digestion unit (CEM) was programmed to ramp to 160 °C at 14 °C min^{-1}, hold for 40 min and then cool for 20 min. The temperature of each digestion vessel was monitored throughout in order to ensure none vented during digestion—this did not occur for any samples in this work.

Analyte concentrations were determined using standard calibration curves generated at the start of each run on the autoanalyser (standard concentrations ranged from 0.005 to 0.500 mg l^{-1} for NH_4^+ and NO_2^- and 0.050–10.000 mg l^{-1} NO_3^- for TON; corresponding persulfate digested standards were used with digested samples). Analytical accuracy was determined as percentage recovery of standards interspersed amongst samples in each run:

$$\% \text{ Recovery} = \frac{(C_m - C_b)}{C_e} \times 100 \qquad (6.1)$$

where C_m is the concentration measured, C_b is the measured concentration of blanks and C_e is the expected concentration of the standard. Mean percentage recoveries across runs over the course of the experiment and their associated SEs are shown in Table 6.1.

Table 6.1 Mean standard percentage recoveries across runs over the course of the experiment

	NH_4^+	TON	NO_3^-
Mean/%	98.0	101.2	99.8
SE (n = 37)	1.2	1.5	0.5

6.3.4.5 Aqueous Sample Hydrogen and Oxygen Isotopic Compositions

The H and O isotopic composition of water samples were initially determined using a Picarro L1102-*i* water isotope analyser. This instrument employs wavelength-scanned cavity ring down spectroscopy (WS-CRDS) to determine the δ^2H and $\delta^{18}O$ values of water samples with an advertised sensitivity and precision of $\delta^2H < 0.5‰$ and $\delta^{18}O < 0.1‰$. Water samples were run in groups of five, bracketed by three house standards of known H and O isotopic composition (previously standardised against the international standards VSMOW2 [δ^2H 0 ± 0.3‰; $\delta^{18}O$ 0 ± 0.02‰], SLAP2 [δ^2H −427.5 ± 0.3‰; $\delta^{18}O$ −55.50 ± 0.02‰] and GISP [δ^2H −189.5 ± 1.2‰; $\delta^{18}O$ −24.76 ± 0.09‰]) which were used to determine sample H and O isotopic compositions by linear regression. Each sample was injected into the instrument 10 times (700 nl per injection) using a CTC Analytics GC Pal autosampler to achieve a water vapour concentration of 20,000 ± 3000 ppm in the cavity. In order to avoid carryover, only the last six of the 10 injections were accepted (and only provided a water vapour of 20,000 ± 3000 ppm was achieved).

Unfortunately, about 12 weeks into the 32-week experiment, however, this instrument failed and the remaining samples were subsequently analysed in the Department of Archaeology at the University of Reading using a Picarro L2120-*i* water isotope analyser. A similar 10 injections per sample procedure was followed, except that only the last five of the 10 injections were accepted and ChemCorrect™ accept/reject flagging software [65] was used. Stretch corrections were carried out using VSMOW2 and SLAP2 standards and drift corrections were applied using within run IAEA standards. In order to check that dissolved constituents, particularly in the soil water samples, did not interfere with δ^2H and $\delta^{18}O$ values measured by WS-CRDS, as is sometimes a concern (e.g. [65–67]), sample δ^2H and $\delta^{18}O$ values were also determined using a Thermo DeltaV Plus IRMS fitted with a SMART Thermo chemical elemental analyser liquid autosampler. Agreement between the two methods was excellent (δ^2H $m = 0.995$, $R^2 = 0.999$; $\delta^{18}O$ $m = 0.990$, $R^2 = 0.980$).

6.4 Results and Discussion

6.4.1 Background Monitoring

6.4.1.1 Rainfall Intensities, Volumes, Nitrogen Concentrations and Nitrogen Isotopic Compositions

In the absence of irrigation, rainfall supplies the infiltrating water that facilitates transport of surface-applied NO_3^- through the surface soil layers. The volume of rainfall that fell in each 24 h period (from 09:00 on 17th October 2014) is shown in Fig. 6.8 alongside the cumulative total volume of rainfall over the course of the field experiment. A minimum of at least 0.2 mm of rainfall was recorded on 69% of

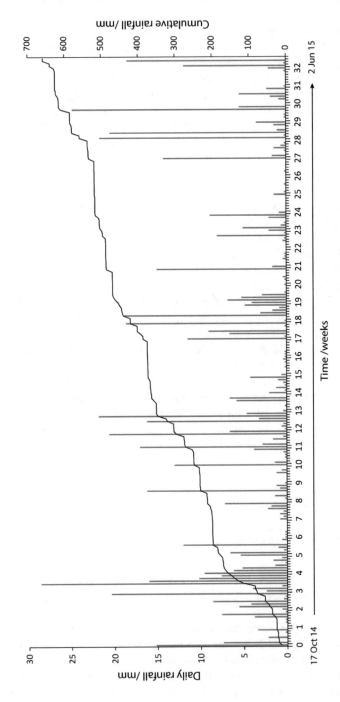

Fig. 6.8 Daily rainfall volumes (24 h periods from 09:00 to 08:59 on each day) and the cumulative rainfall over the course of the 32-week field experiment

days during the field experiment and of these days, the mean daily rainfall is 4.2 mm and the highest, 28.6 mm (on the 10th November 2014, 3–4 weeks after the start of the experiment). The cumulative total volume of rainfall that fell between treatment application and the end of the experiment on 2nd June 2015 is 658.2 mm.

The volume of water available (after evapotranspiration) affects both NO_3^- transport velocities and the dilution of NO_3^- as it travels through the soil. Soil water potentials were not measured directly over the course of the field experiment, but estimates of SMDs were generated by Wessex Water's Catchment Science Team using IRRIGUIDE, a field-scale water balance model developed by ADAS [50–52]. As would be expected, the combination of autumn/winter rainfall and decreasing evapotranspiration leads to decreasing modelled SMDs during the first part of the field experiment (Fig. 6.9a). On 9th November 2014, 3–4 weeks after the start of the experiment, modelled SMDs fall below 0 mm, indicating that the soil has reached

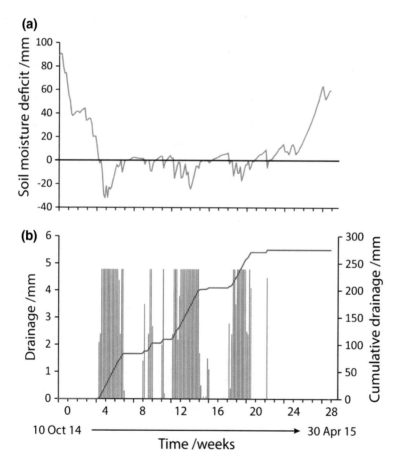

Fig. 6.9 IRRIGUIDE modelled **a** daily SMDs and, **b** daily and cumulative drainage during the field experiment

field capacity. Modelled SMDs then remain at or below 0 mm until early March 2015, *ca.* 20 weeks after the start of the experiment.

The IRRIGUIDE model also estimates daily drainage volumes (Fig. 6.9b). Modelled drainage begins on 9th November 2014 (3–4 weeks after the start of the experiment) and continues, interspersed with periods of no drainage, until 2nd March 2015 (19–20 weeks after the start of the experiment). Modelled SMDs then begin to rise and the model predicts only 4.5 mm of further drainage on 14th March 2015, following a large volume of rainfall on 13th March 2015 (21 weeks after the start of the experiment). In total, the IRRIGUIDE model estimates that cumulative drainage over the course of the field experiment was 275.8 mm.

Until early November 2014 when the site is thought to have reached field capacity, vertical, gravity driven percolation is likely to have dominated $^{15}NO_3^-$ transport. In addition, rainfall intensity is also important in determining how rainwater infiltrates the soil and the subsequent transport velocities of dissolved NO_3^-. During this experiment, the mean hourly rainfall rate was 0.9 mm h^{-1}, the median hourly rainfall rate was 0.4 mm h^{-1} and the highest hourly rainfall rate recorded was 13.6 mm h^{-1} (on 14th May 2015 between 08:00 and 09:00, 29–30 weeks after the start of the experiment). The relatively steady supply of low intensity rainfall is likely to have promoted surface infiltration over surface run-off.

The effect of rainfall N inputs in reducing measured soil ^{15}N enrichments has already been discussed in Sect. 5.5.2 and also applies to roots, foliage and soil water ^{15}N enrichments. Rainfall N concentrations are shown in Fig. 6.10 and are relatively low (0.3–1 mg l^{-1}), with the exception of 2–3 peaks (to maximum TN concentrations of 3.8 mg l^{-1}), which interestingly coincide with periods when dairy

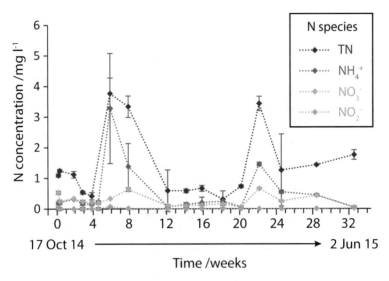

Fig. 6.10 Mean rainfall N concentrations in rainwater collected in the three different rainfall collectors. Error bars are ± SE (n = 3)

cows were grazing in Little Broadheath, surrounding the experimental site. Generally, NH_4^+ predominates and NO_2^- concentrations are lowest (commonly 0.0 mg l^{-1}), but 'organic N' contributes considerably to TN in the latter stages of the experiment. Agriculture, and in particular animal excreta, represents a major global source of atmospheric NH_3 [68, 69] and (intensive) dairy operations are known to lead to the dominance of reduced N in local/downwind deposition [69, 70].

Volume-weighted TN contents indicate that *ca.* 780 mg N m^{-2} was deposited in rainfall over the course of the field experiment. This would be equivalent to *ca.* 1.3 g N m^{-2} year^{-1} of rainfall N deposition, lower than that estimated in Sect. 5.5.2, but still higher than the 2012 CBED wet *and* dry estimate of 1 g N m^{-2} year^{-1} [71] and again, this is thought to be due to the regular close proximity of large numbers of cows to the sampling area. The N isotopic composition, as well as the amount of N deposited in rainfall affects the final 'mixed' N isotopic composition of soil, roots, foliage and soil water, but likely only to a very small extent in this experiment where the effect of the applied highly labelled $^{15}NO_3^-$ will dominate. Rainfall N isotopic compositions were quite variable and $\delta^{15}N$ values ranged from -3.7 to 11.6‰ (mean: 3.5‰). These results fall within the ranges reported for NH_4^+ and NO_3^- in precipitation (Fig. 1.7) and were highest in November and December, matching to some extent the reported pattern of higher cool season rainfall NO_3^- $\delta^{15}N$ values [72].

6.4.1.2 Soil Water Volume Yields and Nitrogen Concentrations

Porous ceramic cup yield volumes range from 0 to 22 ml, generally peaking from November to March, as would be expected based on modelled SMDs (Fig. 6.9a). Particular porous ceramic cups yielded consistently 'high' or 'low' volumes, reflecting the local soil conditions surrounding each porous ceramic cup. Accordingly, there are no patterns in volume yields by row or column, but a graph of mean yield volumes by depth reveals greater yield volumes from shallower porous ceramic cups (Fig. 6.11).

Soil water TN concentrations range from 0.4–123 mg l^{-1} and as expected, are dominated by NO_3^-, while NH_4^+ and NO_2^- concentrations are consistently low, ranging from 0.0 to 1.0 mg l^{-1} and 0.0 to 0.7 mg l^{-1}, respectively. Patterns in TN concentrations over the course of the field experiment vary and peak TN concentrations occur at different times for individual porous ceramic cups. Similarly to yield volumes, particular porous ceramic cups tended to have consistently 'high' or 'low' TN and NO_3^- concentrations relative to one another (and not usually dependent on coinciding 'low' or 'high' volume yields). Volume-weighted TN contents range from 4.2 to 2330 mg and for most porous ceramic cups, patterns in TN concentrations and contents are very similar, except towards the end of the experiment when high TN concentrations are observed in low volume samples for some porous ceramic cups. This effect can be observed in Fig. 6.12 in which TN concentrations and volume-weighted TN contents over the course of the experiment are shown (by depth and distance from the treated area).

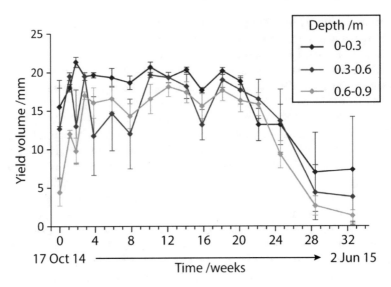

Fig. 6.11 Mean porous ceramic cup volume yields by depth. Error bars are ± SE (n = 3, 4 and 14 for 0–0.3, 0.3–0.6 and 0.6–0.9 m depth ranges, respectively)

   Volume/dilution effects aside, the TN concentration/content of individual porous ceramic cup samples depends not simply on soil water TN concentrations, but on several other factors summarised in Sect. 6.3.2, including soil structure, the origin (radius and pore size) of water sampled for each porous ceramic cup (which will vary with installation depth) and the local environment of the porous ceramic cup in a heterogeneous soil system [40, 41, 43–45, 47]. Thus, differences in TN concentration/content patterns between different porous ceramic cups in the presumably heterogeneous soil of the field experiment are expected. That particular porous ceramic cups yield consistently 'high' or 'low' concentration samples suggests that a constant difference between the ceramic cups, for example, depth or another function of local environment (e.g. surrounding preferential flow pathways) is important.

   Figure 6.12 reveals that soil water TN concentrations/contents (and fluctuations in soil water TN concentrations/contents) generally decrease with increasing sampling depth, which is consistent with other studies (e.g. [73]). It suggests that only a portion of the NO_3^- present at shallower depths ever leaches beyond 0.6 m (probably due to plant uptake) and unsurprisingly, that variability is attenuated (by dispersion and diffusion; [38]) during leaching. Despite variability between individual porous ceramic cups, results grouped as in Fig. 6.12 reveal that (with the exception of the large peak in TN concentrations/contents in the treated area at 4 weeks) peak TN concentrations for shallower porous ceramic cups in each row occur in samples from 27th December 2014, 10 weeks after treatment application. It is estimated, based on IRRIGUIDE modelling, that 110 mm cumulative drainage had occurred by this time. A similar pattern of peak NO_3^- concentrations after *ca.* 100–150 mm cumulative drainage has been observed in previous porous ceramic cup leaching experiments,

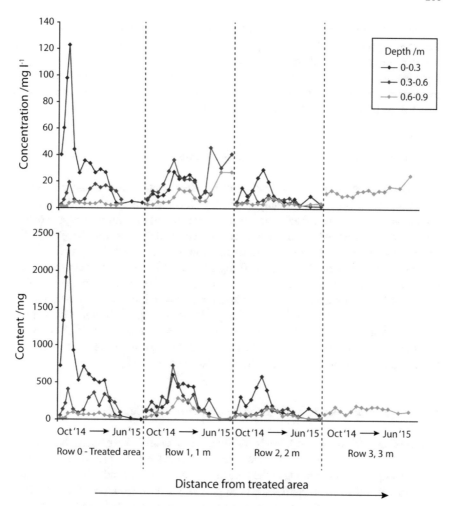

Fig. 6.12 Mean soil water TN concentrations and volume-weighted TN contents over the course of the 32-week field experiment by depth and distance from the treated area (row). The maximum SD of mean TN concentrations for grouped measurements (i.e. from the same depth range and row) is 26.5 mg l^{-1} and of mean TN contents is 396 mg

but soil structure in these experiments was poorer and cups were generally installed at greater depths (>1 m; [40, 46]). In a more analogous experiment involving porous ceramic cups buried at depths of 0.35 and 0.6 m in a structured dual porosity silt loam soil, Wang et al. [74] also recorded similar NO_3^- concentration profiles. As in Fig. 6.12, higher peak concentrations were observed earlier (following *ca.* 90 mm [actual] cumulative drainage) in samples from shallower porous ceramic cups.

The large peak in TN concentrations/contents at 4 weeks for shallower soil water samples from the treated area does not match the pattern of the other rows in this experiment or the literature discussed. Initial flushes of NO_3^- at the onset of drainage are not uncommon (e.g. [75]), but in this case, it is thought to result directly from treatment application (5.5 g ^{15}N-labelled N as K^{15}NO$_3$). The shallowest porous ceramic cup in the treated area is buried only 0.2 m below the surface and is likely to have captured a large proportion of the applied N relatively quickly. This is supported by the fact that this pattern does not occur in the other rows and by very high treated area soil water residue δ^{15}N values at this time (Figs. 6.15 and 6.17).

6.4.1.3 Groundwater Nitrogen Concentrations

Groundwater represents the end reservoir in which water collects and into which lost/excess fertiliser N is eventually leached. Compared with rainfall and soil water, groundwater is a large, mixed body in which water and contaminant inputs over a number of years are averaged. As expected due to its mobile nature, NO$_3^-$ was the major contributor to TN concentrations in groundwater (borehole) samples. Mean TN and NO$_3^-$ concentrations of 11.0 mg l^{-1} and 9.7 mg l^{-1}, respectively, were recorded over the course of the 32-week experiment in the nearest borehole. Ammonium and NO$_2^-$ concentrations were persistently low, with mean concentrations of 0.05 mg l^{-1} and 0.005 mg l^{-1}, respectively. Variation in groundwater N concentrations was also low, but TN and NO$_3^-$ concentrations were generally highest in late winter/early spring, as is consistent with the pattern in groundwater NO$_3^-$ concentrations previously observed at the site and elsewhere [76, 77].

6.4.1.4 Soil, Roots and Foliage Percentage Nitrogen Contents

The imperceptible impact of small N additions on bulk soil % TN contents has already been discussed in Chaps. 3–5 and such measurements therefore provide information on the background functioning of the field plot, rather than the N addition specifically. Bulk soil % TN contents across the field plot vary over the course of the 32-week field experiment, but as in the laboratory incubation experiments, show no overall trend (Fig. 6.13a). Root and foliage % TN contents were determined in the treated area only and display more substantial changes during the experiment, with root % TN contents being relatively stable/rising slightly for the first 20 weeks of the experiment and then declining (Fig. 6.13b). Foliage % TN contents decline almost consistently throughout the experiment (Fig. 6.13c).

The reason for the slow rise and fall in root % TN contents is not definitively known. Root N uptake and storage/transport will be related to some extent to soil N availability [31, 78], which based on bulk soil % TN contents, does not appear to change over the course of the experiment (Fig. 6.13a). Bulk soil % TN contents are not, however, necessarily indicative of 'bioavailable' N concentrations in the soil and the decline in root % TN contents may be due to declining soil available N (as no further fertiliser was applied) during spring when plant N uptake and growth is expected to be increasing [23, 38, 79]. Soil water TN concentrations/contents (Sect. 6.4.1.2), which are dominated by NO$_3^-$ and provide a better indication of

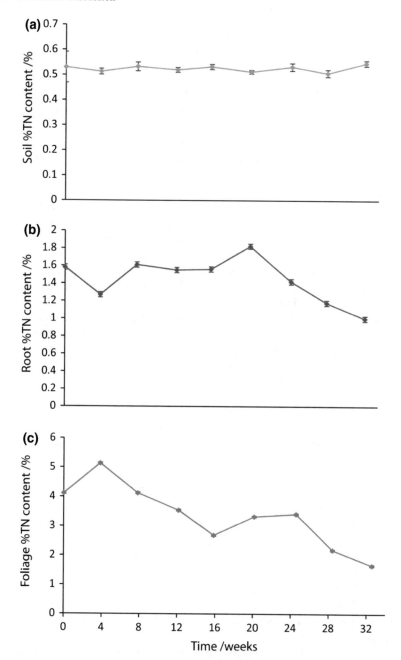

Fig. 6.13 Bulk % TN content: **a** soil, error bars are ± SE of at least three samples at each time point, **b** roots, error bars are the maximum SD of QC and duplicate sample measurements (±0.03%) and, **c** foliage, error bars are the maximum SD of QC and duplicate sample measurements (±0.03%)

bioavailable N, generally rise for the first 10 weeks of the experiment before declining (Fig. 6.12).

Alternatively, increased root growth/extension during spring may have resulted in lower root % TN contents by mass if (even increased) N uptake did not keep pace with root growth. Thus, the total amount of N contained in roots may have risen during this time, but % TN contents by mass will still have declined. As root samples in this experiment were taken from soil cores, rather than sampled specifically, it is difficult to accurately assess changes in root biomass over the course of the experiment (any changes in root extension, for example, would not have been recorded), but there was no obvious increase in root density in soil cores over the course of the experiment.

Finally, root % TN contents may have declined due to increased translocation of N stored in roots to shoots during spring. While declining foliage % TN contents do not support this suggestion, it is still possible as increasing foliage biomass over the course of the experiment and particularly during spring may have masked increases in overall foliage N contents, leading to lower measured % TN contents by mass. It is also possible that declining foliage % TN contents reflect decreasing N availability (as no further fertiliser was applied and soil water N concentrations/contents declined; Fig. 6.12) and turnover of high % TN content foliage over the course of the experiment. The initial rise (to $t = 4$ weeks after treatment application) may be due to foliar absorption of the spray-applied ^{15}NO$_3^-$ [80] and this is supported by very high foliage δ^{15}N values at this time (Fig. 6.13c).

6.4.2 ^{15}N Enrichments and Partitioning in the Treated Area

6.4.2.1 ^{15}N Enrichment of Soil, Roots and Foliage Within the Treated Area Over the Course of the 32-Week Experiment

Following ^{15}NO$_3^-$ application, ^{15}N enrichment was very evident in soil, roots and foliage sampled from the treated area throughout the experiment (Fig. 6.14). The mean N isotopic composition of bulk soil in the treated areas was elevated from pre-application δ^{15}N values of $6.4 \pm 0.2\%$ to $558 \pm 77\%$ at $t = 3$ h, the first sampling point, and maximised at $t = 12$ h to $6440 \pm 2290\%$ (2.66 ± 1.19 atom %; Fig. 6.14a). Bulk soil δ^{15}N values then fell relatively quickly until $t = 16$ days and fluctuated around ca. 200‰ for the rest of the monitoring period. Reasons for the pattern in bulk soil δ^{15}N values over the first 64 days are discussed in detail in Sect. 5.5.2 and thereafter, relatively stable bulk soil δ^{15}N values indicate that the system had reached a temporary equilibrium.

The lower early sampling frequency of roots and foliage prevents comparison of the timing of initial ^{15}N partitioning between soil and roots/foliage, but the earlier peak in bulk foliage δ^{15}N values (to 39,100‰ or 12.9 atom % at $t = 4$ weeks), compared with bulk root δ^{15}N values (to 11,400‰ or 4.35 atom % at $t = 8$ weeks) supports suggestions made in Sects. 5.5.2 and 6.4.1.4 of foliar ^{15}NO$_3^-$ absorption [80]. Following these peaks, both bulk root and foliage δ^{15}N values decline over

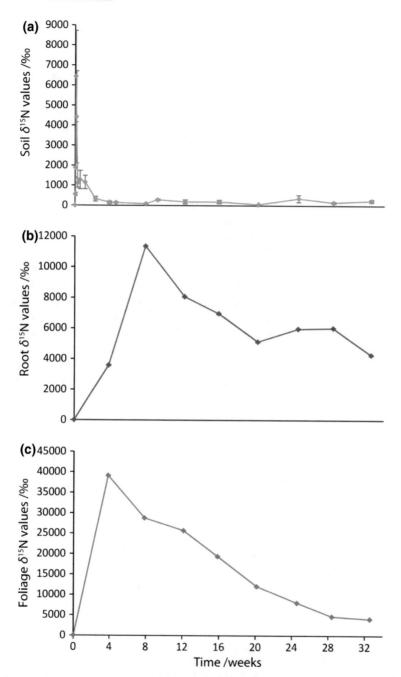

Fig. 6.14 Bulk $\delta^{15}N$ values in the treated area over the course of the 32-week field experiment: **a** soil, error bars are $\pm$ SE of three samples at each time point, **b** roots, error bars are the maximum SD of QC and duplicate sample measurements ($\pm 28.9‰$) and, **c** foliage, error bars are the maximum SD of QC and duplicate sample measurements ($\pm 28.9‰$)

the remainder of the experiment, to 4270 and 4090‰ (1.90 and 1.84 atom %), respectively after $t = 32$ weeks.

Neither bulk root nor foliage δ^{15}N values 'stabilised' like bulk soil δ^{15}N values during the 32 weeks, probably due to the faster, continual turnover of roots and foliage compared with the bulk soil. Maximum recorded bulk δ^{15}N values are highest for foliage and lowest for soil and as values were determined most frequently for soil, this is unlikely to result from differences in sampling frequency (Fig. 6.14). Due to differences in % TN content and pool size, however, this does not confirm that more of the applied ^{15}NO$_3^-$ was partitioned into foliage than soil.

6.4.2.2 ^{15}N Enrichment of Soil Water Residues from Porous Ceramic Cups Underneath the Treated Area Over the Course of the 32-Week Experiment

Bulk soil water residues sampled from the three porous ceramic cups below the treated area following application similarly show considerable ^{15}N enrichment (Fig. 6.15). The δ^{15}N values of individual porous ceramic cup soil water residues are likely to be affected by some of the same factors as soil water TN concentrations/contents (Sect. 6.4.1.2), but also by the interaction of these factors with the applied ^{15}NO$_3^-$ and thus differences in the dilution of the ^{15}N signal with differences in water origin, cup depth and local environment. This appears to have been obviated for the most part through the application of highly labelled ^{15}NO$_3^-$ and clear, single peaks in

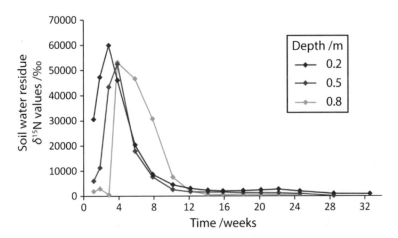

Fig. 6.15 δ^{15}N values of bulk soil water residues from the three porous ceramic cups under the treated area. Analytical accuracy and precision were monitored using a QC standard, the maximum SD of which was 8.0‰, but this is unlikely to be representative of the instrumental uncertainty in bulk soil water residue δ^{15}N values as many of these measurements were necessarily made on very small samples. Overall uncertainty in soil water residue δ^{15}N values is difficult to assess further, as there are no replicate samples

Table 6.2 Approximate ^{15}N leaching rates based the timing of peak soil water residue δ^{15}N values at each depth

	Depth/m		
	0.2	0.5	0.8
Leaching rate/m week^{-1}	0.07	0.13	0.21

soil water residue δ^{15}N values for each porous ceramic cup/depth indicates that the experiment was successful, at least to some extent, in tracing the vertical leaching of the applied ^{15}NO$_3^-$.

Unequivocal ^{15}N enrichment was evident at all depths from the first sampling point, only 1 week after treatment application, indicating that vertical leaching of some of the applied ^{15}NO$_3^-$ to 0.8 m depth occurred rapidly (i.e. leaching at rates greater than 0.7 m week^{-1}; Fig. 6.15). Maximum soil water residue δ^{15}N values of *ca.* 60,000‰ (18 atom %) were recorded at 0.2 m depth only 3 weeks after treatment application and lower peak δ^{15}N values of *ca.* 52,000 and 53,000‰ (16 and 17 atom %) were observed 1 week later at depths of 0.5 and 0.8 m, respectively. Approximate leaching rates based on the time taken for soil water residue δ^{15}N values to peak at each depth are shown in Table 6.2.

The decrease in the maximum ^{15}N enrichment of soil water between 0.2 and 0.5 m is likely due at least partly to the dispersion, diffusion and dilution of the applied ^{15}NO$_3^-$ during the leaching process. It could, however, also be due to differences (other than depth) between the local environment of the two porous ceramic cups and/or non-uniformity of the surface treatment application and finally, is likely partly an artefact of sampling frequency (i.e. the peak in soil water ^{15}N enrichment at 0.5 m was missed). This suggestion is supported by the earlier rise, but then slightly lower maximum ^{15}N enrichment of soil waters at 0.5 m depth compared with at 0.8 m 4 weeks after treatment application. Thus, that maximum δ^{15}N values at 0.5 and 0.8 m were both observed after 4 weeks is likely an artefact of sampling frequency.

That it took 3 weeks for peak δ^{15}N values to be recorded at 0.2 m depth (leaching rate of 0.07 m week^{-1}; Table 6.2) and then only 4 weeks for peak δ^{15}N values to be recorded at both 0.5 and 0.8 m depth (leaching rates of 0.13 and 0.21 m week^{-1}, respectively; Table 6.2), however, also suggests ^{15}N leaching rates were non-linear with depth (e.g. initial infiltration took longer than subsurface leaching) or varied in space (as the three porous ceramic cups have different local environments and are spread across the overall 1 m^2 treated area, across which the treatment may have been imperfectly distributed). Some difference in the vertical transport of ^{15}N past each porous ceramic cup is also supported by the incongruity of peak width with depth. While the peak in soil water residue δ^{15}N values is widest at 0.8 m as expected due to the longer travel time (and thus greater en route diffusion and differences in transit times), the peak at 0.5 m rather than 0.2 m depth is narrowest.

It took 12–14 weeks for the peaks in soil water residue δ^{15}N values to subside. Thus, in this experiment the major flush of autumn applied ^{15}NO$_3^-$ fertiliser passed through the soil in less than 3.5 months and before plant growth re-established in spring. Following the peaks, soil water residue δ^{15}N values indicate that a portion of the applied ^{15}N was retained in the soil/soil water for much longer and values remained high until the end of the sampling period, 32 weeks after treatment application. This 'retained' ^{15}N may well result from the portion of ^{15}NO$_3^-$ cycled through other N forms (i.e. organic N via microbial/plant uptake) in the soil before being released back to soil water either as dissolved organic ^{15}N or ^{15}NH$_4^+$ or converted back to ^{15}NO$_3^-$.

6.4.2.3   Mass-Balanced Partitioning of ^{15}N in and Under the Treated Area

Whilst valuable in evaluating the dynamics of the applied ^{15}N in the field system, the δ^{15}N values of system components (soil, roots, foliage and soil water) do not provide any indication of the quantitative partitioning or fate of the applied ^{15}N. Such assessments must take into account component pool sizes as well as ^{15}N enrichments. It can be challenging, however, to measure or accurately estimate the pool sizes of components in unconfined systems and to take into account differential component turnover rates in such measurements/estimates. In addition, measured N isotopic compositions are not necessarily representative of the entire pool. This is unfortunate, as accurate mass-balanced ^{15}N partitioning calculations are key to understanding the N balance of a system. As a result, even approximate, or 'back-of-the-envelope' calculations to gauge the relative mass-balanced partitioning of ^{15}N between different components of the system are worthwhile. The results of such a calculation for this experiment are shown in Fig. 6.16 as percentage partitionings of the applied ^{15}N into the soil, root, foliage and soil water pools of the field experiment treated area.

The mass-balanced percentages of ^{15}N partitioned into each component were calculated based on the following assumptions and pool size estimates:

(i)   Soil—the mass of soil in the 'system' was calculated based on a 3×0.33 m treated area, 0.1 m sampling depth and a soil density of 1 g cm^{-3}; measured bulk soil δ^{15}N values and % TN contents were used to calculate the excess moles ^{15}N per gram of soil (Sect. 2.6.4); and the total excess moles ^{15}N in the defined mass of soil was then calculated as a percentage of the total moles excess ^{15}N applied.

(ii)   Roots—the mass of roots per gram of soil was used to calculate the total root mass in the defined soil volume; measured bulk root δ^{15}N values and % TN contents were used to calculate the excess moles ^{15}N per gram of roots (Sect. 2.6.4); and the total excess moles ^{15}N in the defined mass of roots was then calculated as a percentage of the total moles excess ^{15}N applied.

(iii)   Foliage—the mass of foliage (grams dry matter) in the 3×0.33 m treated area was estimated from sub-samples collected at each sampling point; measured bulk foliage δ^{15}N values and % TN contents were used to calculate the excess moles ^{15}N per gram of foliage (Sect. 2.6.4); and the total excess moles ^{15}N in foliage at each sampling point was then calculated as a percentage of the total moles excess ^{15}N applied.

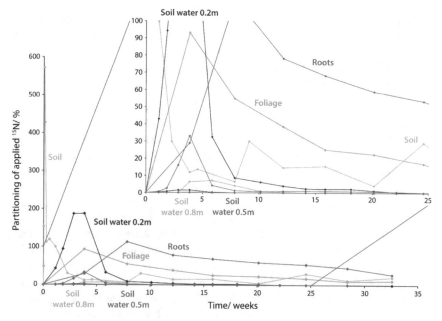

Fig. 6.16 Approximate percentage partitioning of the applied ^{15}N into the soil, root, foliage and soil water pools of the field experiment treated area

(iv) Soil water—since porous ceramic cups sample water from an unknown volume of soil ([41]; Sect. 6.3.2), it is difficult to assess the pool size which soil water samples and associated residue δ^{15}N values represent, accordingly calculations were performed for a sampling radius of 0.1 m (minimum) and 0.5 m (maximum; [41]); the water content of these soil volumes was estimated based on the WHC of the soil at field capacity and modelled SMDs; measured soil water residue δ^{15}N values, % TN contents and residue concentrations per millilitre were used to calculate the excess moles ^{15}N per millilitre of soil water; and the total excess moles ^{15}N in soil water in the minimum and maximum sampling radius at each sampling point was then calculated as a percentage of the total moles excess ^{15}N applied.

It is important to note that Fig. 6.16 shows the percentages of applied ^{15}N in each component pool at the point in time at which the sample was collected. As a result, the calculations only poorly capture the partitioning of ^{15}N into the dynamic soil water pool and do not provide accurate insights into the amount of applied ^{15}N which has been lost by leaching. In addition, it is immediately clear in Fig. 6.16 that the calculations have generated apparently impossible percentages of well over 100% of the applied ^{15}N partitioned into individual component pools at given time points, particularly near the start of the experiment (e.g. 571% of the applied ^{15}N is in the soil pool at $t = 12$ h). This is thought to be due to the uneven distribution of ^{15}N within this pool (i.e. the measured δ^{15}N values are not representative of the system's whole estimated/defined mass of soil; Sect. 5.5.2). As discussed in Sect. 5.5.2, the distribution of ^{15}N in the system is likely to improve with time, but is unlikely to become completely even during the experiment.

Taking into consideration the caveats and assumptions, it is possible to make several interesting observations based on Fig. 6.16:

(i)   The early peak in ^{15}N in soil (to 571% of that applied at $t = 12$ h) is likely due to uneven distribution of the applied ^{15}N in the soil and this is not observed to the same extent in other component pools due to lower sampling frequencies. From 2–32 weeks a mean of $ca.$ 17% of the applied ^{15}N remained in the soil. This is comparable with the percentage of $NO_3^- $-^{15}N recovered in the top 23 cm of soil under a grass-clover ley 48 days after treatment application (25%; [81]) and with the percentage of autumn-applied $NO_3^- $-^{15}N recovered in the top 10 cm of two arable soils by Holbeck et al. [82] over a similar period.

(ii) Partitioning of ^{15}N into roots is slower than that observed for any other pool, but by $t = 8$ weeks, after early (likely distribution-derived) peaks in other component pools have subsided, roots contain the highest proportion of the applied ^{15}N (113%). Partitioning of ^{15}N into roots remains greatest over the remainder of the experiment. This is a positive result and may partly be due to the efficiency of roots in taking up $^{15}NO_3^-$, but could also result from the ability of roots to capture $^{15}NO_3^-$ from up to 1 m depth and transport it back up to the surface, a process which is not considered in the simple calculations used to produce Fig. 6.16.

(iii) Foliage retains the next highest proportion of the applied ^{15}N until 24 weeks after treatment application, when ^{15}N in foliage falls below that retained in soil. This is due to declining foliage % TN contents (Fig. 6.13c) and δ^{15}N values (Fig. 6.14c) and despite increasing foliage dry matter over the course of the experiment.

(iv) Foliage AA extraction and compound-specific δ^{15}N analysis of a temporal transect of samples (at $t = 4$, 20 and 32 weeks after treatment application) indicates that, in addition, a declining percentage of the applied ^{15}N was present in hydrolysable foliage AAs (23.7, 6.8 and 3.3%, respectively) over the course of the experiment. These amounts represent an increasing proportion of the ^{15}N retained in foliage, however (26.2, 33.1 and 37.2% at $t = 4$, 20 and 32 weeks after treatment application, respectively), demonstrating the importance of N nutrition in plant AA and protein production.

(v)   Soil water percentage applied ^{15}N partitioning calculations based on porous ceramic cup sampling radii of 0.1 and 0.5 m are shown in Fig. 6.16 for each depth, but an estimated sampling radius of 0.1 m gives a very small soil water pool and these results are barely visible on the figure. A 0.5 m sampling radius gives a soil volume of just over 0.5 m^3 from which soil water may be sampled and this was used to calculate the soil water pool size (in millilitres) for each porous ceramic cup used in Fig. 6.16. Whether this is indeed a good/fair representation of the size of the soil water pool under the 1 m^2 (total) treated area is debatable and one simple alternative could be to use a 1 m^3 volume to account for all of the soil water up to 1 m depth below the 1 m^2 treated area. This would simply result in a doubling of the 0.5 m radii percentages shown in Fig. 6.16 and does not particularly affect any of the other observations made. In addition, this does not reflect the different depths (and therefore different soil water sample ^{15}N enrichments) of the porous ceramic cups well.

(vi) The percentages of applied ^{15}N partitioned into soil waters decreases with depth, as expected from decreasing soil water N concentrations (Fig. 6.12) and residue $\delta^{15}N$ values (Fig. 6.15) with depth and similarly, there is a lag in peak soil water ^{15}N percentages with depth. In terms of percentages of the applied ^{15}N, larger amounts of the applied ^{15}N are only present in soil waters up until 8 weeks (or 10 weeks at 0.8 m) after treatment application (compared with 12–14 weeks when considering soil water residue $\delta^{15}N$ values; Sect. 6.4.2.2). This suggests that the majority of leaching $^{15}NO_3^-$ has passed through the soil by this time, but as mentioned already, it is not possible to assess total or cumulative $^{15}NO_3^-$ leaching using these calculations or Fig. 6.16.

Alternatively, the total amount of ^{15}N in/passing through the transient soil water component of the system at each depth during the 32-week field experiment can be estimated from the integrals of ^{15}N leaching per week (in mol m^{-2} $week^{-1}$) against time (in weeks) plots (Fig. 6.17; Table 6.3). ^{15}N leaching (mol m^{-2} $week^{-1}$) was estimated by assuming soil water $\delta^{15}N$ values at each depth represent the ^{15}N

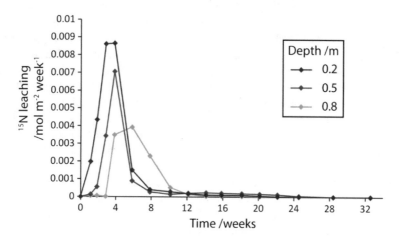

Fig. 6.17 Estimated ^{15}N leaching (mol m^{-2} $week^{-1}$) over the course of the 32-week field experiment

Table 6.3 Moles and percentages of ^{15}N leached over the course of the 32-week field experiment estimated by integration of ^{15}N leaching-time curves

	Depth/m		
	0.2	0.5	0.8
^{15}N leaching/mol m^{-2} $week^{-1}$	0.19	0.14	0.13
% applied ^{15}N/%	109	79.3	76.9

enrichment of NO_3^- leaching past that depth and all of the water in the soil volume above that depth (3×0.33 m by 0.2, 0.5 or 0.8 m depth; water content estimated based on the WHC of the soil at field capacity and modelled SMDs) moves at the rates estimated in Table 6.2. The percentages of applied ^{15}N leaching/present in the soil water component during the experiment are shown in Table 6.3 and are comparable to percentages of applied ^{15}N in other system components in the latter stages of the experiment (Fig. 6.16). These percentages of applied ^{15}N leaching do not accurately represent ^{15}N leaching *losses*, however, because: (i) not all of the ^{15}N present in soils to depths <1 m will leach as this ^{15}N is still physically accessible to plants (and soil microorganisms); and (ii) (estimated) drainage volumes (to >1 m; Fig. 6.9b) are smaller than the soil water volumes used in these calculations.

A simple method used to quantitatively estimate NO_3^- leaching in a field environment involves multiplying drainage volumes (either recorded from soil lysimeters or modelled) by measured respective soil water NO_3^- concentrations (e.g. [40, 46, 83, 84]). Performing this calculation using IRRIGUIDE modelled drainage (Fig. 6.9b) and ^{15}N concentrations (mol l^{-1}) recorded in the porous ceramic cup at 0.8 m gave an estimate of 0.0039 mol ^{15}N in total leaching in drainage over the season. This is equivalent to 2.3% of the total moles ^{15}N applied.

Considering the component partitioning results (Fig. 6.16) in comparison with the shorter term partitioning of ^{15}N into the hydrolysable soil AA pool (Chap. 5), 6.62% of the applied ^{15}N was partitioned into the hydrolysable soil AA pool over a period of approximately 9 weeks (Fig. 5.9; Table 5.4; and it is suggested that a similar level will be maintained for some time beyond this, before declining). This is comparable to the percentage of applied ^{15}N in soil water at around this time. What is particularly interesting about this percentage is that it is directly comparable to that of the WA $^{15}NO_3^-$ laboratory incubation experiment (6.39%; Fig. 4.11; Table 4.21), implying that $^{15}NO_3^-$ assimilation into hydrolysable AAs in the field was not limited by any of the additional competing (loss) processes operating under field conditions (e.g. leaching and plant uptake; Sect. 5.5.4). The percentage partitioning results shown in Fig. 6.16 support this suggestion as the bulk of $^{15}NO_3^-$ leaching appears to have occurred by this point, and has apparently not reduced $^{15}NO_3^-$ incorporation into hydrolysable soil AAs. The same is true of early/direct foliar ^{15}N uptake. It is also interesting that the percentage of applied ^{15}N partitioned into roots maximises after soil microbial $^{15}NO_3^-$ assimilation into hydrolysable AAs has plateaued (Figs. 5.9 and 6.16). While this may be an artefact of different sampling frequencies, it does fit with the short-term out-competition of plants by soil microbes (Sect. 6.1.3).

6.4.3   *Lateral Transport of ^{15}N from the Treated Area*

As well as investigating the dynamics of ^{15}N in different components of the field system, vertical $^{15}NO_3^-$ leaching and ^{15}N partitioning in the treated area, the field experiment was also designed to assess lateral $^{15}NO_3^-$ transport. The most important mechanism by which this is likely to have occurred in this experiment is via lateral soil water movement. Figure 6.18 shows soil water residue $\delta^{15}N$ values by depth and distance from the treated area over the course of the 32-week field experiment. On the logarithmic scale any $\ln(\delta^{15}N)$ value above *ca.* 1.9 may reasonably be attributed to ^{15}N enrichment as a result of the applied $^{15}NO_3^-$. Thus, some level of lateral ^{15}N transport appears to have occurred to all rows. In addition, ^{15}N enrichment is generally greatest at shallower depths.

Soil water residues from beneath the treated area are most ^{15}N-enriched and although this enrichment decreases with depth, it remains above that observed in other rows, suggesting that a greater proportion of the applied $^{15}NO_3^-$ was transported vertically compared with laterally. This is consistent with the low intensity of rainfall following treatment application, which will have promoted infiltration over surface run-off, and below field capacity soil MCs at the start of the experiment, which will have led to the dominance of gravity-driven vertical leaching over lateral flow (Sect. 6.4.1.1).

The soil water residue $\delta^{15}N$ data shown in Fig. 6.18 confirms that some lateral ^{15}N transport did occur, but perhaps surprisingly, soil water residue ^{15}N enrichments do not simply decrease with distance from the treated area. While clear peaks in ^{15}N

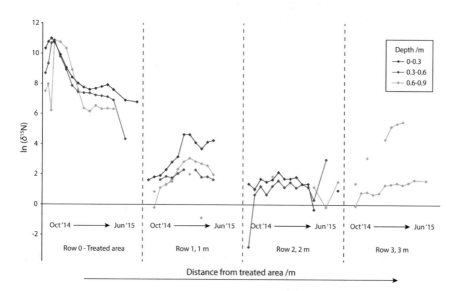

Fig. 6.18 Natural logarithm of $\delta^{15}N$ values of bulk soil water residues from porous ceramic cups in the experimental area over the course of the 32-week field experiment. All data points represent individual samples, with the exception of the lower dataset in Row 3, which represents the mean soil water residue $\delta^{15}N$ values of the seven porous ceramic cups in this row which did not show perceptible ^{15}N enrichment. The maximum SE of the mean for these data is 1.2‰. Analytical accuracy and precision were monitored using a QC standard, the maximum SD of which was 8.0‰, but this is unlikely to be representative of the instrumental uncertainty in bulk soil water residue $\delta^{15}N$ values as many of these measurements were necessarily made on very small samples. Overall uncertainty in soil water residue $\delta^{15}N$ values is further difficult to assess, as there are no replicate samples

enrichment (decreasing with depth) are evident in Row 1, Row 2 only shows a hint of ^{15}N enrichment (with the exception of the last sampling point for the 0.6–0.9 m depth range). In Row 3, on the other hand, soil water residues from one porous ceramic cup from the middle of the row and at 0.9 m depth become highly ^{15}N-enriched, while soil water residue δ^{15}N values from all other porous ceramic cups only rise slightly over the course of the experiment and to insufficient levels to confirm ^{15}N enrichment by the applied ^{15}NO$_3$$^-$. Thus, Row 2 appears to have been 'skipped' somehow. This may result from the experimental design and sampling/analysis strategy—there were no porous ceramic cups below 0.5 m near the middle of this row, the porous ceramic cup at 0.9 m failed shortly after the start of the experiment and samples from the porous ceramic cup at 0.85 m were generally selected for aqueous N concentration analysis. Alternatively, it could result from the operation of preferential lateral flow pathways during the experiment. The latter suggestion is supported by the fact that only one of the porous ceramic cups in Row 3 shows ^{15}N enrichment (all are at 0.9 m depth).

As well as indicating that the extent of lateral ^{15}N transport was less than that vertically, Fig. 6.18 also intimates that the rate of lateral ^{15}N transport was lower. It took approximately 6 weeks for ^{15}N enrichment to be observed in Row 1 (at 0–0.3 m depth), compared with less than 1 week for Row 0, while maximum ^{15}N enrichment in Row 1 occurs 14 weeks after treatment application, compared with 3–4 weeks in Row 0. The ^{15}N-enriched porous ceramic cup in Row 3 therefore shows ^{15}N enrichment unexpectedly quickly, after only 8 weeks. The soil water residue δ^{15}N values of this porous ceramic cup do not maximise equally quickly and instead continue to rise throughout the experiment (i.e. are highest in the last sample, taken 20 weeks after treatment application). This further supports the suggestion that preferential lateral flow pathways operated during the experiment.

It would not be surprising, given the lower rate of lateral transport compared with vertical leaching, that the depth of maximum ^{15}N enrichment should increase with distance from the treated area. Indeed, there is some evidence for such a pattern in Fig. 6.18, but unfortunately, the experimental design and sampling/analysis strategy does not allow this conclusion to be drawn definitively. In addition, any boundary layers in the soil profile with depth are likely to have influenced the depths at which lateral through-flow occurred. The soil-chalk boundary under the experimental area lies at a depth of 0.6–0.7 m. Preferential lateral flow along this boundary could be partly responsible for the observed ^{15}N enrichment in soil water residue δ^{15}N values in the 0.6–0.9 m depth range in Rows 1 and 3.

Based on soil water residue ^{15}N enrichments, the bulk δ^{15}N values of a selection of soil cores (0–0.3 m depth, from the middle of Rows 1–3) and a transect (in time and space) of root and foliage samples were determined in order to further assess lateral ^{15}N transport (Figs. 6.19 and 6.20). There is little conclusive evidence of ^{15}N enrichment (above natural abundance values of *ca.* 6.4‰) in the soil cores, with the possible exception of one sample from Row 2 taken 24 weeks after treatment application (11.3‰), and perhaps two samples from Rows 1 and 3 taken 28 weeks after treatment application (8.5‰). This contrasts with soil water residue data (Fig. 6.19), which shows much clearer ^{15}N enrichment and in the 0–0.3 m range particularly, in

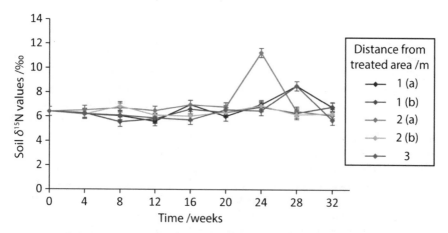

Fig. 6.19 Bulk soil δ^{15}N values for a selection of soil cores from Rows 1–3 (1–3 m from the treated area) over the course of the 32-week field experiment. **a** and **b** represent sets of soil cores sampled from two different areas at that distance from the treated area. Error bars are the maximum SD of QC and duplicate sample measurements (±0.4‰)

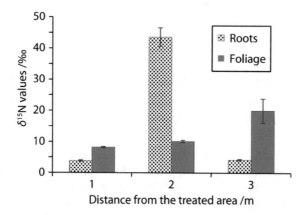

Fig. 6.20 Transect in time (7th March 2015, *ca.* 20 weeks after treatment application) and space (one sample from each of Rows 1–3, 1–3 m from the treated area) of bulk root and foliage δ^{15}N values. Root error bars are the maximum SD of QC and duplicate sample measurements (±0.2‰ for near natural abundance samples and ±3.0‰ for ^{15}N enriched samples) and foliage error bars are the SD of triplicate measurements for each sample

Row 1 at *ca.* 14 weeks. This is likely partly due to the greater dilution of any transported ^{15}N by native soil N. It is interesting, however, that the timing of maximum soil water residue δ^{15}N values do not match the potential maxima in bulk soil δ^{15}N values, reducing confidence that these peaks result from the applied ^{15}N rather than some other variation in the soil.

Both root and foliage samples from the selected transect through the rows on 7th March 2015 (*ca.* 20 weeks after treatment application) show clearer ^{15}N enrichment

(Fig. 6.20) than soil. Root δ^{15}N values are well above natural abundance levels of *ca.* 4‰ in Row 2 on this date, while foliage δ^{15}N values are higher than natural abundance values of *ca.* 1‰ in all rows. As root and foliage % TN contents are higher than that of bulk soil (Fig. 6.13), this cannot be due to lower ^{15}N dilution and suggests that any ^{15}N transported laterally from the treated area is preferentially partitioned into roots and foliage compared with soil. This may be facilitated by the extension of grass roots through the soil profile, enabling access to ^{15}N at depths of up to 1 m. It is surprising that maximum root and foliage ^{15}N enrichments on 7th March 2015 are observed in different rows—Row 2 for roots (43.6‰) and Row 3 for foliage (20.0‰; Fig. 6.20). While root and foliage samples were taken from the same general above- and belowground area, whole plants were not sampled, which could account for this difference. These contrasting results highlight the heterogeneity of the soil-plant system and non-uniformity in the transport of ^{15}N.

Unsurprisingly, groundwater (borehole sample freeze-dried residue) δ^{15}N values did not show any sign of the applied ^{15}N over the course of the 32-week experiment and a mean value of 6.1 $\pm$ 0.1‰ was recorded. Firstly, the applied ^{15}N may not have leached to groundwater in this time and secondly, even if most of the applied ^{15}N is eventually leached to groundwater, it is unlikely that any ^{15}N enrichment will be detectable due to dispersion, diffusion and dilution of the applied ^{15}NO$_3{}^-$ en route and in the groundwater reservoir.

6.4.4 Using Rainfall, Soil Water and Groundwater Hydrogen and Oxygen Isotopic Compositions to Understand Soil Water Mixing and Flow and ^{15}N-nitrate Transport

Relatively recent work has demonstrated the value of rainfall and soil water δ^2H or δ^{18}O values to investigate soil water mixing, flow and nutrient leaching (e.g. [48, 49]). In this experiment, rainfall, soil water and groundwater δ^2H or δ^{18}O values were recorded alongside soil water residue δ^{15}N values in an attempt to use patterns in water isotopes to investigate NO$_3{}^-$ transport in an undisturbed soil. Figure 6.21 shows the variations in rainfall, soil water and groundwater δ^2H and δ^{18}O values over the course of the field experiment.

Rainfall δ^2H and δ^{18}O values are quite variable, as expected from previous research (e.g. [49, 85] but, due to mixing during flow through the soil, this variability is muted by depth through the soil profile [48, 49]). It is difficult, however, to conclusively trace particular rainfall δ^2H and δ^{18}O values/events through the soil water profile based on Fig. 6.21, as only the combined minimum in rainfall δ^2H values and peak in rainwater δ^{18}O values 15–16 weeks after the start of the experiment appears clearly in soil water samples. As no other minima/maxima can be matched, it is also hard to confirm whether this match is an accurate interpretation of the relationship between rainwater and soil water H and O isotopic compositions. A lack of clear rainfall-derived patterns in soil water H and O isotopic compositions with

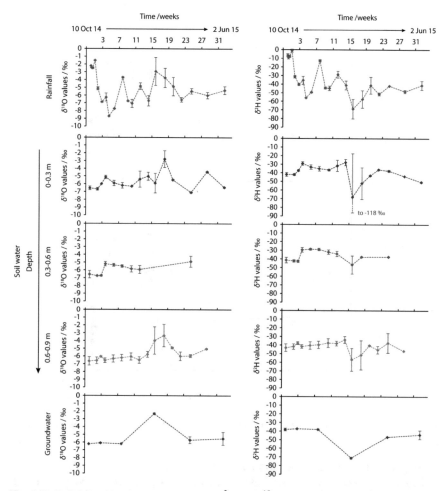

Fig. 6.21 Rainfall, soil water and groundwater δ^2H and δ^{18}O values (in vertical profiles) over the course of the field experiment. Error bars are ± SD of measurements

depth could indicate that uniform percolation or piston flow is not important in this soil [49] but this cannot be confirmed on this basis alone.

Figure 6.22 shows the correlation between depth and the SD of soil water δ^2H and δ^{18}O values, which may be used as an indicator of flowpath length [48]. Although negative, indicating that a longer flowpath resulted in less variability, as expected, the correlations are poor. This indicates that soil water mixing (and attenuation of soil water H and O isotopic variability) has not occurred in proportion to depth and provides further support for the operation of preferential (vertical) flow of water through the soil in the field experiment.

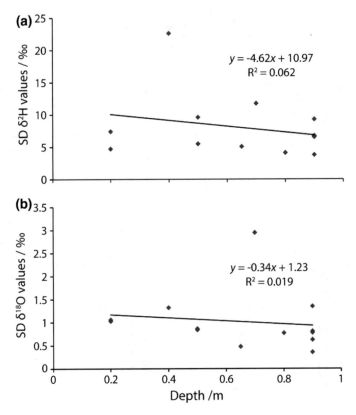

Fig. 6.22 Correlation between depth and the SD of soil water: **a** δ^2H, and **b** δ^{18}O values

It is very unexpected that the minima and maxima in rainwater δ^2H and δ^{18}O values 15–16 weeks after the start of the experiment seem to be reflected in groundwater δ^2H and δ^{18}O values at the same time as rainfall is very unlikely to reach groundwater so quickly and rainfall from one particular period cannot usually influence groundwater H and O isotopic compositions. In fact, the 2 weeks preceding this (which the rainfall sample represents) were quite dry (Fig. 6.8), making both of these occurrences even less likely. This suggests that another factor is responsible for the difference in groundwater δ^2H and δ^{18}O values 15–16 weeks after the start of the experiment, and this factor could also have influenced soil water and possibly even rainwater δ^2H and δ^{18}O values (if related in some way to sample treatment, for example).

It was hoped that the combined analysis of soil water H and O isotopic compositions and residue δ^{15}N values could be used to investigate whether ^{15}NO$_3^-$ was transported in rainfall from recent events or in older, better-mixed soil water. This was not possible, however, as soil water δ^2H and δ^{18}O isotopic patterns could not be conclusively related to rainfall δ^2H and δ^{18}O values.

6.5 Conclusions

The work described in this chapter further investigates the partitioning and (vertical and lateral) transport of $^{15}NO_3^-$ in an undisturbed soil in a unique plot-scale field experiment and in the context of its microbial incorporation (Chap. 5). Weather and the N concentrations/contents of various components of the field system were monitored over the course of the 32-week field ^{15}N tracing study in order to contextualise the ^{15}N tracing results and provide important information about the background N dynamics of the system. The patterns in system component (soil, roots, foliage and soil water) $\delta^{15}N$ values were used to trace the dynamics of the applied ^{15}N in the field system. In all system components, $\delta^{15}N$ values rose to a peak (though at different times) and then fell as ^{15}N was redistributed within the pool and/or lost from that component. Differences in the timing of maximum $\delta^{15}N$ values between system components were used to assess the rate of partitioning between and/or transport to different system components (taking into account the effects of differences in sampling frequency). An earlier peak in foliage $\delta^{15}N$ values compared with roots (Fig. 6.14), was deemed indicative of direct foliar absorption of the spray applied $^{15}NO_3^-$, for example. The clear, single peaks in soil water residue $\delta^{15}N$ values for each sampling depth below the treated areas (Fig. 6.15) confirmed that the experiment was successful, at least to some extent, in tracing the vertical leaching of NO_3^- and indicate that there was a single main leaching flush of the applied $^{15}NO_3^-$. Patterns in the timing (and shape) of these peaks enabled rates of vertical NO_3^- leaching to be estimated (0.07–0.21 m week^{-1}) and indicated that while some uniform percolation seemed to be occurring, leaching rates were either non-linear with depth, or in space. This could be due to preferential flow. It took only 3–4 weeks for soil water residue $\delta^{15}N$ values to peak at all depths and the major flush of leaching autumn applied $^{15}NO_3^-$ fertiliser passed through the soil in less than 3.5 months, before plant growth re-established in spring.

In order to assess the approximate quantitative importance of the ^{15}N enrichments observed in the component pools in terms of the overall partitioning/fate of the applied $^{15}NO_3^-$, a mass-balanced ^{15}N partitioning calculation based on estimated pool sizes was necessary. This calculation emphasised the difficulties involved in estimating pool sizes in the field and the potential for poor/uneven ^{15}N distribution within estimated pools to strongly affect calculation results. Furthermore, accounting for differences in the dynamism and turnover of the pools represents a particular problem with no simple solution. Overall, based on the calculation presented in Fig. 6.16, roots and then foliage represented the quantitatively most important fates of the applied ^{15}N in the medium-term of the experiment. The much shorter residence time of ^{15}N in the dynamic soil water pool is clear, but the quantitative importance of ^{15}N in soil water/leaching could not be determined using this calculation. Instead, the total amount of ^{15}N passing through the soil water pool at each depth was estimated to range from 109 to 76.9% (at 0.2–0.8 m, respectively) of the applied ^{15}N by integrating ^{15}N leaching over time. These estimated percentages probably represent a fairer comparison of ^{15}N in the soil water pool with the ^{15}N stored in less transient components (Fig. 6.16) than the snapshot percentages for soil water shown in Fig. 6.16. ^{15}N leaching losses were estimated to represent 2.3% of the applied ^{15}N over the course of the experiment using IRRIGUIDE modelled drainage (Fig. 6.9b) and soil water ^{15}N concentrations (mol l^{-1}) at 0.8 m.

It is particularly interesting that loss of ^{15}N from the soil water pool occurred before microbial NO_3^--^{15}N incorporation into the hydrolysable soil AA pool reached steady state (Chap. 5), while root ^{15}N uptake peaked after this point. In Chap. 5 it was suggested that microbial $^{15}NO_3^-$ assimilation under field conditions was not limited by ^{15}N losses and this appears to be the case for $^{15}NO_3^-$ leaching, but for plant uptake, out-competition by soil microbes may be responsible for this effect.

The N isotopic compositions of soil waters, soil, roots and foliage from beyond the treated area demonstrated that some lateral transport of $^{15}NO_3^-$ occurred, mainly in soil water, but to a lesser extent and more slowly than vertical transport. These results also provided indications for the operation of preferential flow and emphasised the heterogeneity of the field soil system. It was hoped that rainfall, soil water and groundwater δ^2H and δ^{18}O values could be used to further investigate soil water flow and, in combination with soil water residue δ^{15}N values, improve understanding of NO_3^- transport in soils. This work was only partially successful, however, and provided further indications that uniform percolation and preferential flow are likely to have co-occurred in the soil during the field experiment, but nothing further. It was not possible to determine whether $^{15}NO_3^-$ was transported in rainfall from recent events or in older, better-mixed soil water.

Important, more specific findings, with reference to the hypotheses presented in Sect. 6.2, include:

(i)   The applied $^{15}NO_3^-$ was initially poorly distributed in the field system and it is difficult to assess to what extent this distribution improved over the course of the field experiment. Poor distribution complicates assessment of the quantitative fate of applied ^{15}N.

(ii) System component δ^{15}N values in the treated area initially peaked (though at different times) before declining over the course of the course of the 32-week experiment.

(iii) There was evidence for direct foliar ^{15}N uptake of the spray-applied $^{15}NO_3^-$.

(iv) Vertical $^{15}NO_3^-$ leaching occurred to a greater extent and faster than lateral transport. Overall, soil water flow and NO_3^- leaching in this soil appears to occur via a combination of uniform percolation and preferential flow, as usually occurs in most soils (Sect. 1.2.1), but it has not been possible in this experiment to establish the relative importance of these mechanisms (especially as this will vary spatially and depending on the scale considered).

While the work presented in this chapter provides some interesting and unique further insights into the partitioning and transport of NO_3^- in an undisturbed and unconfined soil, it also serves to highlight why closing the N balance and understanding NO_3^- transport in soils remains a challenge. Two particular disadvantages of the work presented in this chapter relate to the lack of replication (which was simply not possible) and the possible site specificity of some results. While this may reduce the value of the numeric results of the study in relation to other sites, the more general outcomes/insights of the work are likely to be applicable across a range of sites.

References

1. Jenkinson DS (2001) The impact of humans on the nitrogen cycle, with focus on temperate arable agriculture. Plant Soil 228:3–15
2. Addiscott T, Powlson D (1992) Partitioning losses of nitrogen fertilizer between leaching and denitrification. J Agric Sci 118:101–107
3. Näsholm T, Kielland K, Ganeteg U (2009) Uptake of organic nitrogen by plants. New Phytol 182:31–48. https://doi.org/10.1111/j.1469-8137.2008.02751.x
4. Rasmussen J, Kuzyakov Y (2009) Carbon isotopes as proof for plant uptake of organic nitrogen: relevance of inorganic carbon uptake. Soil Biol Biochem 41:1586–1587. https://doi.org/10.1016/j.soilbio.2009.03.006
5. Schimel JP, Bennett J (2004) Nitrogen mineralisation: challenges of a changing paradigm. Ecology 85:591–602
6. Falkengren-Grerup U, Månsson KF, Olsson MO (2000) Uptake capacity of amino acids by ten grasses and forbs in relation to soil acidity and nitrogen availability. Environ Exp Bot 44:207–219
7. Hill PW, Quilliam RS, DeLuca TH, Farrar J, Farrell M, Roberts P, Newsham KK, Hopkins DW, Bardgett RD, Jones DL (2011) Acquisition and assimilation of nitrogen as peptide-bound and D-enantiomers of amino acids by wheat. PLoS ONE 6:e19220. https://doi.org/10.1371/journal.pone.0019220
8. Mérigout P, Lelandais M, Bitten F, Renou J-P, Briand X, Meyer C, Daniel-Vedele F (2008) Physiological and transcriptomic aspects of urea uptake and assimilation in Arabidopsis plants. Plant Physiol 147:1225–1238. https://doi.org/10.1104/pp.108.119339
9. Näsholm T, Ekblad A, Nordin A, Giesler R, Högberg M, Högberg P (1998) Boreal forest plants take up organic nitrogen. Nature 392:914–916
10. Näsholm T, Huss-Danell K, Högberg P (2000) Uptake of organic nitrogen in the field by four agriculturally important plant species. Ecology 81:1155–1161
11. Paungfoo-Lonhienne C, Lonhienne TGA, Rentsch D, Robinson N, Christie M, Webb RI, Gamage HK, Carroll BJ, Schenk PM, Schmidt S (2008) Plants can use protein as a nitrogen source without assistance from other organisms. Proc Natl Acad Sci U S A 105:4524–4529. https://doi.org/10.1073/pnas.0712078105
12. Streeter TC, Bol R, Bardgett RD (2000) Amino acids as a nitrogen source in temperate upland grasslands: the use of dual labelled (^{13}C, ^{15}N) glycine to test for direct uptake by dominant grasses. Rapid Commun Mass Spectrom 14:1351–1355
13. Jones DL, Healey JR, Willett VB, Farrar JF, Hodge A (2005) Dissolved organic nitrogen uptake by plants—an important N uptake pathway? Soil Biol Biochem 37:413–423. https://doi.org/10.1016/j.soilbio.2004.08.008
14. Persson J, Näsholm T (2001) A GC-MS method for determination of amino acid uptake by plants. Physiol Plant 113:352–358
15. Rasmussen J, Sauheitl L, Eriksen J, Kuzyakov Y (2009) Plant uptake of dual-labeled organic N biased by inorganic C uptake: results of a triple labeling study. Soil Biol Biochem 42:524–527. https://doi.org/10.1016/j.soilbio.2009.11.032
16. Warren CR (2012) Post-uptake metabolism affects quantification of amino acid uptake. New Phytol 193:522–531. https://doi.org/10.1111/j.1469-8137.2011.03933.x
17. Jones DL, Kielland K (2002) Soil amino acid turnover dominates the nitrogen flux in permafrost-dominated Taiga forest soils. Soil Biol Biochem 34:209–219

18. Lipson D, Näsholm T (2001) The unexpected versatility of plants: organic nitrogen use and availability in terrestrial ecosystems. Oecologia 128:305–316. https://doi.org/10.1007/s004420100693
19. Chapin FS III, Matson PA, Mooney HA (2002) Principles of terrestrial ecosystem ecology. Springer, New York, USA, pp 185–186, 204
20. Weigelt A, Bol R, Bardgett RD (2005) Preferential uptake of soil nitrogen forms by grassland plant species. Oecologia 142:627–635
21. Gessler A, Schneider S, von Sengbusch D, Weber P, Hanemann U, Huber C, Rothe A, Kreutzer K, Rennenberg H (1998) Field and laboratory experiments on net uptake of nitrate and ammonium by the roots of spruce (*Picea abies*) and beech (*Fagus sylvatica*) trees. New Phytol 138:275–285
22. Warren CR (2009) Why does temperature affect relative uptake rates of nitrate, ammonium and glycine: a test with *Eucalyptus pauciflora*. Soil Biol Biochem 41:778–784. https://doi.org/10.1016/j.soilbio.2009.01.012
23. Gutschick VP (1981) Evolved strategies in nitrogen acquisition by plants. Am Nat 118:607–637
24. Arkoun M, Sarda X, Jannin L, Laîne P, Etienne P, Garcia-Mina J, Yvin J, Ourry A (2012) Hydroponics versus field lysimeter studies of urea, ammonium, and nitrate uptake by oilseed rape (*Brassica napus* L). J Exp Bot 63:5245–5258
25. Britto DT, Kronzucker HJ (2002) NH_4^+ toxicity in higher plants: a critical review. J Plant Physiol 159:567–584
26. Siddiqi MY, Malhotra B, Min X, Glass ADM (2002) Effects of ammonium and inorganic carbon enrichment on growth and yield of a hydroponic tomato crop. J Plant Nutr Soil Sci 165:191–197
27. Bloom AJ, Randall L, Taylor AR, Silk WK (2012) Deposition of ammonium and nitrate in the roots of maize seedlings supplied with different nitrogen salts. J Exp Bot 63:1997–2006. https://doi.org/10.1093/jxb/err410
28. Yara (2011) Pure nutrient nitrate fertiliser. Report available from http://yara.com/products_services/fertilizers/pure_nutrient/pure_nutrient_brochure.aspx. Accessed 26 Mar 2014
29. Jackson LE, Schimel JP, Firestone MK (1989) Short-term partitioning of ammonium and nitrate between plants and microbes in an annual grassland. Soil Biol Biochem 21:409–415
30. Kuzyakov Y, Xu X (2013) Competition between roots and microorganisms for nitrogen: mechanisms and ecological relevance. New Phytol. https://doi.org/10.1111/nph.12235
31. Garnett T, Conn V, Plett D, Conn S, Zanghellini J, Mackenzie N, Enju A, Francis K, Holtham L, Roessner U, Boughton B, Bacic A, Shirley N, Rafalski A, Dhugga K, Tester M, Kaiser BN (2013) The response of the maize nitrate transport system to nitrogen demand and supply across the lifecycle. New Phytol. https://doi.org/10.1111/nph.12166
32. Gazzarrini S, Lejay L, Gojon A, Ninnemann O, Frommer WB, von Wirén N (1999) Three functional transporters for constitutive, diurnally regulated, and starvation-induced uptake of ammonium into Arabidopsis roots. Plant Cell 11:937–947
33. Coruzzi G (2003) Primary N-assimilation into amino acids. In: Arabidopsis, The Arabidopsis Book, p e0010. https://doi.org/10.1199/tab.0010
34. Hodge A, Robinson D, Fitter A (2000) Are microorganisms more effective than plants at competing for nitrogen? Trends Plant Sci 5:304–308
35. Kaštovská E, Šantrůčková H (2011) Comparison of uptake of different N forms by soil microorganisms and two wet-grassland plants: a pot study. Soil Biol Biochem 43:1285–1291. https://doi.org/10.1016/j.soilbio.2011.02.021
36. Neff JC, Chapin FS III, Vitousek PM (2003) Breaks in the cycle: dissolved organic nitrogen in terrestrial ecosystems. Front Ecol Environ 1:205–211
37. Inselsbacher E, Hinko-Najera Umana N, Stange FC, Gorfer M, Schüller E, Ripka K, Zechmeister-Boltenstern S, Hood-Novotny R, Strauss J, Wanek W (2010) Short-term competition between crop plants and soil microbes for inorganic N fertilizer. Soil Biol Biochem 42:360–372. https://doi.org/10.1016/j.soilbio.2009.11.019
38. Cameron KC, Di HJ, Moir JL (2013) Nitrogen losses from the soil/plant system: a review. Ann Appl Biol 126:145–173. https://doi.org/10.1111/aab.12014

39. Redmile-Gordon MA, Armenise E, Hirsch PR, Brookes PC (2014) Biodiesel co-product (BCP) decreases soil nitrogen (N) losses to groundwater. Water Air Soil Pollut 225:1831. https://doi.org/10.1007/s11270-013-1831-7

40. Lord EI, Shepherd MA (1993) Developments in the use of porous ceramic cups for measuring nitrate leaching. J Soil Sci 44:435–449

41. Grossmann J, Udluft P (1991) The extraction of soil water by the suction-cup method: a review. J Soil Sci 42:83–93

42. Briggs L, McCall A (1904) An artificial root for inducing capillary movement of soil moisture. Science 513:566–569

43. Addiscott TM (1996) Measuring and modelling nitrogen leaching: parallel problems. Plant Soil 181:1–6

44. Barbee G, Brown K (1986) Comparison between suction and free-drainage soil solution samplers. Soil Sci 141:149–154

45. Curley EM, O'Flynn MG, McDonnell KP (2011) The use of porous ceramic cups for sampling soil pore water from the unsaturated zone. Int J Soil Sci 6:1–11. https://doi.org/10.3923/ijss.2011.1.11

46. Webster CP, Shepherd MA, Goulding KWT, Lord E (1993) Comparisons of methods for measuring the leaching of mineral nitrogen from arable land. J Soil Sci 44:49–62

47. Hatch DJ, Jarvis SC, Rook AJ, Bristow AW (1997) Ionic contents of leachate from grassland soils: a comparison between ceramic suction cup samples and drainage. Soil Use Manage 13:68–74

48. Asano Y, Compton JE, Robbins Church M (2006) Hydrologic flow paths influence inorganic and organic nutrient leaching in a forest soil. Biochemistry 81:191–204

49. Gazis C, Feng X (2004) A stable isotope study of soil water: evidence for mixing and preferential flow paths. Geoderma 119(1–2):97–111

50. Bailey RJ, Spackman E (1996) A model for estimating soil moisture changes as an aid to irrigation scheduling and crop water-use studies: I. Operation details and description. Soil Use Manage 12:122–128

51. Bailey RJ, Groves SJ, Spackman E (1996) A model for estimating soil moisture changes as an aid to irrigation scheduling and crop water-use studies: II. Field test of the model. Soil Use Manage 12:129–133

52. Silgram M, Hatley D, Gooday R (2007) IRRIGUIDE: a decision support tool for drainage estimation and irrigation scheduling. ADAS UK Ltd

53. Defra (2010) Fertiliser Manual (RB209), pp 60, 65. Report available from http://www.ahdb.org.uk/projects/CropNutrition.aspx. Accessed 26 Mar 2014

54. Scholl M (2006) Precipitation isotope collector designs. U.S. Geological Survey. http://water.usgs.gov/nrp/proj.bib/hawaii/precip_methods.htm. Accessed 1 Oct 2014

55. Spangenberg JE (2012) Caution on the storage of waters and aqueous solutions in plastic containers for hydrogen and oxygen stable isotope analysis. Rapid Commun Mass Spectrom 26:2627–2636. https://doi.org/10.1002/rcm.6386

56. Krom MD (1980) Spectrophotometric determination of ammonia: a study of a modified Berthelot reaction using salicylate and dichloroisocyanurate. The Analyst 105:305–316

57. Searle PL (1984) The Berthelot or indophenol reaction and its use in the analytical chemistry of nitrogen: a review. Analyst 109:549–568

58. Bratton A, Marshall E (1939) A new coupling component for sulfanilamide determination. J Biol Chem 128:537–550

59. Irandoust M, Shariati-Rad M, Haghighi M (2013) Nitrite determination in water samples based on a modified Griess reaction and central composite design. Anal Methods 5:5977–5982. https://doi.org/10.1039/c3ay40913a

60. Tsikas D (2007) Analysis of nitrite and nitrate in biological fluids by assays based on the Griess reaction: appraisal of the Griess reaction in the l-arginine/nitric oxide area of research. J Chromatogr B 851(1–2):51–70

61. Gal C, Frenzel W, Möller J (2004) Re-examination of the cadmium reduction method and optimisation of conditions for the determination of nitrate by flow injection analysis. Microchim Acta 146:155–164. https://doi.org/10.1007/s00604-004-0193-7

62. Henriksen A, Selmer-Olsen AR (1970) Automatic methods for determining nitrate and nitrite in water and soil extracts. Analyst 95:514–518
63. Patton CJ, Kryskalla JR (2011) Colorimetric determination of nitrate plus nitrite in water by enzymatic reduction, automated discrete analyzer methods, In: Techniques and methods, Section B, Methods of the National Water Quality Laboratory Book 5, Laboratory Analysis, U.S. Geological Survey, Reston, Virginia, USA
64. Johnes PJ, Heathwaite AL (1992) A procedure for the simultaneous determination of total nitrogen and total phosphorus in fresh water samples using persulphate microwave digestion. Water Res 26:1281–1287
65. Picarro (2010) ChemCorrect^{TM}—Solving the problem of chemical contaminants in H_2O stable isotope research. Picarro, Inc., Sunnyvale, California, USA
66. Brand WA, Geilmann H, Crosson ER, Rella CW (2009) Cavity ring-down spectroscopy versus high-temperature conversion isotope ratio mass spectrometry; a case study on δ^2H and $\delta^{18}O$ of pure water samples and alcohol/water mixtures. Rapid Commun Mass Spectrom 23:1879–1884
67. West AG, Goldsmith GR, Brooks PD, Dawson TE (2010) Discrepancies between isotope ratio infrared spectroscopy and isotope ratio mass spectrometry for the stable isotope analysis of plant and soil waters. Rapid Commun Mass Spectrom 24:1948–1954. https://doi.org/10.1002/rcm.4597
68. Bouwman AF, Boumans JM, Batjes NH (2002) Estimation of global NH_3 volatilization loss from synthetic fertilizers and animal manure applied to arable lands and grasslands. Global Biogeochem Cycles 16:1024. https://doi.org/10.1029/2000GB001389
69. Hojito M, Hayashi K, Matsuura S (2010) Ammonia exchange on grasslands in an intensive dairying region in central Japan. Soil Sci Plant Nutr 56:503–511. https://doi.org/10.1111/j.1747-0765.2010.00466.x
70. Fenn ME, Poth MA, Schilling SL, Grainger DB (2000) Throughfall and fog deposition of nitrogen and sulfur at an N-limited and N-saturated site in the San Bernardino Mountains, Southern California. Can J For Res 30:1476–1488
71. CEH (2016) Nitrogen atmospheric Concentration Based Estimated Deposition (CBED) data for the UK 2012. Accessed 18 Apr 2016
72. Coplen TB, Hopple JA, Böhlke JK, Peiser HS, Rieder SE, Krouse HR, Rosman KJR, Ding T, Vocke RD Jr, Révész KM, Lamberty A, Taylor P, De Bièvre P (2002) Compilation of minimum and maximum isotope ratios of selected elements in naturally occurring terrestrial materials and reagents. U.S. Geological Survey Water Resources Investigations Report 01-4222, U.S. Geological Survey, Denver, USA
73. Nila Rekha P, Kanwar RS, Nayak AK, Hoang CK, Pederson CH (2011) Nitrate leaching to shallow groundwater systems from agricultural fields with different management practices. J Environ Monit 13:2550–2558
74. Wang Q, Cameron K, Buchan G, Zhao L, Zhang EH, Smith N, Carrick S (2012) Comparison of lysimeters and porous ceramic cups for measuring nitrate leaching in different soil types. N Z J Agric Res 55:333–345. https://doi.org/10.1080/00288233.2012.706224
75. Granger SJ, Heaton THE, Bol R, Bilotta GS, Bulter P, Haygarth PM, Owens PN (2008) Using $\delta^{15}N$ and $\delta^{18}O$ to evaluate the sources and pathways of NO_3^- in rainfall event discharge from drained agricultural grassland lysimeters at high temporal resolutions. Rapid Commun Mass Spectrom 22:1681–1689
76. Stuart ME, Chilton PJ, Kinniburgh DG, Cooper DM (2007) Screening for long-term trends in groundwater nitrate monitoring data. Q J Eng Geol Hydrogeol 40:361–376
77. Stuart ME, Chilton PJ, Butcher AS (2008) Nitrate fluctuations in groundwater: review of potential mechanisms and application to case studies. British Geological Survey, Groundwater Science Programme, Open Report OR/08/046
78. Yuan Z, Liu W, Niu S, Wan S (2007) Plant nitrogen dynamics and nitrogen-use strategies under altered nitrogen seasonality and competition. Ann Bot 100:821–830. https://doi.org/10.1093/aob/mcm178
79. Jarvis SC, Macduff JM (1989) Nitrate nutrition of grasses from steady-state supplies in flowing solution culture following nitrate deprivation and/or defoliation. J Exp Bot 40:965–975

80. Bowman CB, Paul JI (1992) Foliar absorption of urea, ammonium, and nitrate by perennial ryegrass turf. J Am Soc Hortic Sci 117:75–79
81. Goulding KWT, Webster CP, Powlson DS, Poulton PR (1993) Denitrification losses of nitrogen fertiliser applied to winter wheat following ley and arable rotations as estimated by acetylene inhibition and ^{15}N balance. J Soil Sci 44:63–72
82. Holbeck B, Amelung W, Wolf A, Südekum K-H, Schloter M, Welp G (2013) Recoveries of ^{15}N-labelled fertilisers (chicken manure, mushroom compost and potassium nitrate) in arable topsoil after autumn application to winter cover crops. Soil Tillage Res 130:120–127
83. Sieling K, Kage H (2006) N balance as an indicator of N leaching in an oilseed rape—winter wheat—winter barley rotation. Agric Ecosyst Environ 115:261–269
84. Goulding KWT, Poulton PR, Webster CP, Howe MT (2000) Nitrate leaching from the Broadbalk Wheat Experiment, Rothamsted, UK, as influenced by fertilizer and manure inputs and the weather. Soil Use Manag 16(4):244–250
85. Darling WG, Talbot JC (2003) The O & H stable isotopic composition of fresh waters in the British Isles. 1. Rainfall. Hydrol Earth Syst Sci 7:163–181

Chapter 7
Overview and Future Work

7.1 Overview

The overriding aim of the work conducted in this thesis was to improve understanding of soil N cycling and NO_3^- partitioning and transport in soils in order to contribute knowledge vital in developing better soil management practices to increase NUE and reduce NO_3^- leaching. In particular, the lack of information surrounding the cycling of organic N in soils, the transfer of N between inorganic and organic forms and the N supplying capacity of soils were identified as key research targets necessary for developing better soil management practices. A major reason for this lack of information is the complexity and heterogeneity of soil organic N, which prevents thorough characterisation and thus investigation of soil organic N formation, degradation and turnover.

It is possible, however, to identify some of the organic N compounds in soils— AAs, for example, which are the constituents of proteins, ubiquitous biomolecules which regulate the reactions essential to life. In fact, proteins are so important that proteinaceous matter generally comprises 20–50% of total soil N [1–4] and are likely to be major 'organic products' of microbial N assimilation/activity. These factors were used alongside the compound-specific GC-C-IRMS ^{15}N-SIP AA N fate work of Knowles [5] to develop a compound-specific AA ^{15}N-SIP method to assess N assimilation by the soil microbial biomass (Chap. 3; Fig. 7.1). The wide utility of the method was demonstrated through laboratory incubation experiments applying $^{15}NH_4^+$, $^{15}NO_3^-$ and ^{15}N-Glu. Its advantages, compared with existing techniques, and the range of potential insights/applications available are well summarised in the conclusions of Charteris et al. [6] and represented in Chap. 3. Amongst other things, the approach provides valuable insights regarding the transfer of N from the inorganic pool into a quantitatively important and active pool of soil organic N, which is likely to have considerable influence on the N supplying capacity of the soil.

Somewhat related to the lack of understanding surrounding the transfer of N between the inorganic and organic N pools in soils is a lack of information about

© Springer Nature Switzerland AG 2019
A. F. Charteris, *^{15}N Tracing of Microbial Assimilation, Partitioning and Transport of Fertilisers in Grassland Soils*, Springer Theses,
https://doi.org/10.1007/978-3-030-31057-8_7

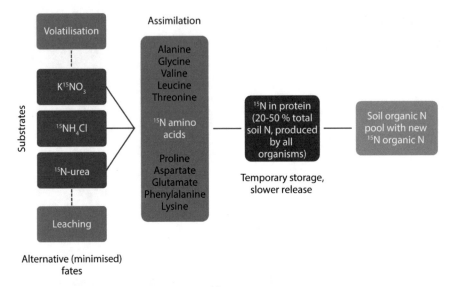

Fig. 7.1 Diagrammatic summary of the AA ^{15}N-SIP approach

the detailed processing of fertiliser N compounds in soils at a molecular level. This has implications not only for the short-term availability of the applied fertiliser N to plants and for leaching, but also for the longer-term release of initially immobilised fertiliser N, and therefore the N supplying capacity of the soil. Accordingly, compound-specific AA ^{15}N-SIP was applied in laboratory incubation experiments to assess the microbial assimilation and transfer into the organic soil protein pool of important fertiliser N compounds, [15]NO$_3^-$, [15]NH$_4^+$ and ^{15}N-urea in two grassland soils (Chap. 4). This work was also extended to investigate the effects of co-application of the ^{15}N fertilisers with manure and with glucose.

Insights were gained into the biochemical routing of the applied ^{15}N and the work enabled quantitative comparisons of the rates and extents of the soil microbial assimilation of different fertiliser N forms applied at environmentally relevant concentrations to two different soils. The extents of N incorporation into the total hydrolysable AA or soil protein pool were ordered as predicted and in accordance with N source preferences and energy and C assimilation requirements—i.e. [15]NO$_3^-$ < ^{15}N-urea < [15]NH$_4^+$ (Fig. 7.2). The generally low overall percentage ^{15}N incorporations of these substrates was, however, quite surprising. Co-addition of ^{15}N with slurry reduced NO$_3^-$-^{15}N incorporation, presumably due to preferential slurry-NH$_4^+$ assimilation, but increased NH$_4^+$-^{15}N incorporation, possibly as a result of the additional C supplied and/or stimulation of the microbial biomass. Co-additions with glucose increased ^{15}N assimilation, as anticipated.

There were no clear differences in the patterns of ^{15}N assimilation into individual hydrolysable AAs between the two soils, due, presumably, to the ubiquity of the primary metabolic pathways that operate in all soil microorganisms. Interesting differences were observed, however, in the rates and extents of assimilation of

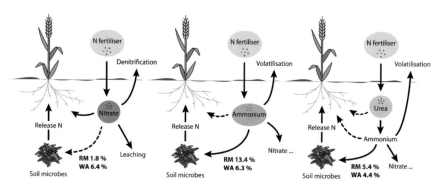

Fig. 7.2 Conceptual diagram of the different N cycling of three N fertilisers in different soils, including the percentages of applied ^{15}N incorporated into the total hydrolysable soil AA pool (microbially assimilated) for each substrate in each soil

each ^{15}N-labelled substrate between the two soils (Fig. 7.2). This could result from inherent differences between the RM and WA soils (e.g. non-calcareous vs. calcareous, differences in the NH_4^+ fixing capacity of the soils' clays and/or differences in microbial biomass size/activity/community structure and as a result, soil N dynamics). Alternatively, or in addition, it could be due to the soils' different management histories (e.g. long-term grassland vs. recently converted from arable to grassland, cattle depositions only vs. cattle depositions and synthetic fertiliser), which are likely to have affected their properties (e.g. soil structure, nutrient contents) and microbial biomass sizes/activities/community structures and N dynamics.

This work represents the first time the biomolecular fate of commonly applied fertiliser N compounds has been traced through microbially-mediated transformations into the total hydrolysable AA or soil protein pool. The results fit well with the literature, both in terms of the biochemistry of N assimilation (Sect. 4.1.2) and previous studies assessing fertiliser N immobilisation based on bulk measurements (Sect. 4.1.3), but the vital mechanistic link which this work has provided between these two different areas of research (theoretical/pure culture derived biochemical expectations vs. bulk level fertiliser immobilisation studies) at very different scales is extremely valuable. In addition, the work constitutes an important step toward greater understanding of the microbially-mediated transformations of fertiliser N to organic N and ultimately contributes to the development of a more complete picture of soil N cycling in response to fertiliser N applications. Finally, the quantitative data regarding these transformations generated through the time course incubation experiments will be of use in soil N cycling models.

A common concern associated with the use of mesocosm laboratory experiments to improve understanding of processes operating in the environment is that the scale reduction and system simplifications reduces the representativeness and thus value of the experiments. Accordingly, it was considered extremely important to assess whether the microbial N assimilation results generated in the laboratory incubation experiments are applicable to the process occurring in field. The results from the

$^{15}NO_3{}^-$ mesocosm laboratory experiment on the WA soil were therefore compared with those of a field experiment at the same site, in which a 1 m² strip of land was treated with $^{15}NO_3{}^-$ and the incubation replicated as far as possible, but under field conditions (Chap. 5).

Differences in bulk soil results between the laboratory and the field were generally easily explained by much greater ^{15}N losses from the soil under field conditions, which were anticipated. There were almost no differences in the patterns of ^{15}N assimilation into individual hydrolysable AAs between the laboratory and the field, despite the potential influence of field conditions on microbial activity (e.g. diurnal and seasonal temperature variation) and additional processes (e.g. plant uptake and leaching). This is reassuring as the fundamental biochemical reactions for life should not change. Microbial activity and the rates and extents of microbial ^{15}N assimilation, on the other hand, were expected to be sensitive to varying environmental conditions (e.g. soil temperature and moisture) and the greater decline in ^{15}N availability in the field. As a result, the remarkably small differences in the overall rates and extents of ^{15}N incorporation into the soil protein pool between the laboratory and the field experiments (6.39 and 6.62% of the applied ^{15}N, respectively), were surprising. This suggests that microbial $^{15}NO_3{}^-$ assimilation was not affected by the different conditions and additional variability and processes in the field, but that a relatively constant proportion of applied $^{15}NO_3{}^-$ is incorporated into the total hydrolysable AA pool. Overall, this work suggests that laboratory incubation experiments can provide a reasonable indication of ^{15}N incorporation into hydrolysable AAs in the field, at least for $^{15}NO_3{}^-$.

Since the microbial assimilation of $^{15}NO_3{}^-$ was found to be low under both laboratory and field conditions, other processes and fates are presumably more important for $NO_3{}^-$ cycling in the soil system. Indeed, the propensity of $NO_3{}^-$ to leaching is well known (Sect. 1.2) and $NO_3{}^-$ is an important plant N source (Sects. 6.1.2 and 6.1.3). Despite considerable research, however, establishing a complete inventory of N fates remains and challenge [7] and, in the case of $NO_3{}^-$ specifically, uncertainty also remains in the mechanisms and rates of $NO_3{}^-$ leaching, particularly in undisturbed and heterogeneous soil systems (Sect. 1.2).

Accordingly, the physical partitioning and (vertical and lateral) transport of $^{15}NO_3{}^-$ in an undisturbed and unconfined soil was investigated in the plot-scale field experiment (Chap. 6). The patterns in system component $\delta^{15}N$ values (soil, roots, foliage and soil water) were used to trace the dynamics of the applied ^{15}N in the field plot system. In all system components, $\delta^{15}N$ values rose to a peak (though at different times) and then fell as ^{15}N was redistributed within the pool and/or lost from that component. Differences in the timing of peak $\delta^{15}N$ values between system components were used to assess the rate of partitioning between and/or transport to different system components (taking into account the effects of differences in sampling frequency). An earlier peak in foliage $\delta^{15}N$ values compared with roots, was deemed indicative of direct foliar absorption of the spray applied $^{15}NO_3{}^-$, for example.

The experiment was successful, at least to some extent, in tracing the vertical leaching of $NO_3{}^-$ and clear, single peaks in soil water residue $\delta^{15}N$ values indicated that there was a single main leaching flush of the applied $^{15}NO_3^-$. Patterns in the timing (and shape) of these peaks enabled rates of vertical NO_3^- leaching to be estimated and indicated that while some uniform percolation seemed to be occurring, leaching rates were either non-linear with depth, or in space. This could be due to preferential flow. It took only 3–4 weeks for soil water residue $\delta^{15}N$ values to peak at all depths and the major flush of leaching autumn applied $^{15}NO_3^-$ fertiliser passed through the soil in less than 3.5 months, before plant growth re-established in spring.

In order to assess the approximate quantitative importance of the ^{15}N enrichments observed in the component pools in terms of the overall partitioning/fate of the applied $^{15}NO_3^-$, a mass-balanced ^{15}N partitioning calculation based on estimated pool sizes was necessary. This calculation emphasised the difficulties involved in estimating pool sizes in the field and the potential for poor/uneven ^{15}N distribution within estimated pools to strongly affect calculation results. Furthermore, accounting for differences in the dynamism and turnover of the pools represents a particular problem with no simple solution. Overall, roots and then foliage represented the quantitatively most important fates of the applied ^{15}N in the medium-term of the experiment. The much shorter residence time of ^{15}N in the dynamic soil water pool was clear, but the quantitative importance of leaching could not be determined using this calculation. Instead, this was estimated to represent 2.3% of the applied ^{15}N over the course of the experiment using IRRIGUIDE modelled drainage and soil water ^{15}N concentrations (mol l^{-1}) at 0.8 m. It is interesting that loss of ^{15}N from the soil water pool occurred before microbial NO_3^--^{15}N incorporation into the hydrolysable soil AA pool reached steady state (Chap. 5), while root ^{15}N uptake peaked after this point.

The N isotopic compositions of soil waters, soil, roots and foliage from beyond the treated area demonstrated that some lateral transport of $^{15}NO_3^-$ occurred, mainly in soil water, but to a lesser extent and more slowly than vertical transport. These results also provided indications for the operation of preferential flow and emphasised the heterogeneity of the field soil system. It was hoped that rainfall, soil water and groundwater δ^2H and $\delta^{18}O$ values could be used to further investigate soil water flow and, in combination with soil water residue $\delta^{15}N$ values, improve understanding of NO_3^- transport in soils. This work was only partially successful, however, and provided further indications that uniform percolation and preferential flow are likely to have co-occurred in the soil during the field experiment, but nothing further. It was not possible to determine whether $^{15}NO_3^-$ was transported in rainfall from recent events or in older, better-mixed soil water.

The work presented in this thesis has thus addressed the four specific objectives given in the Introduction (Sect. 1.5) and contributed further scientific understanding to fertiliser N processing in soils, which could ultimately be of use in improving NUE and reducing NO_3^- pollution. One of the main achievements of the project has been the development, validation/demonstration and use of compound-specific AA ^{15}N-SIP to investigate microbial assimilation and the organic N pool in both mesocosm laboratory experiments and the field. The approach enabled differences to be revealed in the molecular level processing of different fertiliser N forms in different soils for the first time, providing a vital mechanistic link between molecular-scale processes and the observations of field-scale N fertiliser immobilisation studies.

It will be extremely valuable in further work aimed at developing better fertiliser recommendations and soil management practices. Since further investigation is required to develop improved soil management practices based on the findings of this research, however, the main implications of the work lie in directing future research.

7.2 Future Recommendations

There is still much that can be done to improve understanding of soil N cycling and NO_3^- partitioning and transport in soils with a view to adapting land management practices to reduce NO_3^- leaching and pollution. This includes extending the work done in existing experiments (by additional sample analyses and further sampling) and conducting further experiments designed to tackle some of the questions raised by the work presented in this thesis. Some of these questions and potentially useful extensions to the work presented in each chapter are highlighted briefly in the next section and then two further experiments which would be particularly valuable are then described in more detail.

7.2.1 Further Questions and Extensions to Existing Experiments

7.2.1.1 Chapter 3: Compound-Specific ^{15}N Stable Isotope Probing to Assess Microbial Nitrogen Assimilation

As is often the case with the development of a new approach, many opportunities exist for further extension and application of the approach to different questions/areas of research. Accordingly, the potential for extending and refining the AA ^{15}N-SIP method to assess microbial N assimilation presented in Chap. 3 has already been highlighted as a valuable aspect of the approach (Sect. 3.5). Two interesting examples of further applications/extensions include:

(i) For the investigation of N_2-fixation in soils by incubation under a $^{15}N_2$ atmosphere. Nitrogen-fixation represents an important natural process of soil N fertilisation and it is hoped that the process could eventually be manipulated to improve the N supplying capacity of soils and reduce fertilisation rates. This can already be achieved to some extent by inter-cropping with N_2-fixing plants, the water fern Azolla and its cyanobacterial microsymbiont, *Anabaena Azollae*, has historically been used in rice production, for example [8]. Results in terms of increasing productivity and reducing N pollution are, however, variable and further work is required to investigate the use of free-living N_2-fixing microorganisms in soils, many of which are unculturable and thus particularly difficult to study. In this case, the approach also has a major additional advantage over current assay methods (e.g. acetylene reduction; [9]) in that N_2-fixation would be determined directly.

(ii) Adapting the method to investigate N cycling into other compound-specific N-containing biochemical pools in order to further understand soil N cycling, the transfer of inorganic N into the organic pool and the biomolecular fate of applied N. This could require a considerable amount of analytical method development, however, to establish appropriate extraction, isolation and derivatisation protocols for the compound(s). As the next largest identifiable pool of organic N in soils (after AAs; [2]), amino sugars represent an obvious target. Challenges remain, however, in their compound-specific C isotopic analysis by GC-C-IRMS [10] and N isotopic analyses are well known for being even more challenging [11, 12].

7.2.1.2 Chapter 4: Biosynthetic Routing, Rates and Extents of Microbial Fertiliser Nitrogen Assimilation in Two Pasture Soils

As might be expected from the first application of a novel approach to assess microbial N assimilation into an organic N pool, the results of the laboratory incubation experiments raise several further questions, including:

(i) In what chemical form(s) is the remainder of the applied ^{15}N that is not converted into AAs for each substrate? Does most remain in the N form applied, or is it transformed? Does this vary between ^{15}N substrates?

(ii) How long does assimilated ^{15}N remain in the hydrolysable soil AA pool? What is the fate of ^{15}N released from the hydrolysable AA pool?

(iii) What are the major controlling factors for the different ^{15}N incorporation responses of the two soils to a particular N source? Can the percentage ^{15}N incorporation of a substrate be altered by manipulating these controls and what range of responses can be achieved (across different soils)?

(iv) Are the results for arable soils comparable? How do the results change with soil type/land use/climatic regime etc.? Can the results of a systematic study of different soils be extrapolated to predict the behaviour of untested soils?

(v) What are the implications of the differences between N sources and soils for plant N nutrition in the short and longer-term? Are there differences in this for different plants? Is productivity/yield affected? Is it necessary to calculate application rates taking microbial assimilation into account?

(vi) What are the implications of the differences between N sources and soils for N losses (leaching and gaseous) in the short and longer-term?

Most of these questions require further experiments to be designed and conducted. It would certainly be interesting, however, to analyse some of the soil samples remaining from the incubation experiments for their NO_3^- and NH_4^+ concentrations and N isotopic compositions in order to further establish in what form the rest of the applied ^{15}N remains in the soil. Soil NO_3^- and NH_4^+ concentrations are commonly determined by extraction with 2 M potassium chloride (KCl) and their δ^{15}N values by NH_4^+ diffusion [13, 14], although other techniques do exist (e.g. [15]).

7.2.1.3   Chapter 5: Microbial Nitrate Assimilation Under Field Conditions—Testing the Representativeness of the Laboratory Incubation Experiments

It would be valuable to confirm that the conclusion reached in Chap. 5 that laboratory incubation experiments are able to represent N cycling into soil AAs well holds true under different conditions (e.g. in different seasons and with more extreme weather), for other sites/in other soils and for other N forms. This could be achieved by repeating the field incubation experiment at different times of the year, in other areas of the same field and in different fields and soils and conducting the same suite of experiments for other ^{15}N-labelled substrates. True replication cannot be achieved in field experiments due to spatial variability and such an assessment may yield a range of results in terms the representativeness of the laboratory experiments of the field. It would be hoped, however, that in general, the results of the laboratory incubation experiments represent N cycling into hydrolysable soil AAs in the field well.

7.2.1.4   Chapter 6: ^{15}N Tracing of the Partitioning and Transport of Nitrate Under Field Conditions

Many additional soil, root and foliage samples were collected from the wider experimental area over the course of the 32-week field experiment. None of the wider experimental area soil samples analysed in Chap. 6 were unambiguously ^{15}N-enriched and given that the samples selected were those considered most likely to be ^{15}N-enriched, it seems unlikely that conclusive ^{15}N enrichment would be revealed in any of the soil samples yet to be analysed. The small transect of root and foliage samples that was analysed, gave interesting and variable ^{15}N enrichment results, on the other hand. It could therefore be valuable to determine the N isotopic composition of further root and foliage samples from the wider experimental area over the course of the 32-week experiment in order to further assess the spatial and temporal variability of ^{15}N enrichment and by extension, ^{15}N transport.

Finally, as identified in Sect. 7.2.1.2, one important question arising from this research is the longer term turnover/fate of ^{15}N incorporated into the hydrolysable soil AA pool. It would therefore be very valuable to determine the δ^{15}N values of AAs extracted from later soil samples from the treated area to investigate the rate of depletion of ^{15}N from the hydrolysable soil AA pool. This may be via transformation into more resistant forms of organic N or via re-mineralisation to $^{15}NH_4^+$ and has implications for the longer-term supply of the applied ^{15}N to plants or for leaching and other loss pathways. It would be useful to combine this analysis with that of soil NO_3^- and NH_4^+ concentrations and N isotopic compositions (as suggested in Sect. 7.2.1.2) in order to further assess the longer term cycling of the applied ^{15}N. The treated area of the site is likely to remain ^{15}N-enriched for years to come (e.g. [16]) and continued monitoring of the partitioning of ^{15}N into different system components (including the hydrolysable soil AA pool) over the longer term could prove very interesting.

7.2.2 Two Selected, Suggested Further Experiments

7.2.2.1 Determining the Major Factors Controlling the Different Responses of Soils to Nitrogen Fertilisation

It would be extremely interesting and useful to determine the major factors controlling the ^{15}N incorporation response of the two soils used in this work (and other soils) and the extent to which this can be used to manipulate microbial N assimilation. Such an experiment would involve selecting a range of potential contributing differences between soils to be assessed (e.g. soil type, clay content and management history), obtaining appropriate soils and repeating the incubation experiments described in Chap. 4 with a selection of ^{15}N-labelled substrates. In particular, there has been recent interest in so-called 'home-field effects' and the ability of 'tuned' microbial communities to more efficiently process substrates which they regularly receive. An analogous home field advantage (HFA) effect is well-recognised for plant litter decomposition [17]. It would be very beneficial in terms of developing soil management strategies to assess whether home field effects exist for microbially-mediated N cycling using compound-specific ^{15}N-SIP of microbial N assimilation in a range of soils with different management histories. This could be combined with soil microbial community profiling and possibly even DNA ^{15}N-SIP in order to assess whether any home field effect is due to differences in community structure or soil function. Finally, if microbial tuning/home field effects are confirmed, it would be necessary to investigate the implications of this for plant productivity and N losses. A well-tuned microbial community may, for example, reduce N losses, but also N availability to plants and yields by more efficient N assimilation.

7.2.2.2 Further Assessment of the Interaction Between Microbial Nitrogen Assimilation and Release and Hydrological Processes

Through the work conducted in Chaps. 4–6, it appeared that microbial NO_3^- assimilation was low and unaffected by the variable environmental conditions in the field and the additional competing processes that result in N losses and transport. This indicates that microbial NO_3^- assimilation is potentially unaffected by hydrological drivers of N fluxes within the soil system. This counterintuitive observation could have significant implications for understanding soil N cycling and dynamics, especially in the context of farming practices and water quality management. In order to further understand the interaction of microbial processing, which supplies NO_3^- for leaching and the hydrological processes which drive leaching and N transport, it would be valuable to conduct lysimeter-scale leaching experiments assessing microbial ^{15}N assimilation, ^{15}N release and $^{15}NO_3^-$ leaching for a range of different ^{15}N-labelled substrates under differently timed rainfall events at different intensities and durations. It would then be useful to repeat and extend these experiments to include plants (of differing groundcover and rooting structure/depth) in order to assess their contribution to the dynamics of NO_3^- leaching.

7.3   Concluding Remarks

Overall, the work presented in this thesis develops, validates, demonstrates and applies a novel compound-specific AA ^{15}N-SIP approach to investigate N cycling in complex ecosystems. The new (rate and partitioning) measurements available and the new knowledge gained from them will be valuable in ^{15}N tracing soil N cycling models (e.g. [18]) and particularly in any models that involve differentially active sub-pools of organic N (labile and resistant; e.g. [19]). A disconnect currently exists between the empirical finding that NO_3^- derived from the mineralisation and nitrification of SOM is the most important contributor to groundwater NO_3^- pollution and the fact that many catchment-scale NO_3^- leaching models do not include soil microbial processes. Integrating process-based soil N cycling models into catchment-scale nutrient balance models in order to understand the contribution of N cycling to NO_3^- leaching (and other N_r losses) under varying conditions and improve predictions of N availability for plants/loss to aid better catchment management represents a major future challenge. Detailed and quantitative information regarding the rates and extents of underlying N cycling processes under varying conditions is required for the construction of such models and this information will be provided by new experimental approaches to investigate soil N cycling, such as the one presented in this thesis.

References

1. Stevenson FJ (1982) Organic forms of soil nitrogen. In: Stevenson FJ (ed) Nitrogen in agricultural soils. American Society of Agronomy, Madison, Wisconsin, USA, pp 67–74
2. Schulten H-R, Schnitzer M (1998) The chemistry of soil organic nitrogen: a review. Biol Fertil Soils 26:1–15
3. Friedel JK, Scheller E (2002) Composition of hydrolysable amino acids in soil organic matter and soil microbial biomass. Soil Biol Biochem 34:315–325
4. Roberts P, Jones DL (2008) Critical evaluation of methods for determining total protein in soil solution. Soil Biol Biochem 40:1485–1495. https://doi.org/10.1016/j.soilbio.2008.01.001

5. Knowles TDJ (2009) Following the fate of proteinaceous material in soil using a compound-specific ^{13}C- and ^{15}N-labelled tracer approach. Unpublished Ph.D. Thesis, University of Bristol, Bristol, UK

6. Charteris AF, Knowles TDJ, Michaelides K, Evershed RP (2016) Compound-specific amino acid ^{15}N stable isotope probing of nitrogen assimilation by the soil microbial biomass using gas chromatography/combustion/isotope ratio mass spectrometry. Rapid Commun Mass Spectrom 30:1846–1856. https://doi.org/10.1002/rcm.7612

7. Jenkinson DS (2001) The impact of humans on the nitrogen cycle, with focus on temperate arable agriculture. Plant Soil 228:3–15

8. Paul EA (2007) Soil microbiology, ecology and biochemistry, 3rd edn. Academic Press, Oxford, UK

9. Vessey JK (1994) Measurement of nitrogenase activity in legume root nodules: in defense of the acetylene reduction assay. Plant Soil 158:151–162

10. Decock C, Denef K, Bodé S, Six J, Boeckx P (2009) Critical assessment of the applicability of gas chromatography-combustion-isotope ratio mass spectrometry to determine amino sugar dynamics in soil. Rapid Commun Mass Spectrom 23:1201–1211

11. Brand WA, Tegtmeyer AR, Hilkert A (1994) Compound-specific isotope analysis: extending toward ^{15}N/^{14}N and ^{18}O/^{16}O. Org Geochem 21:585–594

12. Brenna JT (1994) High-precision gas isotope ratio mass spectrometry: recent advances in instrumentation and biomedical applications. Acc Chem Res 27:340–346

13. Stark JM, Hart SC (1996) Diffusion technique for preparing salt solutions, Kjeldahl digests, and persulfate digests for nitrogen-15 analysis. Soil Sci Soc Am J 60:1846–1855

14. Sebilo M, Mayer B, Grably M, Billiou D, Mariotti A (2004) The use of the 'ammonium diffusion' method for δ^{15}N-NH$_4{}^+$ and δ^{15}N-NO$_3{}^-$ measurements: comparison with other techniques. Environ Chem 1:99–103. https://doi.org/10.1071/EN04037

15. Lachouani P, Frank AH, Wanek W (2010) A suite of sensitive chemical methods to determine the δ^{15}N of ammonium, nitrate and total dissolved N in soil extracts. Rapid Commun Mass Spectrom 24:3615–3623. https://doi.org/10.1002/rcm.4798

16. Sebilo M, Mayer B, Nicolardot B, Pinay G, Mariotti A (2013) Long-term fate of nitrate fertilizer in agricultural soils. Proc Natl Acad Sci USA 110:18185–18189. https://doi.org/10.1073/pnas.1305372110

17. Austin AT, Vivanco L, González-Arzac A, Pérez LI (2014) There's no place like home? An exploration of the mechanisms behind plant litter-decomposer affinity in terrestrial ecosystems. New Phytol 204:307–314. https://doi.org/10.1111/nph.12959

18. Müller C, Stevens RJ, Laughlin RJ (2004) A ^{15}N tracing model to analyse N transformations in old grassland soil. Soil Biol Biochem 36:619–632

19. Inselsbacher E, Wanek W, Strauss J, Zechmeister-Boltenstern S, Müller C (2013) A novel ^{15}N tracer model reveals: plant nitrate uptake governs nitrogen transformation rates in agricultural soils. Soil Biol Biochem 57:301–310

Appendices

Appendix A1

Tables of amino acid concentrations and percentage contributions of total hydrolysable amino acid nitrogen to the total nitrogen pool for the laboratory incubation experiments presented in Chap. 4 (Tables A1.1, A1.2, A1.3, A1.4, A1.5, A1.6, A1.7, A1.8, A1.9, A1.10, A1.11).

© Springer Nature Switzerland AG 2019
A. F. Charteris, ^{15}N *Tracing of Microbial Assimilation, Partitioning and Transport of Fertilisers in Grassland Soils*, Springer Theses,
https://doi.org/10.1007/978-3-030-31057-8

Table A1.1 Amino acid concentrations and the percentage contribution of total hydrolysable AA (THAA) N to the TN pool for the RM $^{15}NO_3^-$ incubation experiment at each time point analysed

		Time (days)											Mean	SE
		0	0.0625	0.125	0.25	0.5	1	2	4	8	16	32		
Mean concentration (mg g^{-1})	Alanine	2.38	1.86	1.93	1.97	1.93	2.11	1.97	2.18	1.81	1.85	2.32	2.06	0.0558
	Aspartate	1.61	1.92	1.99	2.02	2.01	1.83	2.19	1.82	1.55	1.56	1.15	1.77	0.0660
	Glutamate	1.59	1.78	1.85	1.73	1.81	1.83	1.92	1.80	1.51	1.57	1.13	1.68	0.0551
	Glycine	1.83	1.27	1.39	1.38	1.36	1.53	1.42	1.42	1.20	1.47	1.82	1.49	0.0503
	Hydroxyproline	0.12	0.12	0.12	0.12	0.12	0.14	0.12	0.13	0.11	0.10	0.12	0.12	0.0028
	Isoleucine	0.38	0.49	0.49	0.39	0.35	0.25	0.34	0.39	0.34	0.26	0.37	0.37	0.015
	Leucine	1.02	1.07	1.11	1.02	0.97	0.95	0.95	1.07	0.85	0.91	0.87	0.98	0.017
	Lysine	0.48	0.34	0.41	0.57	0.46	0.39	0.40	0.46	0.53	0.39	0.60	0.44	0.025
	Methionine	0.07	0.13	0.12	0.12	0.10	0.09	0.10	0.07	0.06	0.08	0.06	0.09	0.004
	Phenylalanine	0.48	0.58	0.59	0.59	0.54	0.56	0.43	0.51	0.37	0.49	0.41	0.50	0.015
	Proline	1.23	1.11	1.14	1.08	1.10	1.23	1.12	1.22	1.01	0.97	1.23	1.14	0.0261
	Serine	0.89	0.89	0.95	1.12	1.05	0.90	1.00	0.98	0.87	0.81	0.85	0.93	0.026
	Threonine	0.73	0.82	0.90	0.97	0.88	0.68	0.87	0.86	0.77	0.67	0.65	0.80	0.027
	Tyrosine	0.22	0.34	0.31	0.37	0.34	0.34	0.23	0.26	0.19	0.25	0.21	0.27	0.012
	Valine	0.75	0.72	0.77	0.73	0.63	0.44	0.63	0.76	0.67	0.52	0.62	0.67	0.026
	THAA	13.8	13.4	14.1	14.2	13.6	13.3	13.7	13.9	11.7	11.9	12.4	13.3	0.245
	% THAA N of TN	28.8	25.0	27.1	26.6	26.3	26.1	25.8	27.0	22.6	22.6	24.2	25.9	0.523

Table A1.2 Amino acid concentrations and the percentage contribution of THAA N to the TN pool for the RM $^{15}NH_4^+$ incubation experiment at each time point analysed

		Time (days)											Mean	SE
		0	0.0625	0.125	0.25	0.5	1	2	4	8	16	32		
Mean concentration (mg g^{-1})	Alanine	2.38	1.92	1.79	1.84	1.91	1.87	1.93	1.90	2.02	2.11	2.09	2.03	0.0453
	Aspartate	1.61	2.21	1.80	1.82	2.01	1.94	2.10	2.00	2.11	2.08	2.03	1.95	0.0512
	Glutamate	1.59	2.02	1.81	1.82	1.88	1.81	1.89	1.83	1.92	1.89	1.88	1.83	0.0440
	Glycine	1.83	1.26	1.14	1.41	1.34	1.33	1.52	1.37	1.49	1.49	1.44	1.46	0.0435
	Hydroxyproline	0.12	0.12	0.11	0.12	0.12	0.11	0.12	0.11	0.13	0.13	0.13	0.12	0.0024
	Isoleucine	0.38	0.61	0.54	0.43	0.46	0.52	0.55	0.45	0.48	0.48	0.47	0.48	0.014
	Leucine	1.02	1.16	1.15	1.05	1.07	1.05	1.09	1.03	1.09	1.13	1.12	1.09	0.0119
	Lysine	0.48	0.41	0.27	0.27	0.25	0.34	0.48	0.40	0.51	0.59	0.55	0.42	0.029
	Methionine	0.07	0.12	0.12	0.09	0.13	0.10	0.08	0.13	0.12	0.11	0.13	0.1	0.005
	Phenylalanine	0.48	0.55	0.58	0.45	0.46	0.46	0.48	0.54	0.60	0.59	0.64	0.53	0.014
	Proline	1.23	1.14	1.10	1.19	1.14	1.09	1.14	1.07	1.14	1.18	1.20	1.16	0.0197
	Serine	0.89	1.02	0.85	0.84	0.94	0.92	1.04	0.95	1.07	1.12	1.05	0.97	0.023
	Threonine	0.73	1.07	0.89	0.87	0.90	0.86	0.96	0.92	0.99	1.01	0.93	0.90	0.025
	Tyrosine	0.22	0.29	0.29	0.23	0.24	0.24	0.26	0.30	0.36	0.35	0.35	0.28	0.0099
	Valine	0.75	0.95	0.80	0.76	0.72	0.83	0.90	0.78	0.82	0.83	0.78	0.81	0.022
	THAA	13.8	14.9	13.2	13.2	13.6	13.5	14.5	13.8	14.8	15.1	14.8	14.1	0.197
	% THAA N of TN	28.8	30.5	26.4	26.5	27.5	30.3	33.5	25.9	27.2	29.9	26.8	28.5	0.555

Table A1.3 Amino acid concentrations and the percentage contribution of THAA N to the TN pool for RM $^{15}NO_3^- + 2$ incubation experiment at each time point analysed

		Time (days)											Mean	SE
		0	0.125	0.25	0.5	1	2	4	8	16	32	64		
Mean concentration (mg g^{-1})	Alanine	1.14	1.51	1.33	1.43	1.39	1.23	1.44	1.05	1.35	1.34	1.39	1.33	0.0332
	Aspartate	1.60	2.41	1.66	1.89	2.12	1.60	1.89	1.42	2.11	1.82	1.91	1.86	0.0664
	Glutamate	1.84	2.44	1.87	1.98	2.14	1.73	2.01	1.50	2.24	1.95	2.13	1.99	0.0581
	Glycine	0.92	1.69	1.38	1.36	1.34	1.21	1.30	1.03	1.49	1.35	1.64	1.34	0.048
	Hydroxyproline	0.10	0.15	0.13	0.13	0.13	0.11	0.13	0.09	0.12	0.12	0.12	0.12	0.004
	Isoleucine	0.47	0.35	0.28	0.35	0.33	0.26	0.37	0.39	0.31	0.37	0.26	0.34	0.015
	Leucine	0.97	0.98	0.90	0.95	0.94	0.81	0.93	0.85	0.93	0.95	0.91	0.92	0.015
	Lysine	0.21	0.78	0.80	0.74	0.76	0.26	0.28	0.32	0.71	0.48	0.71	0.55	0.048
	Phenylalanine	0.54	0.44	0.47	0.50	0.49	0.39	0.40	0.38	0.39	0.45	0.43	0.44	0.011
	Proline	0.95	1.24	1.08	1.10	1.11	0.96	1.18	0.85	1.16	1.07	1.18	1.08	0.027
	Serine	0.71	1.19	0.89	0.95	1.01	0.76	0.91	0.66	0.98	0.87	0.94	0.90	0.034
	Threonine	1.06	1.28	0.99	1.12	1.09	0.84	1.06	0.88	1.15	1.09	1.05	1.06	0.037
	Tyrosine	0.24	0.36	0.38	0.33	0.35	0.31	0.26	0.22	0.25	0.25	0.28	0.29	0.011
	Valine	0.94	0.78	0.63	0.77	0.74	0.58	0.79	0.74	0.75	0.78	0.62	0.74	0.025
	THAA	11.7	15.6	12.8	13.6	14.0	11.1	12.9	10.4	13.9	12.9	13.6	12.9	0.348
	% THAA N of TN	22.2	28.5	23.5	25.1	26.3	20.1	22.9	19.1	25.9	24.1	25.5	23.9	0.663

Table A1.4 Amino acid concentrations and the percentage contribution of THAA N to the TN pool for RM ^{15}N-U incubation experiment at each time point analysed

		Time (days)					Mean	SE
		0	0.125	2	16	32		
Mean concentration (mg g^{-1})	Alanine	1.14	1.42	1.34	1.32	1.34	1.31	0.0340
	Aspartate	1.60	1.55	1.42	1.42	1.63	1.53	0.0479
	Glutamate	1.84	1.83	1.72	1.66	1.84	1.78	0.0381
	Glycine	0.92	1.43	1.32	1.29	1.31	1.25	0.056
	Hydroxyproline	0.10	0.14	0.13	0.13	0.13	0.13	0.0048
	Isoleucine	0.47	0.22	0.25	0.22	0.25	0.28	0.027
	Leucine	0.97	0.85	0.83	0.83	0.85	0.87	0.017
	Lysine	0.21	0.37	0.35	0.32	0.37	0.32	0.019
	Phenylalanine	0.54	0.44	0.45	0.45	0.48	0.47	0.011
	Proline	0.95	1.20	1.16	1.11	1.10	1.10	0.034
	Serine	0.71	0.77	0.71	0.71	0.79	0.74	0.022
	Threonine	1.06	0.81	0.78	0.78	0.88	0.86	0.039
	Tyrosine	0.24	0.24	0.25	0.23	0.25	0.24	0.0037
	Valine	0.94	0.49	0.53	0.48	0.54	0.60	0.049
	THAA	11.7	11.8	11.2	10.9	11.8	11.5	0.214
	% THAA N of TN	22.2	22.9	22.0	20.6	21.8	21.9	0.387

Table A1.5 Amino acid concentrations and the percentage contribution of THAA N to the TN pool for WA $^{15}NO_3^-$ incubation experiment at each time point analysed

		Time (days)							Mean	SE
		0	0.125	0.5	2	4	16	32		
Mean concentration (mg g^{-1})	Alanine	0.86	1.00	0.91	0.84	0.96	0.86	0.86	0.89	0.019
	Aspartate	1.25	2.28	2.32	1.84	2.34	2.30	2.19	1.97	0.108
	Glutamate	1.23	1.72	1.72	1.49	1.76	1.58	1.53	1.53	0.0500
	Glycine	1.01	1.18	1.17	1.14	1.29	0.96	1.02	1.10	0.030
	Hydroxyproline	0.09	0.13	0.14	0.10	0.13	0.12	0.11	0.11	0.004
	Isoleucine	0.21	0.23	0.24	0.18	0.20	0.29	0.23	0.22	0.012
	Leucine	0.53	0.64	0.56	0.59	0.59	0.61	0.58	0.58	0.0094
	Lysine	0.23	0.37	0.36	0.30	0.34	0.47	0.40	0.34	0.025
	Phenylalanine	0.25	0.27	0.25	0.24	0.24	0.35	0.33	0.27	0.010
	Proline	0.73	0.84	0.76	0.72	0.78	0.71	0.68	0.74	0.016
	Serine	0.36	0.68	0.62	0.49	0.64	0.67	0.64	0.56	0.031
	Threonine	0.40	0.73	0.60	0.50	0.70	0.89	0.76	0.62	0.041
	Tyrosine	0.09	0.11	0.12	0.10	0.11	0.14	0.13	0.11	0.004
	Valine	0.38	0.50	0.44	0.33	0.39	0.67	0.54	0.45	0.028
	THAA	7.6	10.7	10.2	8.9	10.5	10.6	10.0	9.5	0.91
	% THAA N of TN	21.6	27.5	26.3	23.1	28.0	27.3	26.9	25.3	0.719

Table A1.6 Amino acid concentrations and the percentage contribution of THAA N to the TN pool for WA $^{15}NH_4^+$ incubation experiment at each time point analysed

		Time (days)							Mean	SE
		0	0.0625	0.125	0.5	2	8	32		
Mean concentration (mg g^{-1})	Alanine	0.86	0.78	0.85	0.93	0.84	0.87	0.76	0.84	0.015
	Aspartate	1.25	2.13	1.96	2.31	2.06	2.42	2.04	1.93	0.0954
	Glutamate	1.23	1.53	1.45	1.69	1.50	1.72	1.54	1.49	0.0421
	Glycine	1.01	0.96	1.04	1.15	1.03	1.08	0.93	1.03	0.019
	Hydroxyproline	0.09	0.11	0.11	0.13	0.11	0.12	0.11	0.11	0.003
	Isoleucine	0.21	0.20	0.17	0.26	0.22	0.24	0.22	0.21	0.0089
	Leucine	0.53	0.57	0.55	0.62	0.58	0.62	0.59	0.57	0.0084
	Lysine	0.23	0.31	0.26	0.38	0.31	0.35	0.31	0.30	0.013
	Phenylalanine	0.25	0.26	0.24	0.33	0.25	0.35	0.37	0.29	0.010
	Proline	0.73	0.68	0.68	0.77	0.69	0.72	0.67	0.71	0.012
	Serine	0.36	0.61	0.58	0.70	0.64	0.71	0.60	0.57	0.029
	Threonine	0.40	0.69	0.63	0.88	0.77	0.85	0.73	0.67	0.039
	Tyrosine	0.09	0.09	0.09	0.14	0.12	0.14	0.15	0.11	0.006
	Valine	0.38	0.46	0.42	0.62	0.51	0.55	0.49	0.48	0.020
	THAA	7.6	9.4	9.1	10.9	9.6	10.7	9.5	9.3	0.27
	% THAA N of TN	21.6	23.7	22.8	29.1	24.4	28.1	23.7	24.4	0.720

Table A1.7 Amino acid concentrations and the percentage contribution of THAA N to the TN pool for WA $^{15}NO_3^- + 2$ incubation experiment at each time point analysed

		Time (days)				Mean	SE
		0	2	32	64		
Mean concentration (mg g^{-1})	Alanine	0.72	0.85	0.73	0.69	0.75	0.026
	Aspartate	1.86	1.92	1.78	1.53	1.77	0.0833
	Glutamate	1.38	1.51	1.42	1.24	1.39	0.0470
	Glycine	0.73	1.02	0.94	0.87	0.89	0.038
	Hydroxyproline	0.11	0.12	0.11	0.10	0.11	0.0036
	Isoleucine	0.28	0.19	0.17	0.18	0.21	0.017
	Leucine	0.57	0.54	0.59	0.63	0.58	0.014
	Lysine	0.27	0.31	0.37	0.31	0.32	0.022
	Phenylalanine	0.35	0.27	0.25	0.23	0.28	0.016
	Proline	0.58	0.66	0.64	0.59	0.62	0.016
	Serine	0.53	0.56	0.52	0.46	0.52	0.021
	Threonine	0.70	0.60	0.54	0.47	0.58	0.040
	Tyrosine	0.12	0.13	0.17	0.13	0.14	0.0074
	Valine	0.62	0.56	0.32	0.32	0.45	0.045
	THAA	8.8	9.3	8.5	7.7	8.6	0.29
	% THAA N of TN	23.9	24.8	23.2	21.4	23.3	0.690

Table A1.8 Amino acid concentrations and the percentage contribution of THAA N to the TN pool for WA $^{15}NH_4^+ + 2$ incubation experiment at each time point analysed

		Time (days)				Mean	SE
		0	2	32	64		
Mean concentration (mg g^{-1})	Alanine	0.72	0.71	0.77	0.74	0.74	0.019
	Aspartate	1.86	1.65	1.81	1.53	1.72	0.0699
	Glutamate	1.38	1.32	1.44	1.23	1.34	0.0426
	Glycine	0.73	0.87	1.02	0.90	0.88	0.038
	Hydroxyproline	0.11	0.10	0.10	0.10	0.11	0.0026
	Isoleucine	0.28	0.19	0.18	0.15	0.20	0.016
	Leucine	0.57	0.52	0.57	0.53	0.55	0.0094
	Lysine	0.27	0.26	0.39	0.27	0.29	0.025
	Phenylalanine	0.35	0.25	0.26	0.21	0.27	0.017
	Proline	0.58	0.60	0.65	0.59	0.61	0.015
	Serine	0.53	0.47	0.54	0.45	0.50	0.020
	Threonine	0.70	0.51	0.56	0.45	0.56	0.038
	Tyrosine	0.12	0.11	0.17	0.12	0.13	0.0086
	Valine	0.62	0.53	0.35	0.30	0.45	0.041
	THAA	8.8	8.1	8.8	7.6	8.3	0.27
	% THAA N of TN	23.9	22.4	24.7	20.8	22.9	0.814

Table A1.9 Amino acid concentrations and the percentage contribution of THAA N to the TN pool for WA $^{15}NO_3^- + G$ incubation experiment at each time point analysed

		Time (days)		Mean	SE
		0	32		
Mean concentration (mg g^{-1})	Alanine	0.72	0.76	0.74	0.022
	Aspartate	1.86	2.14	2.00	0.158
	Glutamate	1.38	1.56	1.47	0.0865
	Glycine	0.73	0.94	0.84	0.052
	Hydroxyproline	0.11	0.11	0.11	0.0040
	Isoleucine	0.28	0.24	0.26	0.021
	Leucine	0.57	0.55	0.56	0.016
	Lysine	0.27	0.31	0.29	0.022
	Phenylalanine	0.35	0.26	0.30	0.026
	Proline	0.58	0.65	0.61	0.021
	Serine	0.53	0.63	0.58	0.041
	Threonine	0.70	0.77	0.73	0.059
	Tyrosine	0.12	0.14	0.13	0.0073
	Valine	0.62	0.55	0.59	0.039
	THAA	8.8	10.3	9.6	0.61
	% THAA N of TN	23.9	27.1	25.5	1.62

Table A1.10 Amino acid concentrations and the percentage contribution of THAA N to the TN pool for WA $^{15}NH_4^+$ + G incubation experiment at each time point analysed

		Time (days)		Mean	SE
		0	32		
Mean concentration (mg g^{-1})	Alanine	1.08	0.96	1.02	0.0475
	Aspartate	1.57	1.13	1.35	0.124
	Glutamate	1.54	1.19	1.37	0.114
	Glycine	1.03	0.85	0.94	0.065
	Hydroxyproline	0.15	0.13	0.14	0.0066
	Isoleucine	0.25	0.20	0.23	0.015
	Leucine	0.76	0.69	0.73	0.025
	Lysine	0.40	0.18	0.29	0.069
	Phenylalanine	0.41	0.30	0.35	0.028
	Proline	0.82	0.74	0.78	0.037
	Serine	0.74	0.52	0.63	0.051
	Threonine	0.78	0.56	0.67	0.055
	Tyrosine	0.20	0.17	0.19	0.010
	Valine	0.53	0.43	0.48	0.027
	THAA	11.1	8.1	9.6	0.77
	% THAA N of TN	23.0	17.6	20.3	1.69

Table A1.11 Amino acid concentrations and the percentage contribution of THAA N to the TN pool for WA ^{15}N-U incubation experiment at each time point analysed

		Time (days)					Mean	SE
		0	0.125	2	16	32		
Mean concentration (mg g^{-1})	Alanine	1.08	1.17	1.15	1.20	1.14	1.15	0.0241
	Aspartate	1.57	1.36	1.72	1.25	1.60	1.50	0.0730
	Glutamate	1.54	1.51	1.77	1.45	1.69	1.59	0.0591
	Glycine	1.03	1.15	1.16	1.19	1.11	1.13	0.0290
	Hydroxyproline	0.15	0.17	0.18	0.19	0.19	0.18	0.0061
	Isoleucine	0.25	0.20	0.24	0.17	0.23	0.22	0.011
	Leucine	0.76	0.72	0.76	0.70	0.75	0.74	0.013
	Lysine	0.40	0.29	0.33	0.27	0.23	0.30	0.027
	Phenylalanine	0.41	0.37	0.39	0.37	0.36	0.38	0.0086
	Proline	0.82	0.94	0.91	1.03	0.99	0.94	0.030
	Serine	0.74	0.58	0.73	0.60	0.72	0.68	0.028
	Threonine	0.78	0.63	0.79	0.57	0.77	0.71	0.036
	Tyrosine	0.20	0.18	0.21	0.19	0.20	0.20	0.0044
	Valine	0.53	0.43	0.52	0.39	0.50	0.47	0.020
	THAA	11.1	10.3	11.5	10.1	10.9	10.8	0.297
	% THAA N of TN	23.0	21.0	23.6	21.1	23.2	22.4	0.644

Bibliography

BADC. Accessed 21 Jan 2015

Högberg P (1991) Development of ^{15}N enrichment in a nitrogen-fertilised forest soil-plant system. Soil Biol Biochem 23:335–338

Orchard VA, Cook FJ (1983) Relationship between soil respiration and soil moisture. Soil Biol Biochem 15:447–455

Ratkowsky DA, Olley J, McMeekin TA, Ball A (1982) Relationship between temperature and growth rate of bacterial cultures. J Bacteriol 149:1–5

© Springer Nature Switzerland AG 2019
A. F. Charteris, *^{15}N Tracing of Microbial Assimilation, Partitioning and Transport of Fertilisers in Grassland Soils*, Springer Theses, https://doi.org/10.1007/978-3-030-31057-8